58th Annual Loss Prevention Symposium (LPS 2024)

Held at the 2024 AIChE Spring Meeting and 20th Global Congress on Process Safety

New Orleans, Louisiana, USA
24-28 March 2024

ISBN: 979-8-3313-1700-3

58th Annual Loss Prevention Symposium (LPS 2024)

Held at the 2024 AIChE Spring Meeting and 20th Global Congress on Process Safety

New Orleans, Louisiana, USA
24-28 March 2024

AIChE
The Global Home of Chemical Engineers

Printed from e-media with permission by:

Curran Associates, Inc.
57 Morehouse Lane
Red Hook, NY 12571

Some format issues inherent in the e-media version may also appear in this print version.

Printed with permission by Curran Associates, Inc. (2025)

For permission requests, please contact AIChE
at the address below.

AIChE
120 Wall Street, FL 23
New York, NY 10005-4020

Phone: (800) 242-4363
Fax: (203) 775-5177

www.aiche.org

Additional copies of this publication are available from:

Curran Associates, Inc.
57 Morehouse Lane
Red Hook, NY 12571 USA
Phone: 845-758-0400
Fax: 845-758-2633
Email: curran@proceedings.com
Web: www.proceedings.com

TABLE OF CONTENTS

Strategic Emergency Planning – Case Study for a Gas Separation Plant

Stuart Chia
Sherpa Consulting Pty Ltd
Suite 1, Level 8, North Tower, 1-5 Railway Street, Chatswood, NSW 2067, Australia
Stuart.Chia@sherpaconsulting.com.au

Dusit Pitanupong, Thakatkaew Saingam and Polkrit Chouyyoung
Trans Thai-Malaysia (Thailand) Ltd, Songkhla, Thailand

Keywords: barrier and control management, bowties, pre-incident plan

Abstract

Trans Thai-Malaysia (TTM), a joint venture between PTT (Thailand) and Petronas (Malaysia), has been enhancing their process safety management (PSM) framework for major hazard control across their energy assets.

Major Accident Events (MAE) bowties have been used to identify and strengthen barriers (controls) to prevent and mitigate against a loss of primary containment. Application of barrier analysis found weaknesses in critical controls including emergency response. As the final mitigation barrier, TTM recognized that effective emergency planning and pre-incident plans (PIP) were required to provide a measured response against MAEs.

TTM sought guidance from the CCPS PSM Emergency Management principles and industry best practices to improve MAE PIPs. This has resulted in a better understanding of the potential consequences from both credible and escalated events, a timeline with required response activities, simplified decision tree, establishment of hazard zones and understanding whether safe access to fixed firefighting equipment is possible. The PIPs are supported by a consequence mapping tool using Google Earth.

The next steps will involve PIP awareness training, followed by desktop reviews and field trials. These activities will further refine the PIPs, improve training programs and function test critical firefighting equipment against performance standards. Key experiences gained during this strategic planning phase are shared.

1 TTM Process Safety Journey

1.1 Background

Trans Thai-Malaysia (TTM), a joint venture between PTT (Thailand) and Petronas (Malaysia), has embarked upon initiatives to improve their process safety management (PSM) framework for major hazard control across their offshore and onshore energy assets.

As part of corporate operational excellence requirements, TTM has adapted industry PSM models including the Center of Chemical Process Safety (CCPS) and guidelines to enhance their PSM framework. A cornerstone of the multi-year PSM program (starting in 2020) has been the implementation of a barrier management program to deliver assurance to stakeholders that TTM has in place effective operational barriers to ensure its major hazard risks (known as Major Accident Events, MAE) are acceptable.

MAE bowties have been used to identify and strengthen barriers to prevent and mitigate against a loss of primary containment. Application of barrier analysis including validation found weaknesses in critical controls including emergency response. As the final mitigation barrier, TTM recognized that effective emergency planning and pre-incident plans (PIP) were required to provide a measured response to control MAEs. Emergency management in turn has interfaces with PSM elements including hazard identification, operational knowledge, training and competence as well as emergency system reliability.

1.2 TTM Business

As part of the Joint Development Area (JDA) between Malaysia and Thailand, TTM operates a major Gas Separation Plant (GSP) near Songkhla, fed by an offshore pipeline from the Block A18 central processing platforms. The GSP processes the natural gas feed into products consisting of methane, liquified petroleum gas (LPG) and natural gas liquids (NGL). These products are delivered to Petronas and PTT via a network of onshore transmission pipelines and road tankers.

At the Board level, understanding material risks to the organization is mandatory. The TTM QSHE Policy has a specific focus upon process safety and loss prevention. Concerned with the range and impact of LOPC events occurring globally at similar facilities and nationally, the JDA partners challenged the QSHE team to provide assurance of 'how do we know that we are operating safely?' and 'how do we demonstrate safe operations?'

1.3 TTM Barrier Management

During this period, PTT was also developing their major hazard management program framed upon barrier thinking concepts and bowtie construction guidance outlined in the Chemical Center of Process Safety (CCPS) *Bow Ties in Risk Management: A Concept Book for Process Safety* [1]. As a member of the PTT PSM working group, TTM determined this approach, and the group guidelines could be adapted with some modification to their Songkhla operations.

The set of TTM Guidelines are in two parts with the first document outlining in a step-by-step fashion, the MAE bowtie process to demonstrate effectiveness of barriers, relationship with safety critical controls and site management systems. The second document covers the site validation process with an emphasis on safety critical barriers meeting performance standards and identifying potential shortfalls.

Application of the barrier management process has allowed TTM to address fundamental process safety questions [2]:

1. Do we **understand** what can go wrong in our operations?
2. Do we know **what barriers** there are to **prevent and mitigate** this from happening?
3. Do we have the **information** to **know (assurance)** they are **working effectively**?

In consultation with the site workforce, desktop workshops (2020-2023) have resulted in the development including barrier analysis of MAE bowties covering the GSP and pipeline operations. In 2023, TTM initiated the site barrier validation reviews.

2 Requirement to improve Emergency Management

2.1 Bowtie Findings

A review of the GSP MAE bowtie mitigation barriers and initial barrier validity found:

- Emergency response as 'the last line of defense' appeared in all bowties. Underpinning this barrier was the emergency response plan.
- The site response team (comprised of a dedicated fire crew and shift team personnel) were responsible for a) applying cooling water from fixed and mobile fire equipment b) activating the water deluge in the field c) conducting remote isolations and/or d) search and rescue. Underpinning this barrier was the pre-incident plan (PIP). This is illustrated in Figure 1.

The workshop team considered these mitigation barriers as 'effective, 'independent' and 'auditable' and assigned the control as 'Valid' as per CCPS guidance [1]. Interestingly the validity rating was premised upon a compliance viewpoint, for example:

- PIP covers this process area? [if 'yes', barrier considered effective]
- Fire monitors tested as per schedule? [if 'yes', barrier considered effective and auditable]
- Site team trained and up to date? [if 'yes', barrier considered effective and auditable]

Figure 1. GSP Bowtie – Depropanizer (illustrative example)

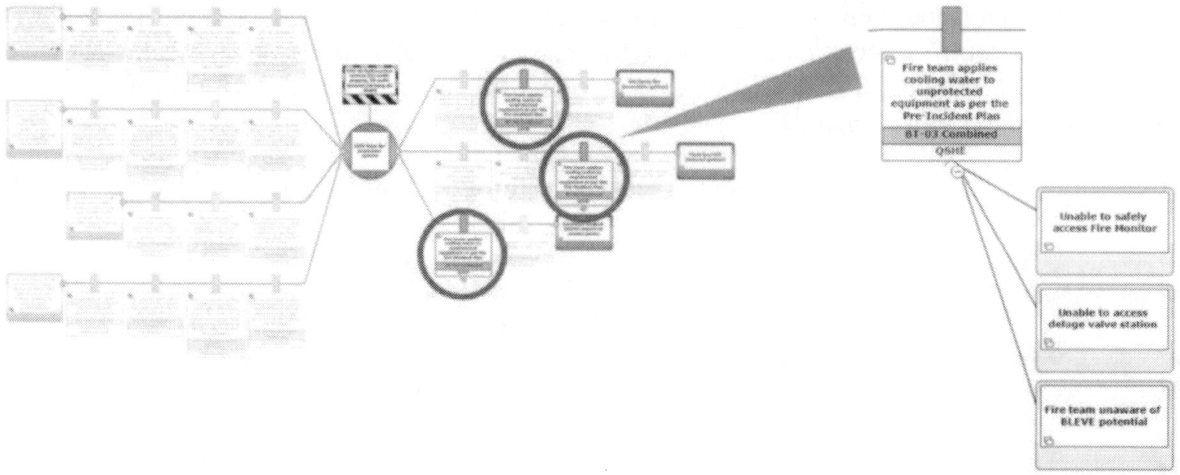

Whilst the classification appeared sound, the PSM team realized the validity testing was too high level resulting in potentially unrevealed barrier weaknesses. The PSM team was mindful of the 'watermelon effect' where despite the barrier validity indicating good health (ie green color as control is in place and functioning), there was sufficient underlying barrier degradation resulting in poor health (red color). Such effects could compromise major hazard control management.

2.2 Application of CCPS expectations to improve Emergency Management expectations

To improve the barrier assessment and validation process, TTM applied the CCPS PSM definition of emergency management to identify expectations to allow for a deeper review in particular PIPs. Examples of expectations applied to pre-incident planning are shown in Table 1.

These expectations will be added to the emergency element within the PSM Framework and referenced in the site Barrier Validation guideline.

Table 1. Expectations for PIP (modelled after CCPS PSM Emergency Management)

No.	CCPS Emergency Management Principle	TTM Expectations – Pre-Incident Plans
1	Maintain a dependable practice	MAE PIPs cover the range of credible LOPC events, recognizing the material hazards present.
2	Prepare for emergencies	MAE PIPs cover the potential range of consequences and impacts to the gas plant as well as the emergency response team including escalated events. MAE PIPs accounts for required response, safe access to isolation systems, fixed firefighting equipment and limitations. All emergency response team personnel are trained in the use and application of MAE PIPs.
3	Periodically Test the Adequacy of Plans and Level of Preparedness	Periodic exercises (desktop and field) are based upon MAE PIPs to test and improve emergency response, with formal feedback to update the PIPs.

Note: Within the TTM Emergency Management PSM element, there are other expectations.

2.3 Shortfall Assessment of Existing PIP and Required Improvements

Based upon Table 1, TTM applied the PIP expectations against those barriers related to emergency management resulting in a re-classification to 'Partially Valid'. This was a result of:

- PIPs were general in nature and did not cover the potential MAE and hazard zones. TTM required an understanding of the hazard zones (ie hot, warm, cold) that could result from ignited and unignited LOPC (ie how big and how bad).

- PIPs assumed the emergency response team could safely access a) firefighting equipment such as fire monitors b) deluge valve stations. TTM required an understanding of which firefighting appliances could be safely accessed under the range of MAEs.

- PIPs did not account for escalated incident potential especially process plant and storages with pressurized fluids such as propane, LPG (BLEVE potential). TTM needed to recognize the potential for an escalated incident and when to withdraw from the incident to minimize endangerment of the emergency response team. The 2022 Cuban Matanzas oil storage facility explosion [3] and 2015 Tianjin port explosion [4] had resulted in multiple fatalities mostly to emergency crews caught unaware in an escalated event.

Lessons learnt from published incident reports were sought by TTM to identify additional improvements for emergency management [3]. Common findings from analysis of emergency plan failures of relevance to TTM included:

- Poor emergency response procedures with unclear roles and responsibilities.
- Lack of inadequate training for emergency responders and emergency exercises.
- Emergency procedures not covering MAEs due to their perceived low probability.
- Late evacuation caused additional threats to personnel and the surrounding population.

Finally, the PIPs had to summarize critical information necessary to manage a potential MAE.

3 The Results so far

3.1 Enhanced MAE Pre-Incident Plan

The enhanced MAE PIPs covering the GSP has been completed with the offshore and onshore pipelines scheduled for early 2024. Key features of the MAE PIPs are described in Table 2 and the PIP format shown in Figure 2.

Table 2. Enhanced TTM MAE PIP Format

Section	Title	Description
A	MAE Description	Provides a summary of the material hazard, physical properties, process units and potential range of consequences.
B	Location Map	Provide the site layout showing the MAE location together with the assembly points, control room and fire station.
C	Incident Event flowchart and Hazard Zones	Simplified incident response flowchart from 'Incident Declared' through 'All Clear' with consideration of escalation potential A summary table of the Hazard Zones allows the on -scene commander to establish the appropriate response zones. The PHAST consequence software package was used to model releases for immediate and escalated events.
D	Timeline and Actions	Summarizes the key activities as a timeline of the Response Team both in the control room and in the field.
E	Response and Incident Map	A site map with the MAE incident superimposed showing locations of fire monitors, fire hydrants and deluge valve stations.

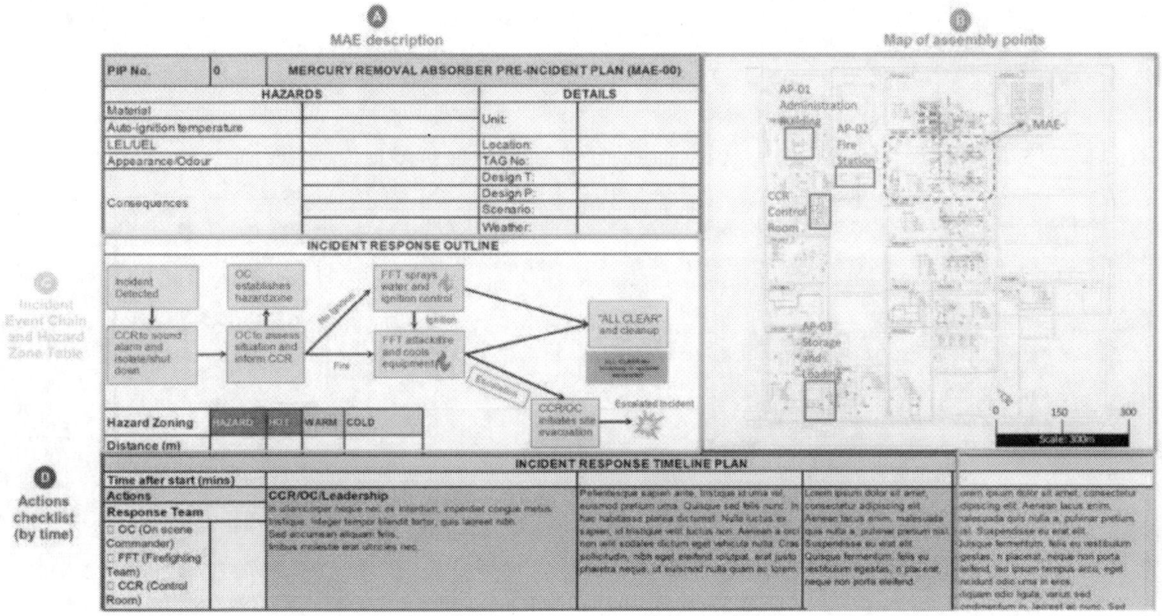

Figure 2. Excerpt of the TTM PIP

3.2 PIP Google Earth Application

An application is currently in development with input from the QSHE group. Together with the bowties and fire consequence analysis, these have been mapped with the site firefighting system using Google Earth.

For directional releases (eg jet fires), the application allows the user to point the incident towards the plant to consider escalation potential or towards fire systems where access to fire protection systems (eg fire monitors) may not be possible.

The application has also been layered with the site fire zones, locations of critical infrastructure (eg control room) as well as the assembly points.

This will allow TTM to have a tool that can be used to:

- Train personnel in hazard knowledge.
- Undertake emergency planning.
- Practice responses against the MAE PIPs and identify improvements.

A screen shot of PIP Google Earth application is shown in Figure 3.

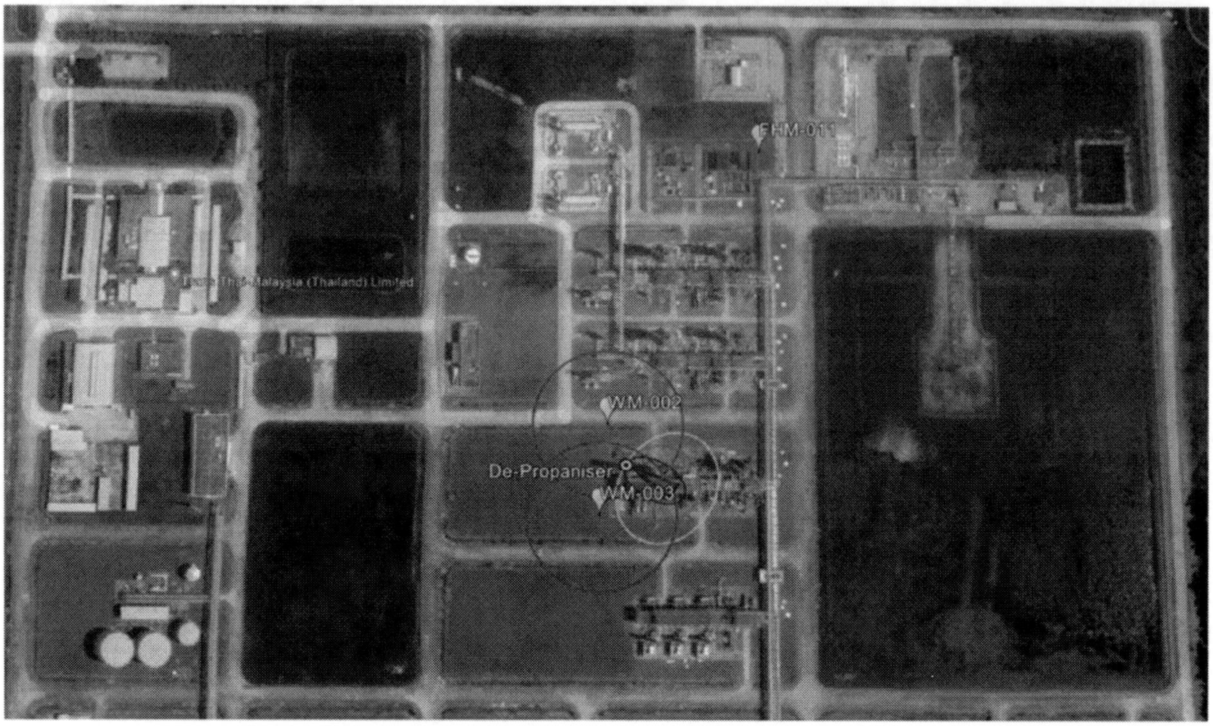

Figure 3. Excerpt of the TTM PIP Google Earth Application

4 The Next Step on the Journey

The next step in the TTM PIP transformation project is the implementation phase that covers:

1. [PSM Competence element / Operational Knowledge element] PIP awareness training: A specific PIP training course with targeted learning outcomes to the emergency response team and site personnel will be developed. The learning outcomes will focus upon a) understanding the different potential consequences of Major Accident Events b) understanding how the MAE consequences were used to develop the PIPs and c) how to use the MAE PIPs.

2. [PSM Competence element] PIP desktop reviews and field trials: A desktop review will be undertaken using the Google Earth application to allow the team to understand better the use of the PIPs. A simulation exercise based upon a selected MAE will be used to stress test the PIP and compare it against performance standards. These reviews and trials will be used to further enhance the PIPs.

3. [PSM Reliability element/ Operational Knowledge element] PIP field trials: Critical equipment (eg fire monitors) required for emergency response will be tested during the field trial against performance standards.

5 Conclusion

TTM has implemented a barrier management program to demonstrate that effective operational barriers are in place to ensure its major hazard risks are acceptable. Bowties have been used to identify and strengthen barriers to prevent and mitigate against a loss of primary containment. Application of barrier analysis found weaknesses in the emergency response control especially the Pre-Incident Plans (PIPs). The intent of the site PIPs is to provide a measured response to control MAEs.

TTM has embarked upon a transformation project to enhance the PIPs. The first step was the strategic response planning to identify improvements against expectations derived from the CCPS PSM Emergency Management principles. Engagement with the site emergency response team and operations personnel involved in emergency response has resulted in PIPs focused upon responding to GSP MAEs. The PIPs have been designed to be simple to use, providing a basis for informed decision making to respond to credible and escalated incidents. The PIPs are supported by a consequence mapping tool using Google Earth to allow the response team to consider emergency responses to the various MAEs.

With completion of the PIPs, TTM will now enter the execution step commencing with targeted training and desktop reviews. The PIPs will be stress tested through field trials and compared against performance markers for personnel and emergency response facilities.

6 References

[1] CCPS, “Bowtie in Risk Management, a Concept Book for Process Safety”, in association with the Energy Institute (2018).

[2] Do you know on a day-to-day basis if you are in control of your operations? Available on https://www.cgerisk.com/solutions/understand-your-risks/3-basic-process-safety-questions/ Accessed on February 16, 2024.

[3] Copernicus Emergency Management Service,” Directorate Space, Security and Migration, European Commission Joint Research Centre (EC JRC), “EMSR622: Industrial accident in Matanzas", Cuba”, https://www.emergency.copernicus.eu/ accessed February 16, 2024.

[4] Z. Gyenes, “Learning from emergency response in the process industries’, IChemE Hazards 28, Symposium Series No. 163, 2018.

Digitizing Emergency Response: Risks and opportunities of emerging technology for incident and emergency responders

Gareth Black
Senior Crisis Consultant, CHEMTREC
gblack@chemtrec.com

Keywords: crisis, emergency, response, incident, digitization.

Abstract: In this session we will explore the traditional challenges of emergency response and crisis management and ask whether technological solutions provide the answers. We will introduce and explore CHEMTREC's Digital Emergency Response ecosystem pilot project, showing how technology can offer significant benefits to responders; enhancing situational awareness, supporting real-time communications, streamlining reporting and improving coordination between organizations and other responders such as L2/3 clean up providers. However we'll also strike a note of caution, and discuss the complexities of crisis and emergency situations, the complexities of human decision making and why AI and other digital solutions are not a panacea to our challenges and could in fact create barriers to effective decision making under pressure.

Introduction

The world we live in is increasingly volatile, organizations are now 35% more likely to suffer from a natural disaster than they were 30 years ago. We have seen the impacts of this over the last decade, with a global pandemic, increased severe weather events and major global geo-political instability. Between 2017 and 2022 the US suffered an average of 18 climate related disasters each causing in excess of $1billion in damages (Gusnor, 2023). While these challenges impact society in general, they also pose specific risks to the chemical sector. They pose a direct threat, with an estimated 746 chemical facilities at risk of being affected by a category 4 or 5 hurricane (Samora, 2023). They also pose an indirect threat, with the wider impacts of COVID-19 on the international supply chain well documented.

Outside the global and environmental dynamic, the chemical industry remains high risk. Despite innovations and a constant strive for safety, over 270 hazardous chemical incidents occurred in the US in 2023 alone (Nelms, 2023), with notable high profile incidents including the East Palestine Rail Derailment. While it is right and proper that the industry continues to focus on safety and reducing the likelihood of error, it must also look at ways it can enhance it's emergency response provision. Effective emergency and crisis management are crucial to reducing the scale, duration and impact of any event. This paper examines how the adoption of digital solutions can enhance emergency management provision across the sector.

Section 1: The Challenges of Effective Crisis and Emergency Management:

Crisis and emergency responses are by their nature complex. We are faced with high risk situations, often with little time to respond and very little information upon which to make decisions. Indeed communication and the availability of information is the single biggest challenge in incident, emergency and crisis management the world over. The challenge is illustrated in the figure 1 (MacFarlane & Leigh, 2014)

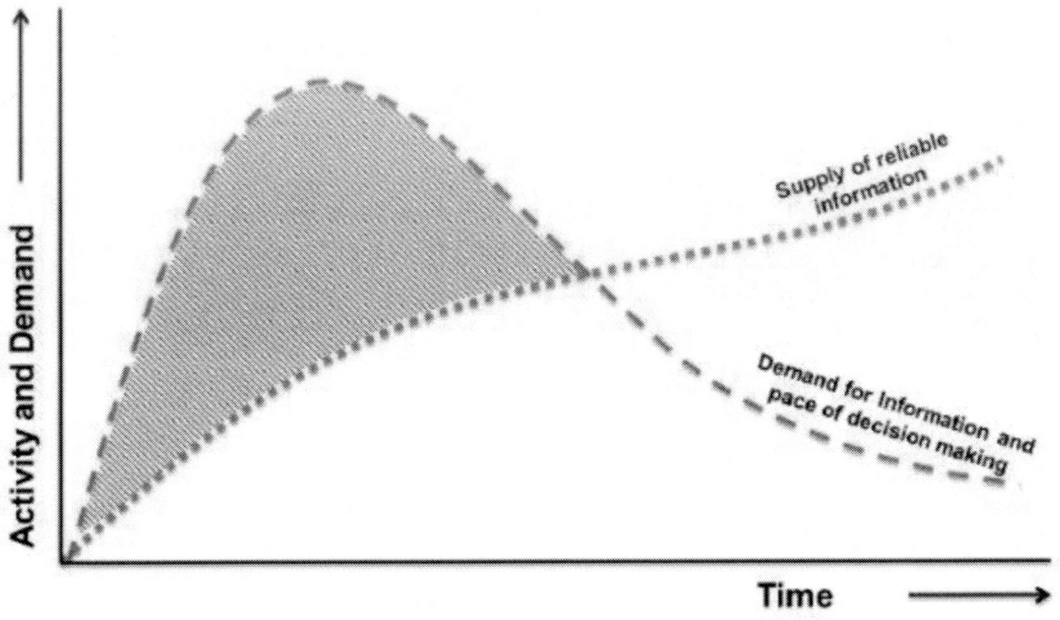

Figure 1: Evidence gap in crisis management (MacFarlane, R and M. Leigh, 2014).

The demand for information required to make critical decisions rarely, if ever keeps pace with the supply of reliable information. Crisis and emergency plans and structures are designed to address this challenge as far as possible. The use of command structures, a more directive leadership styles, and limited spans of control are key processes in ensuring information flows quickly from the scene of the incident to those in other parts of the response. Figure 2 below shows the importance of this; communication with 3 people, results in 3 lines of communication, communication with 14 people results in 91 lines of communication.

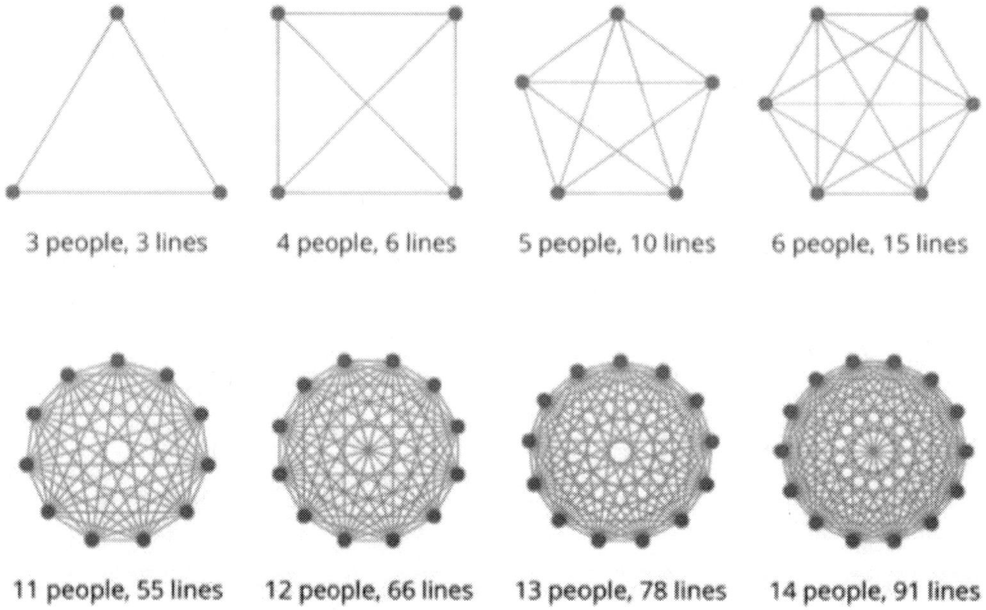

Figure 2: Lines of communication

This is the purpose of the command structures we establish as part of any crisis plan, they aim to reduce lines of communication and enhance communication flows. In the Incident Command System, the system widely adopted throughout the world, 1 person should have a span of control between 3 and 7. If they are in charge of less than 3 people, there's no point having a structure, more than 7 and they need to build a hierarchy to improve communications flows. Crisis and emergency plans also aim to increase the flow of communication through standardized reporting templates, ensuring the correct information is shared with other responders. Despite these systems and structures, the information gap remains a challenge.

Obtaining information about what is happening is also only one part of the challenge. Good crisis and emergency management requires a proactive approach. It is therefore vital that responders do not just consider what has happened and what is happening, but also what may happen next. Building up a picture of these three elements forms part of our situational awareness. Dr Mica Endsley's (1995) model of situational awareness provides an overview of this process. We first take in data from the world around us, we perceive it. We then being to make sense of it, what does this data mean in context. We turn data into meaningful information using our knowledge and experience. Finally, we make a projection. We predict what may happen next as a result of the information we have.

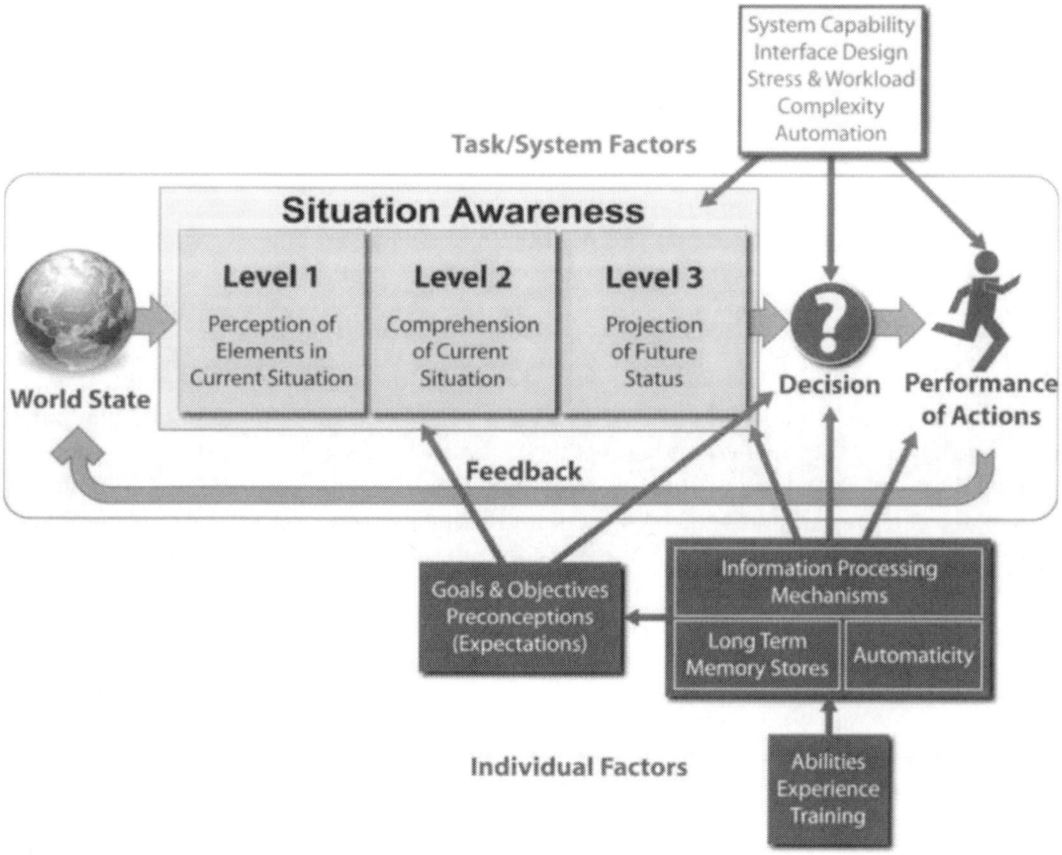

Figure 3: Model of Situation Awareness in Dynamic Decision Making (Endsley, 1995)

Our ability to work through this process is reliant on a number of factors. The availability of data and information is the first challenge. While we outlined some of this above, there are also contextual elements which we need to feed in. At a local level this may include elements like the weather forecast and receptors, at a strategic level this may include media interest and the political and financial context. Once we have the information required, or indeed have the information we can gather in the time available, we make a decision. Again the availability of information poses some critical challenges here. Do we act based on incomplete information and risk making a wrong decision, or the situation worse, or do we wait for more information before making a decision and risk the situation escalating.

Lets put some of this in context through an example case study which highlights the challenges we see when working with a typical chemical company. Company A is a multi-national chemical company. It has sites in several countries, each site has an Emergency Action Plan (EAP) and at a corporate level the organization has a crisis management plan, it ships material by road, rail and ship.

An incident occurs 1 mile from a site in South America, a road traffic collision involving one of the company's trucks resulting in a spillage. The incident is reported to the site and they activate their local response plan. The site now has to coordinate with a wide range of stakeholders. This requires several phone calls and emails to each to both share and gather data, so that they can build their situational awareness and take effective decisions. Stakeholders include:

- The driver of the vehicle
- The site response team
- The wider site
- Local responders
- The company's corporate crisis management team and media team
- The media
- Their contracted L2/3 clean up provider

Not only do the site have to liaise with all of these individuals, they also have to funnel information received from each into a cohesive overview of the current situation. The manual process is time consuming, fragmented and slows the organization's ability to close the information gap. Even the activation of the company's corporate crisis team is slow. The site makes a single call to an on-call manager who then emails the corporate team, or make a series of manual phone calls to activate them. As we discussed at the start of this paper, the role of effective emergency and crisis management is to minimize the scale, duration and impact of any event. As we have explored so far in our challenges, any slow down in communication, slows down the building of situational awareness and so slows down decision making at all levels of the response.

Section II: Digitization of Emergency Response

The first and most obvious thing to say about technology and crisis and emergency management is that it's only useful if it's available. COVID saw many companies accelerate their digital transformation program and crisis and emergency management was no different. COVID was the driving force behind a revolution in business continuity and crisis and emergency management across many organizations, it was a living case study of the need for such arrangements. With the restrictions of home working, it also presented the perfect opportunity for organizations to digitize their response. Utilizing platforms which were emerging or increasing in adoption across their business many organizations built their crisis management plans into platforms like Microsoft teams, leveraging its coordination and tasking capabilities. These are excellent capabilities where available and address many of the challenges outlined earlier from an information collection and dissemination perspective, however many organizations have become overly reliant on these functionalities. The very nature of crisis and emergency management dictates that organizations are dealing with disruption, and with increasing frequency, that disruption is cyber related. It is foolish for organizations to build their crisis and emergency management programs upon their primary systems which are likely to be affected and potentially unavailable during a cyber event. This is where external crisis and emergency management SAAS software solutions have huge benefits.

There are a wide range of SAAS platforms now available addressing the crisis and emergency management market. All share a range of similar features which serve to address the challenges described above. Let's revisit Company A and the incident it has suffered.

The company now has an incident management software platform for use throughout the organization. Upon crashing, the driver is instructed to call and report the incident to the organizations level 1 emergency response provider. This provides a central, single point of contact for reporting all incidents and events. The call taker takes all relevant information from the driver and then the call is triaged and recognized as meeting the threshold for a "level 2 activation" in the organizations crisis plan. This means activation of the nearest site, the organizations crisis management team, and the nearest level 2/3 clean up provider. Through the SAAS platform all relevant team members are activated instantly via sms, phone, email and push notifications and taken to a relevant page in the platform where they can receive instant updates on the latest situation and collaborate with other key stakeholders in the response. The team can meet through meeting functionality in the platform, and communicate key decisions made to other teams through the platform. Emergency warnings can also be sent to the wider organization, with mass communications used to keep employees up to date on the incident and any wider implications. Mass communication tools like this are important for emergency alerting, the average person receives 121 emails a day and 32% of employees never use the company intranet – breaking through with direct messaging is critical in emergency situations (AlertMedia, 2023).

Mass communications systems include features which facilitate swift and fast communication to multiple users (1000s) instantly via multiple means and channels (i.e. text, email, phone, app), includes the option for 2 way communication, offer secure communication using appropriate encryption and security and include messages which can be pre-populated to again increase the speed of any alert. Wider incident coordination platforms build on these features and also offer a library in which to host crisis and emergency management plans, the ability to directly task individuals with specific tasks from action cards, these are automatically sent to role leads upon activation of the crisis management team and updates are sent as tasks are completed. New tasks can also be created and allocated as the incident progresses. While providing a constant stream of information, platforms also allow for the creation of more formal situation reports which may be shared with those not involved in the platform or directly involved in the response such as shareholders. The main situational awareness element of any platform is a dashboard providing live task tracking, incident updates and audit trail of decisions made.

The benefits of having a digital ecosystem which underpins any emergency response are clear. Rapid activation of teams, rapid communication throughout the response structure and seamless coordination with other stakeholders. However it is important to note that the digital ecosystem is not a panacea. The systems only work when they are built upon effective crisis and emergency management structures which are underpinned by a program of training and exercising. If the driver fails to follow the correct process, the system fails. If the call taker fails to properly triage the call, the system fails. As with any automation in any industry, the system offers huge benefits, but is still at the mercy of human error.

Section III: The role of AI in Emergency Management: Pros and Cons

If this is the potential of software solutions, then what of the new frontier in computing, the use of AI in emergency response. In 2022 over 50% of companies were leveraging AI in some capacity. Cheng et al., (2020) argue that the capacity of AI to process large amount of data in real time, offers huge potential for emergency response teams, providing them with timely and accurate information. Ahmed et al., (2021) argues that the ability of AI to analyse data from a wide range of sources can offer realtime insights and facilitate automated decision making. Indeed research projects in the Netherlands have shown that AI has the potential to increase performance in emergency dispatch decisions. The system increased performance in emergency dispatch of ambulances by 0.77 percent the equivalent of adding more than 7 weekly ambulance shifts. It did this by following dispatchers decision trees and examining how dispatchers had historically used this (Leete, 2023). This is where AI has the potential to excel, it offers excellent capabilities for basic, repeatable decision making processes and where there is a depth of evidence upon which to base those decisions. AI also has the ability to provide additional context and information to responders, GIS as an example can provide real time information on the impacts of any incident by analyzing infrastructure and key demographic data into one map. This is something we have been attempting to do for some time, utilise the information gathered during a transportation risk assessment, to support the response to an incident. GIS data has the potential to do this assessment in realtime.

AI also offers excellent predictive capabilities, and can be used to effectively forecast the potential for incidents, again based on historical data, current weather conditions and other critical information. It is being used in California to do so effectively by the California Department of Forests where image recognition technology is used to identify wildfires before humans are able to. Similar systems could be deployed in process industries providing early warning systems for incidents or potential incidents. AI tools such as LSTM (Long Short-Term Memory) and BiLSTM (Bilateral Long Short Memory) are able to leverage data sources like satellite imagery and maps to support hazard monitoring efforts. They are able to offer real-time monitoring of risks and support incident forecasting (Munno et al., 2024). AI systems like GRUs (Gated Recurrent Units), offer further support from a predictive standpoint, using historical data such as river levels and rainfalls amounts to identify emergency situations.

Whilst AI's ability to analyze and provide a wide range of data is a strength, it also presents challenges for emergency responders. AI offers the potential to significantly reduce the information gap discussed at the start of this study by offering a wide range of facts about the incidents, or predictions based on analysis to fill the information gap. However it does not possess the capabilities to analyze this information in the wider context of the incident. Nor is it likely to have experienced many of the contextual issues which make up a crisis situation. Crisis situations are crisis situations because they are outliers. They are events which have rarely happened before and which were unexpected. The nature of AI means it has little experience base upon which to base its assumptions. While it is able to analyze vast, it is unable to place this in context in these outlier scenarios. The other danger associated with AI and emergency response, is the capacity of responders to deal with the information provided. The answer to the information gap described at the start of this paper, is not an infinite amount of data. There is only so much information an emergency responder can process, and this processing capacity is limited further by the stresses of an incident response.

That is not to say AI does not offer any benefits, but it is not the solution to all problems. When combined with human judgement and used a decision support tool it has huge benefits. Where it can be used to analyse data and provide meaningful information and insights, taking away much of the processing associated with the comprehension and projection stage of the human situational awareness process, it can enhance situational awareness and decision-making. The key with AI is for it to provide decision prompts, and increase a responders capacity to respond, rather than overload them with data. This is a careful balance to strike.

1 References.

Ahmed, S., Hasan, M. S., & Vasilakos, A. V. (2021). Artificial intelligence for disaster management: A review. IEEE Transactions on Engineering Management, 68(1), 7-22.

AlertMedia (2023) *Employee communications software: Key benefits and features*, *AlertMedia*. Available at: https://www.alertmedia.com/employee-communications-software/#:~:text=Mass%20communication%20software%20will%20enable,being%20received%20loud%20and%20clear. (Accessed: 23 February 2024).

Cheng, H., Zhang, Y., & Wu, Z. (2020). Artificial intelligence in disaster management: Progress, challenges and prospects. Engineering, 6(8), 930-941.32–64. doi:10.1518/001872095779049543.

Gusner, P. (2023) *Natural disaster facts and statistics 2024*, *Forbes*. Available at: https://www.forbes.com/advisor/homeowners-insurance/natural-disaster-statistics/ (Accessed: 23 February 2024).

Leete, A. (2023) *Ai in public safety: The future for emergency services?*, *EENA*. Available at: https://eena.org/blog/ai-in-public-safety-the-future-for-emergency-services/ (Accessed: 23 February 2024).

MacFarlane, R. and Leigh, M. (2014) *Information management and shared situational awareness*, *Emergency Planning Society*. Available at: https://www.epcresilience.com/application/files/3816/7146/9178/Information_Management_and_Shared_Situational_Awareness.pdf (Accessed: 23 February 2024).

Munno, C., Proto, I. and Trancu, P. (2024) *Ai and disaster management: Potential and applications - about resilience*, *About Resilience -*. Available at: https://www.aboutresilience.com/ai-and-disaster-management-potential-and-applications/ (Accessed: 23 February 2024).

Nelms, D. (2023) *Over 270 hazardous chemical incidents have occurred in the United States this year alone, highlighting need for stronger chemical safety regulations*, *PreventionWeb*. Available at: https://www.preventionweb.net/news/over-270-hazardous-chemical-incidents-have-occurred-united-states-year-alone-highlighting-need (Accessed: 23 February 2024).

Samora, S. (2023) *Why chemical companies should prepare for hurricane season*, *Manufacturing Dive*. Available at: https://www.manufacturingdive.com/news/csb-calls-on-chemical-cos-noaa-prepare-above-normal-hurricane-season/692881/ (Accessed: 23 February 2024).

Fire Safety Design Philosophy (FSDP): Guideline for Oil, Gas and Petrochemical Industries

M Rosli A Seman
P.Eng. (BEM, MIEM), Malaysia
Menara 3 PETRONAS, Kuala Lumpur, Malaysia
rosli.abuseman@petronas.com.my

N Izzah Yaakub (Presenter)

Shaiful Khalid[1], Rizal Harris Wong[1], M Shaik Fareed M Naja Buddin[1], N Farhanah Nadhirah M Adam[1]

Keywords: Fire Safety Design Philosophy (FSDP), Roles, Work Process, Content, Checklist.

Abstract

The employers or occupiers of high-risk industrial facilities are required to prepare and submit a Fire Safety Design Philosophy (FSDP) to Jabatan Bomba & Penyelamat Malaysia (JBPM) also known as Fire and Rescue Department of Malaysia. This document must be reviewed and endorsed based on fire and explosion risk assessment prior to the submission of architectural and M&E drawings.

JBPM has identified discrepancies in the contents, fire protection requirements and overall quality of the reports submitted by high-risk industrial facilities including those in the Oil, Gas and Petrochemical Facilities. These reports often lack of essential information or supporting documentation, necessitating numerous clarifications and follow-up before they were finally accepted as complete and satisfactory by JBPM.

With this new guideline, the owner and stakeholders of relevant installations and parties responsible for the preparation and review of the report will be aware of the requirements and will be able to produce a comprehensive report meeting the requirement of the Regulations and expectation of JBPM. This will also ensure consistency on the report contents and fire protection requirement, applying relevant Code and Standard across the industries.

1 Introduction

This Guideline produced in collaboration with JBPM and PETRONAS. This initiative was realized due to unavailability of Standard Guideline for industries and authority for reference.

This paper presents the key points of "Guideline for the Preparation And Updating The Fire Safety Design Philosophy (FSDP) For Oil, Gas And Petrochemical Facilities in Malaysia." The essential points include the work process for document submission to authority and the requirements of preparation FSDP report to meet the expectation of JBPM. The guideline also describes the roles and responsibilities of key personnel involved in the planning and preparation of the FSDP report to ensure compliance with the requirements set out in the report.

2 Objectives of Guideline

The objectives of the Guidelines are to address the pain points faced by the industries. This Guideline is developed as guidance to ensure:-

a. Clear work process of FSDP development for report submission;
b. Consistency of contents and quality of FSDP to avoid resubmission; and
c. Adequate fire safety system design and installation based on fire risk analysis.

3 Scope of Guideline

This Guideline covers the Oil, Gas and Petrochemical facilities as listed below:-

a. Refinery industries;
b. Chemical and Petrochemical industries;
c. Chemical Warehouse and Storage facilities; and
d. Building with unique and special design or industries with special hazards, which will be determined by Director General of JBPM.

This scope of this Guideline is in line with Clause 236 in UBBL 1984 for special hazards which require FSDP submission.

This Guideline should also refer to the Fire Safety Systems Guidelines for Oil, Gas and Petrochemical Facilities in Malaysia.

4 Roles and Responsibilities in Preparation FSDP Report

Several key personnel are involved in the planning and preparation of the FSDP report, and their roles are crucial to ensure compliance to JBPM requirement and subsequently to obtain approval of the report. These key personnel are Facility Owner/ Project Manager, Fire Safety Consultant (FSC), Principal Submitting Person (PSP) and Submitting Person (SP). Figure 1 summarizes the roles and responsibilities of the key personnel.

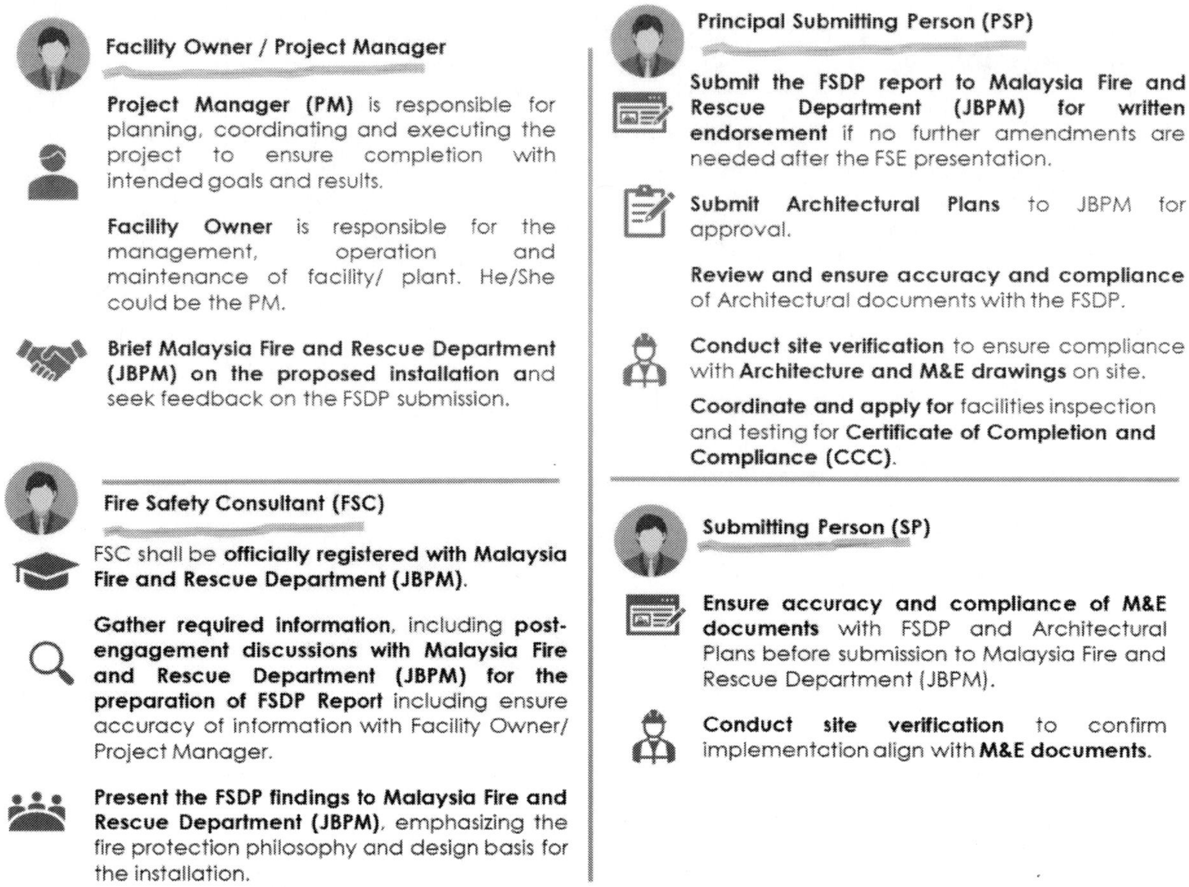

Figure 1. Roles and Responsibilities for FSDP Report

4.0 Work Process for Submission of FSDP

The work process for submission of FSDP is carefully synchronized with the established approach mandated by JBPM. The sequential steps involved in the meticulous planning and preparation of FSDP is visually represented in Figure 2, providing a clear and comprehensive overview of the entire process.

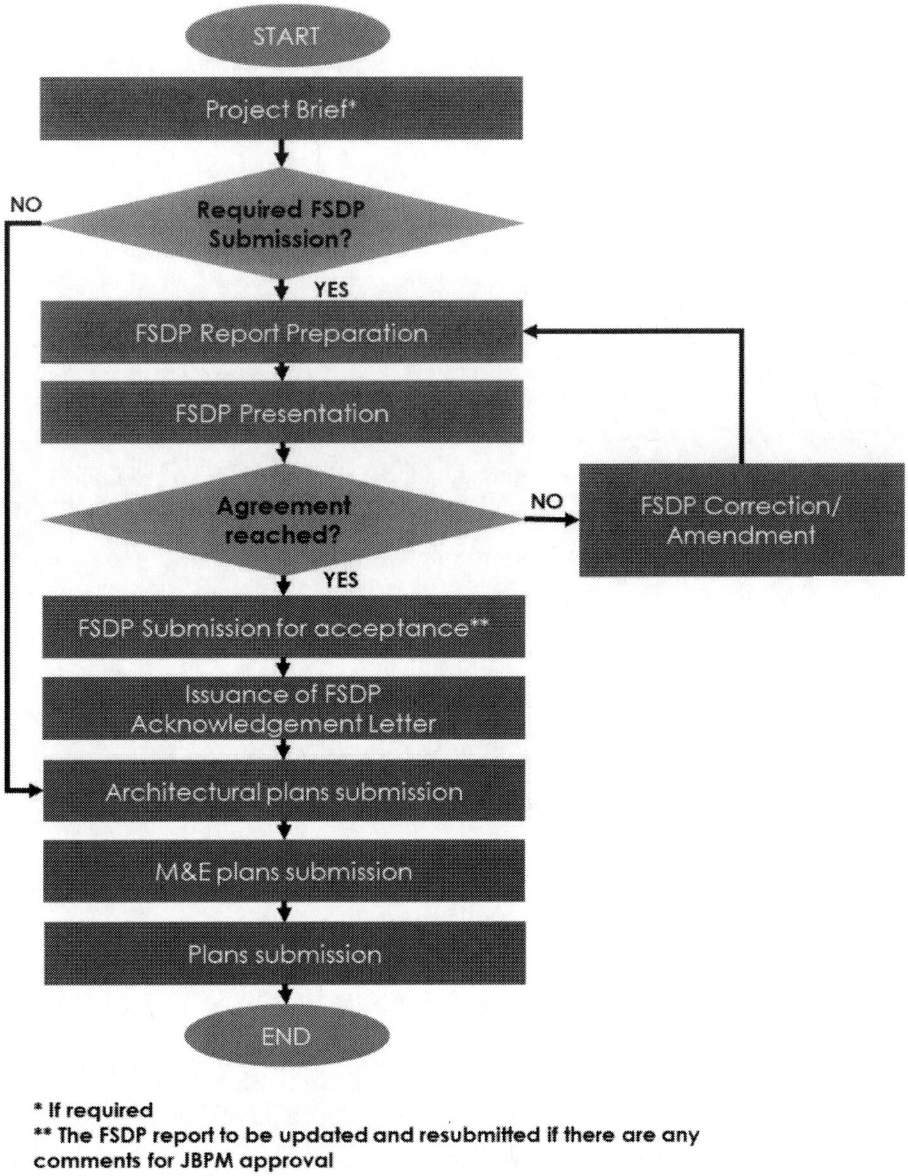

Figure 2. Workflow for FSDP approval by JBPM

6.0 Content of FSDP Report

The FSDP Report is divided into seven (7) parts and structured as follows:

i. Front Page
ii. Document Authorization and Document Revision History
iii. Table of Content
iv. List of Acronym and Definition
v. Section 1.0: Introduction of FSDP
vi. Section 2.0: Standards and Codes

vii. Section 3.0: Facility Descriptions
viii. Section 4.0: Risk Assessment
ix. Section 5.0:Prevention Management
x. Section 6.0: Fire Protection Management
xi. Section 7.0: People Safety Management

Table 1 describes the minimum information required to be included in each of the main sections. The comprehensive information for each section is elaborated further in the Guideline.

Table 1. Structures of for FSDP Report

Section No.	Section Title	Description
1.0	Introduction of FSDP	This section provides the purpose of the FSDP development, objectives and stockholders involved in the preparation of the report. • Purpose of Submission • Objective • Scope
2.0	Standards and Codes	This section shall provide the list of applicable Standards and Codes for the Project or Facility as follows: • Applicable Malaysian Laws and Regulations • Other applicable Codes and Standards i.e. Malaysia Standards (MS), British Standards (BS, AS), American Standards (NFPA, API, ASME) Note: Standards other than those conforming to the standards listed above shall be approved by Director General of JBPM.
3.0	Facility Descriptions	This section shall provide description of the facilities as follows: • Facilities site location, layout and Surrounding Area • Plant capacity • Facility descriptions e.g., processing units, structures, buildings, and material/ finished products. • Process block diagram/ flow diagram • Stakeholders
4.0	Risk Assessment	This section shall provide information related to the facility hazards and list of safety studies to be undertaken to identify and mitigate the hazards. • Hazard Identification • Summary of Hazards

Section No.	Section Title	Description
		• Risk and Evaluation Assessment
5.0	Prevention Management	An installation should be designed to avoid or minimize the risk of hazardous events such as fire, explosion or toxic release. Inherent safer concepts should be applied at the design stage and shall be explained in the FSDP. • Inherently Safer Design Concept • LOPC Containment • Overpressure/ Vacuum Protection and Disposal System • Ignition Control
6.0	Fire Protection Management	Fire protection is a concept and/or engineering design put in place to mitigate the consequence of a fire scenario from escalation and to ensure that the fire can extinguish quickly and with minimal consequential damage. This section describes the Fire Protection Philosophy for the Project or Facility. Fire protection equipment in the Project or Facility shall be in accordance with Malaysia Standards and certified by certification/ testing body e.g., SIRIM and recognized by JBPM. • Active Fire Protection (AFP) • Passive Fire Protection (PFP) • Fire and Gas Detection and Communication System
7.0	People Safety Management	This section describes the arrangements made to ensure a safe, timely and effective response during emergency events. This shall include arrangement, communication and coordination within the facility and external parties such as the local authority. • Escape Route and Muster Point • Emergency Response

5 Checklist

The Guideline includes a comprehensive checklist to assist stakeholders in systematically reviewing and ensuring compliance with all necessary requirements before submitting the document to the relevant authority. This checklist serves as a guidance to ensure all the essential information in the FSDP are meeting the expectation and requirement set out by JBPM.

6 Conclusion

In conclusion, the Guideline aims to establish standardization by providing structured frameworks that ensures consistency and conformity in planning and preparation practices of FSDP report for Oil, Gas and Petrochemical Facilities in Malaysia. With this new guideline, the owner and stakeholders of relevant installations and parties responsible for the preparation and review of the report will be aware of the requirements and will be able to produce a comprehensive report meeting the requirement of the Regulations and expectation of JBPM.

7 References

[1] Malaysia Fire and Rescue Department (JBPM), Prosedur Perakaun Pelan Teknikal and Sokongan Pengeluaran Siap & Pematuhan (OSC & CCC).

[2] Institute of Fire Engineers (UK) Malaysia Branch and Malaysia Fire and Rescue Department (JBPM), Guide to Fire Protection in Malaysia.

[3] NFPA 551, Guide for the Evaluation of Fire Risk Assessment.

[4] BS 9999:2008 - Code of practice for fire safety in the design, management and use of buildings

[5] API PUBLICATION 2510A - Fire-Protection Considerations for the Design and Operation on Liquefied Petroleum Gas (LPG) Storage Facilities

[6] API RECOMMENDED PRACTICE - Fire Protection in Refineries

[7] LASTFIRE 2018 – Risk Reduction and Option

Using Human Factors to Improve Cybersecurity

Patrick O'Brien
exida
80 N Main St
Sellersville, PA 18960
pobrien@exida.com

Keywords: Cybersecurity, Human Factors

Abstract

Managing human factors has long been recognized as a critical step for reducing the rates of incidents with the potential for negative impacts on personnel safety, the environment, equipment damage, and unplanned downtime. As cybersecurity incidents for industrial control systems become a growing concern, it is critical to apply the same lens of human factors to improving cybersecurity hygiene. A 2018 study by the Ponemon Institute, including over 150 companies from 13 industries, found that roughly 80% of all insider cybersecurity incidents were accidental. With the advent of sophisticated control system-specific malware variants, this means that accidentally plugging in an infected USB could not only lead to equipment downtime but could lead to a potential process safety incident as well. This paper will evaluate how existing practices and techniques for human performance improvement can be effectively leveraged to improve cybersecurity readiness.

1 Introduction

Often, cybersecurity for both Information Technology (IT) and Operational Technology (OT) systems focuses on the technical aspects of cybersecurity such as the structure of malware or specific controls and security settings for each device type. Although the technical aspects of cybersecurity are important, at its core, cybersecurity is a "people problem." Not only are people actively attempting to "hack" or compromise weakened computer systems, but people are also on the other side of those systems working hard to secure them and keep them up to date as the current list of vulnerabilities and cybersecurity threats continuously evolves.

In many cases, despite best efforts, it is also the people using OT systems who inadvertently invite cybersecurity incidents into their environment. A 2018 study by the Ponemon Institute, including over 150 companies from 13 industries, found that roughly 80% of all insider cybersecurity incidents were accidental [1]. This means that the likelihood of an employee or contractor unintentionally triggering a cybersecurity incident was roughly 4 times greater than the likelihood of an insider intentionally manipulating the system. If companies could reduce the rate at which mistakes lead to cybersecurity incidents, this could lead to significant improvements in operational up time as well as safety for process plants.

Human factors can be defined as the "discipline concerned with designing machines, operations, and work environments so that they match human capabilities, limitations, and needs [2]." Although the challenges of human performance related to cybersecurity concerns are relatively new, the discipline of human factors has been in place for several decades and has undergone a number a series of improvements specifically for use in process plants. Leveraging existing models for understanding and improving human factors in process plants could offer a significant benefit to OT organizations looking to improve their cybersecurity resiliency.

This paper will look to answer three key questions related to human factors and cybersecurity:

1. What is "human factors?"
2. Why is human factors relevant to cybersecurity?
3. How can human factors improve cybersecurity?

2 What is "Human Factors?"

Human factors is the study of human performance in operator-machine systems considering a range of topics such as training, work planning, and fatigue. Human factors provides insights into factors that influence human performance [3]. These factors that can influence human performance to either strengthen it or weaken it are sometimes called Performance Influencing Factors (PIFs); several common PIFs are identified in Figure 1.

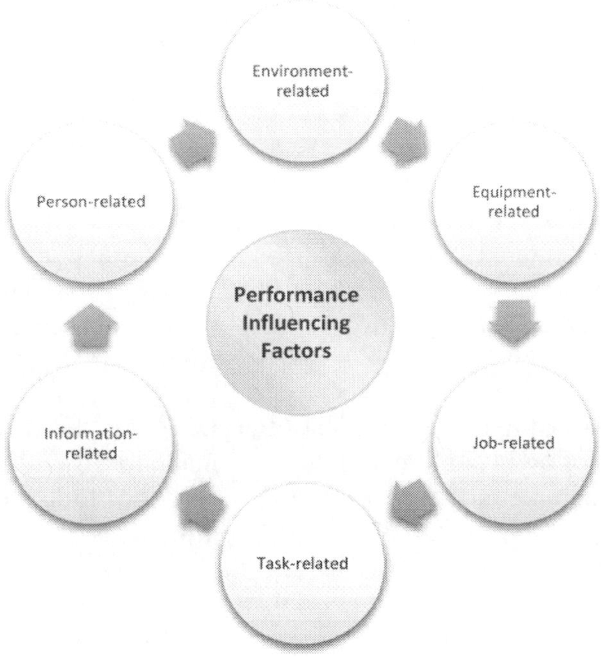

Figure 1 Performance Influencing Factors Adapted from Human Factors Handbook for Process Plant Operations [3]

Table 1 lists each performance influencing factor with both a typical process operations/safety example and an OT cybersecurity example. In some cases, such as an environment-related PIF, the same factors may impact both operations and security; in others, the specific causes may differ but have similar impacts on performance.

Table 1 PIFs with Process and Cybersecurity Examples

PIF	Process Safety Example	Cybersecurity Example
Environment-related	High noise levels and ambient temperature leads to difficulty focusing on complex tasks.	High noise levels and ambient temperature makes it difficult to download and install the correct configuration for a smart field device.
Equipment-related	Ineffective Human Machine Interface (HMI) layout requires additional steps to check all process areas leading to less frequent monitoring of certain units.	Configuration of security settings makes it difficult for operators and maintenance personnel to complete daily tasks, resulting in a greater potential for security settings to be bypassed.

PIF	Process Safety Example	Cybersecurity Example
Job-related	Excess workload leads to operators being more likely to skip steps in a procedure to save time.	Excess workload for controls engineer leads to greater likelihood that a change to a computer-system is not evaluated for cyber impact.
Task-related	Alarm flood overwhelms operator with multiple input sources, increasing the complexity of responding the upset condition.	Large numbers of system logs, system faults, and cybersecurity-related messages make it difficult for security personnel to discern threat activity from normal behavior.
Information-related	Unclear operating procedures leads to difficulty following and recording each step completed.	Unclear cybersecurity procedures make it difficult to know if current activities follow the suggested guidelines.
Person-related	Lack of on-the-job training makes it difficult to identify abnormal conditions.	Lack of cybersecurity awareness training makes it difficult to recognize potential signs of cybersecurity incidents.

Understanding what factors lead to lower levels of human performance allows organizations to better train and support operators and other personnel to reduce the likelihood of human performance issues and resulting incidents. Ultimately, it is the practice of consistently identifying the demands of specific tasks and providing the required support that leads to high standards of human performance [3].

Different tasks require different levels of attention and knowledge to be performed successfully. Understanding how the demands of a task change the way in which people perform it is needed to identify the appropriate support measures. For example, the knowledge required to quickly diagnose and respond to an abnormal site condition may be very different from the skills required to quickly execute a series of documented steps in the control system. One common way of differentiating tasks is the Skill, Rule and Knowledge-based model (SRK), which presents familiarity with the task on the y-axis and required attention level on the x-axis [4] [5]. Using this method, tasks can be identified as three different types of performance:

- Skilled: high familiarity, low attention level required
- Rule-based: moderate familiarity, moderate attention level required
- Knowledge-based: low familiarity, high attention level required

The most effective strategies for supporting human performance depend on the type of task as highlighted in Figure 2.

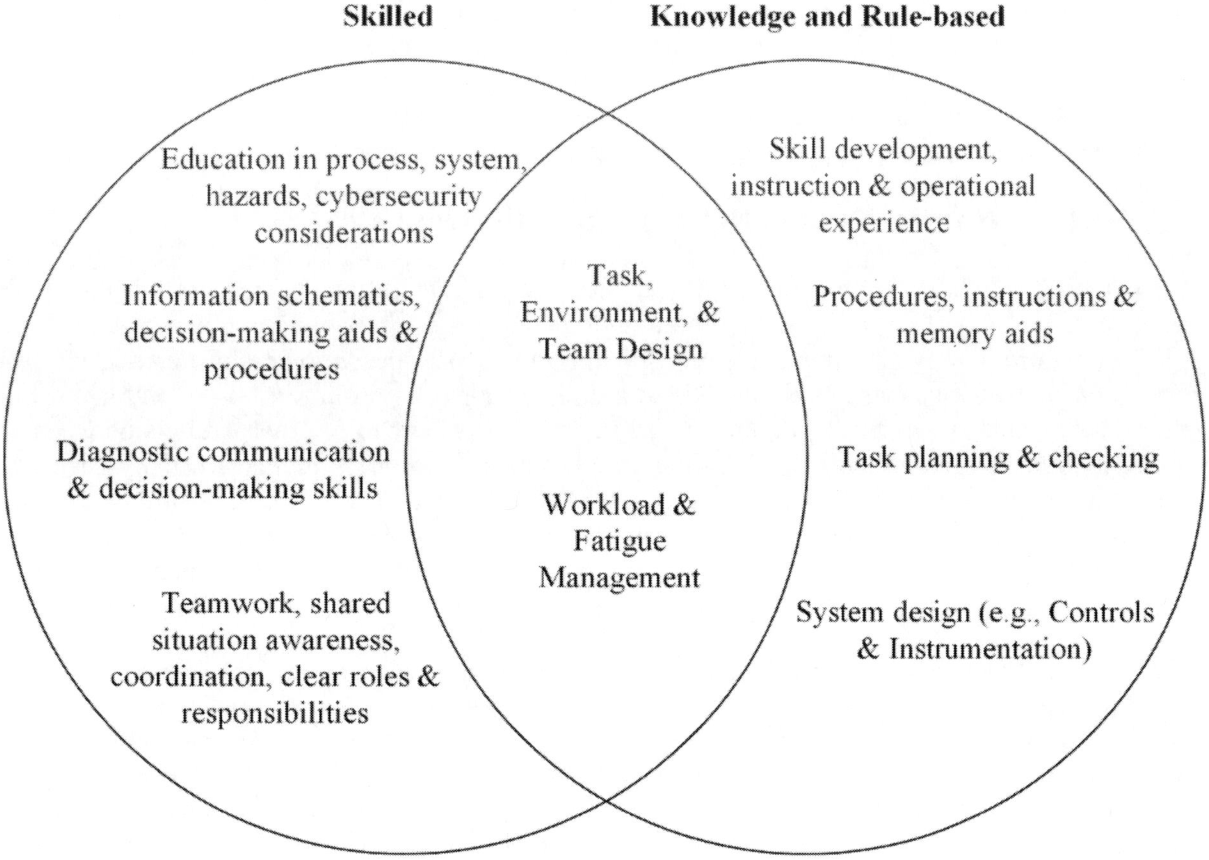

Figure 2 Common Strategies for Supporting Human Performance by Task Type Adapted from Human Factors Handbook for Process Plant Operations [3]

Providing an environment clear from distractions and managing overall workload and fatigue are pre-requisites for supporting human performance for all tasks. For tasks requiring skill-based performance, a deeper understanding of the potential hazards with the system can lead to increased attention for related tasks. For example, in a process with highly exothermic reactions, the difference between a setpoint of 10°C and 100°C could mean the difference between safe operation and a runaway reaction with the potential for an overpressure and loss of containment. The understanding of the critical nature of the setpoint may significantly change the way operators approach the process. Building on this foundation with informative decision-making aids (e.g., decision trees, process flow charts), diagnostic skills (e.g., practice exercises to identify abnormal conditions), and shared awareness of team members can further improve the team's performance and help to better identify potential concerns before they arise.

For knowledge and rule-based performance, operational experience is a key determiner of success since these types of tasks require people to respond to new situations that they have not seen before or see infrequently. On-the-job training, combined with focused skill development training, can help personnel become more comfortable with responding to these non-routine situations. Task planning is another important factor for success because it helps to prepare possible responses to different situations that may occur during the activity. One example of task planning in control systems is the use of backout plans (a contingency plan for how to restore the previous version of the system) when making a configuration or software update. These plans help team members to quickly respond in case the system demonstrates unexpected behavior after the update or the update is unsuccessful. In addition to training and task planning, developing support procedures and ensuring that the system design is fit for purpose can improve proactive planning and help with effective decision making in the event of unexpected conditions.

3 Why are Human Factors Relevant to Cybersecurity?

OT cybersecurity is the "prevention of illegal or unwanted penetration of, or interference with, the intended operation of an industrial automation and control system using computer-based systems [6]" - or more simply put, the prevention of *intentional* or *unintentional* interference with the operation of computer-based systems used in an operational environment. If "operator-machine interface" is substituted for "computer-based system," it is easy to see the parallels between the study of human factors and implementing cybersecurity in process plants. In many cases, the unintentional interference with OT systems can be directly tied back to PIFs. Additionally, many intentional cybersecurity incidents, such as the recent attack on a Florida water treatment facility, originate from a vulnerability or weakness in the system resulting from a human performance issue.

3.1 Florida Water System and Human Factors in Cybersecurity

Background: In February 2021, hackers remotely connected to the control system of the Oldsmar Water Treatment Facility in Pinellas County Florida [7]. After connecting, the hackers were able to modify the concentration of water treatment chemicals and increase the amount of sodium hydroxide (lye) injected into the system by a factor of 100. Had this higher concentration been allowed to continue, it had the potential to result in serious illness to the public as well as lead to significant corrosion and pipe damage [8].

Effective Human Performance: Luckily, the hazardous level of sodium hydroxide was quickly identified by operators for the site when they saw the concentration outside of the normal range. They were able to reset the concentration to the correct value before there was any negative impact downstream [7]. The awareness of the operator and their response to the abnormal condition provided the final barrier in between the hackers and their final objective and was effective in mitigating the impact of the incident.

Ineffective Human Performance: As security researchers looked into the incident to understand how the hackers gained initial access, they found that a compromised TeamViewer (remote access software) account had been used to connect remotely. TeamViewer had originally been installed on the system to allow for remote troubleshooting and status checks on operational issues and alarms, but it was no longer being used by the site and had not been used for at least 6 months prior to the incident [7]. Although compromising a remote access software may seem like a technology-based attack, the opportunity for attackers to take advantage of TeamViewer was due to a lack of cybersecurity procedures related to allowed software as well as the requirements for removing unused software. A more robust approach to supporting human performance in relation to the maintenance of the computer systems for the site could have potentially prevented this attack before it occurred.

3.2 Characterizing People Impacts on Cybersecurity

There are many different ways of characterizing cybersecurity attacks based on the type of attacker executing the attack, the method of the attack, and the desired outcome. One of the simplest ways to differentiate between attacks is by determining if the incident resulted from an external agent attempting to gain access (outsider) or from an internal employee or contractor using existing access (insider). Because insider cybersecurity incidents directly involve the interaction of site personnel with computer systems, it is easy to see the correlation with human factors.

There are many different roles with access to control systems including engineers, technicians, supervisors, operators, contractors, vendors, and system administrators. Insider cybersecurity threats can be particularly difficult to protect against because the source of the risk is already allowed to access the system. These insider-based incidents can be further categorized by considering if there is an intent or motive to cause harm and/or whether a decision to act inappropriately is made. By answering these questions, insider incidents can be grouped into three categories as shown in Table 2 adapted from *Insider Threats* [9].

Table 2 Sub-Groups of Insider Cyber Activities [9], [6]

	Accidental	**Negligent**	**Malicious**
Decision to act inappropriately?	No	Yes	Yes
Motive to cause harm?	No	No	Yes

Malicious insider incidents arise when an intent to cause harm is paired with an intentional decision to act outside of normally acceptable behavior. This intentional insider incident is commonly referred to as a "disgruntled employee" scenario. The motives for disgruntled employees vary but often stem from a workplace disagreement, failure to achieve a desired promotion, unfavorable termination, outside influence, or personal stresses. Although it may never be possible to fully eliminate the potential for disgruntled insiders, effective Human Resources policies for screening new hires and monitoring employee engagement and well-being over time, as well as providing a supportive environment with workload management, may be effective in reducing the frequency of occurrence for intentional incidents.

Unintentional sources of insider risk can result from accidents or negligent behavior, depending on the situation. Unintentional insider incidents are often overlooked but play a key role in the security of OT systems. A 2018 study by the Ponemon Institute of over 150 companies across 13 industries found that nearly 80% of all insider cybersecurity incidents were unintentional [1]. An analysis of this set of unintentional insider incidents indicated that the most common sources were employee or contractor negligence (64%) and credential theft (e.g., accidentally sharing passwords, or using credentials that have already been leaked) (13%) [1].

Unintentional incidents are not driven by a formal motive. Instead, they are often the result of accidental or negligent behavior, both of which are heavily dependent on PIFs. Common examples of insider unintentional incidents include:

- Connecting a previously infected USB or device to the control network. Note: studies of infected USBs have shown that ransomware like WannaCry and OT-specific malware such as Stuxnet are still commonly found today [10].
- Configuration mistakes of network or control devices.
- Failing to follow cybersecurity practices/ procedures (e.g., not complying with password policies, using unapproved devices).
- Failing to follow good cybersecurity hygiene (e.g., leaving an unused account active or leaving unused software installed on the control system).

In addition to incidents resulting from unintentional actions, insiders can also provide opportunities for outside hackers to gain access by clicking on malicious links from phishing emails, accidentally sharing or losing credentials, or visiting infected websites. (This is often a concern for laptops or other devices used to download configurations or software that are then connected directly to OT systems.)

4 How Can Applying Human Factors Improve Cybersecurity?

When looking at all of the factors impacting the cybersecurity for OT systems, a conclusion is drawn that, due to all of the opportunities for insider unintentional incidents or the potential for people to bypass or limit the effectiveness of cybersecurity controls, people are the "weakest link" in the security equation. However, if methods to address human factors are leveraged to improve human performance in cybersecurity, there is the potential to completely rewrite this equation.

When calculating the risk of a scenario for either process safety or cybersecurity, the risk is equal to the consequence severity multiplied by the frequency of occurrence or likelihood (Equation 1).

$$R = C x F \qquad \textbf{Equation 1}$$

Where:

R is risk

C is the consequence

F is the frequency

In process safety, a third factor - operational discipline - has been identified to have a significant impact on process risk. Operational discipline is an outcome of an effective conduct of operations program that covers the procedures, tools, and training needed to guide personnel's behaviors in the desired direction. Operational discipline supports personnel in conducting work activities for effective human performance. By incorporating operational discipline into the risk equation in Equation 2, it is possible to visualize how improved human performance can directly reduce the risk in both process safety and cybersecurity applications.

$$R = \frac{C \: x \: F}{f\,(Operational\ Discipline)} \qquad \textbf{Equation 2}$$

Where:

R is risk

C is the consequence

F is the frequency

f (Operational Discipline) is a function based on the implementation of the operational discipline program

Just as operational discipline has a similar impact on both process and cybersecurity risk, other strategies for supporting human performance can be adapted from process safety to support cybersecurity activities. Common human performance support strategies are listed along with related cybersecurity tasks and examples of implementation in Table 3.

Table 3 Implementing Strategies for Supporting Human Performance in Cybersecurity

Strategy for Supporting Human Performance	**Related Cybersecurity Activities**	**Example Implementation**
Task, environment, & team design	Defining cybersecurity roles and responsibilities for OT systems	Clearly defining the roles and responsibilities for all personnel related to security including ongoing activities and the reporting of suspected cyber-attacks in a formal RACI (Responsible, Accountable, Consulted, Informed) chart helps to avoid any confusion on ownership of different cybersecurity tasks. This helps to maintain accountability and drive continuous improvement.
Workload & fatigue management	Defining responsibilities for control systems engineers	Although there are many activities required to maintain high availability of controls systems and security posture over time, assigning too many tasks can quickly lead to overwork and increased fatigue. Prioritizing the most important tasks for operations and security can help to support a manageable workload. Additionally, if it is identified that the required activities require multiple personnel to support, they should be clearly delegated and defined.

Strategy for Supporting Human Performance	Related Cybersecurity Activities	Example Implementation
Education in process, system, hazards, cybersecurity considerations	Cybersecurity awareness training (for all personnel)	Lack of cybersecurity awareness training makes it difficult to recognize potential signs of cybersecurity incidents and makes it difficult for personnel to understand their potential to impact the system. Effective awareness training helps to clearly define the impact that people have on the cybersecurity of the OT system and provides easy actions for improving cyber hygiene.
Information schematics, decision-making aids & procedures	Responding to potential cybersecurity incidents	Discerning potential cybersecurity incidents from normal control system behavior can be difficult. Providing clear guidance in incident response procedures can help by identifying common indications of cyberattacks, providing support for making key decisions, and identifying the reporting structure can help to greatly increase the likelihood of an attack being caught quickly.
Diagnostic communication & decision-making skills	Responding to cybersecurity alerts	Identifying cybersecurity alerts and requirements for periodic reviews of security logs enables diagnosing potential cybersecurity concerns more quickly. Providing procedures for incident handling and reporting then allow these concerns to be escalated quickly and effectively to the required personnel, which can provide another important input to cybersecurity incident and response plans.
Teamwork, shared situation awareness, coordination, clear roles & responsibilities	Cybersecurity tabletop exercises	Provide a safe environment for employees to respond to cybersecurity incidents and practice following cybersecurity procedures. Practicing communicating with personnel during a tabletop can support faster response times during an incident and also help identify potential roadblocks in current procedures so that they can be updated prior to an incident.
Skill development, instruction & operational experience	OT cybersecurity skill training	In addition to general awareness training, it is important that personnel with roles for implementing and maintaining cybersecurity (e.g., controls engineers, process safety engineers, maintenance supervisors) are provided the training and mentorship to develop the skills necessary to complete their responsibilities. Cybersecurity is a relatively new and fast-moving area, which often requires focused training on topics like cybersecurity risk assessments and effective cybersecurity controls.

Strategy for Supporting Human Performance	Related Cybersecurity Activities	Example Implementation
Procedures, instructions & memory aids	Cybersecurity policies and procedures for acceptable use	Providing clear cybersecurity guidelines in the form of policies and procedures helps to define the expected cybersecurity hygiene requirements for all personnel. This guidance, combined with memory aid/awareness posters, can significantly improve the performance of insiders against phishing attacks and the use of unsecured USB devices.
Task planning & checking	Deploying a software or configuration update	Develop procedures for management of change (including cybersecurity), patch management, and cybersecurity work instructions for system updates. For each software update or configuration change, it is important to have step-by-step instructions as well as a backout plan in case the system has an unexpected response to the change so that the operational disruption can be minimized.
System design (e.g., Controls & Instrumentation)	Designing cybersecurity controls for OT systems	If security settings or controls make it so that personnel cannot complete their job responsibilities, it is very likely that they will be bypassed. Including personnel from multiple disciplines in the evaluation of cybersecurity risks and the implementation of recommendations is helpful in finding the necessary balance between availability and security of key systems and providing the best possible design of security controls for a given site.

The total number of activities that can be implemented to support human performance in cybersecurity is quite large. However, significant improvement can be achieved by focusing on providing the necessary training to all personnel so that they can identify the ways they impact the security of the system and providing clear guidelines for identifying and responding to potential cybersecurity incidents.

Once this baseline is met, implementing a robust cybersecurity management system - with the policies, procedures, and practices to support cybersecurity activities like risk assessment and the implementation of security controls - can help to deploy a risk-based approach to cybersecurity much the same way as it is deployed in process safety [6].

5 Conclusion

At its core, cybersecurity is really about people and how they can either protect or compromise computer-based systems (intentionally or unintentionally). Finding ways to better support the people protecting OT systems will reduce the likelihood of people unintentionally compromising or exposing OT systems to threats and could significantly reduce the cybersecurity risk facing many process sites.

Although the focus on cybersecurity in process sites is relatively new in the last 10-15 years, the study of human factors in the process industry has been around for significantly longer. By leveraging existing strategies for reducing the negative impact of Performance Influencing Factors and providing the key steps to support human performance, it is possible to identify the most critical steps for improving the human performance element of cybersecurity.

A robust cybersecurity program needs to encompass more than just policies and procedures. Supporting these written requirements with awareness training, skill-based training, and clearly defined roles can help to significantly improve human performance. Supporting human performance helps to both reduce the likelihood of an unintentional cybersecurity incident and increase the effectiveness of the barriers against cybersecurity incidents that rely on human capabilities (e.g., account management, response to cybersecurity alerts, configuration of security settings). For a human factors-driven approach to be fully implemented in process plants, a shift is needed in the way that people are viewed in the cybersecurity risk equation, moving people from the cause of cybersecurity incidents to being the most important barrier against cybersecurity intrusions. This can be used to adopt a fundamentally new mantra for cybersecurity *"People are the solution, not the problem* [3]*."*

6 References

[1] Ponemon Institute, "2018 Cost of Insider Threats: Global," Ponemon Institute, Traverse City, MI USA, 2018.

[2] CCPS, "CCPS Process Safety Glossary," Center for Chemical Process Safety, 2024. [Online]. Available: www.aiche.org/ccps.

[3] CCPS, Human Factors Handbook for Process Plant Operations, Improving Process Safety and System Performance, Hoboken, NJ USA: John Wiley & Sons, 2022.

[4] J. Rasmussen, Information Processing and Human-Machine Interaction: An Approach to Cognitive Engineering, New York: Elsevier Science Inc, 1986.

[5] J. Reason, Human Error, Cambridge: Cambridge University Press, 1990.

[6] CCPS, Managing Cybersecurity in the Process Industry, A Risk-based Approach, Hoboken, NJ USA: John WIley & Sons, 2022.

[7] A. Greenberg, "A Hacker Tried to Poison a Florida City's Water Supply, Officials Say," Wired, San Francisco, CA USA, 2021.

[8] J. Peiser, "A Hacker Broke into a Florida Town's Water Supply and Tried to Poison it with Lye," Washington Post, Washington, D.C. USA, 2021.

[9] Imperva, "Insider Threat," Imperva, Redwood Shores, CA, 2020.

[10] Kaspersky, "USB Threats from Malware to Miners," Secure List, Moscow, Russia, 2018.

OT Cybersecurity: What is this all about?

Blake Benson
Randal Montgomery
Scott Sandler

ABS Consulting
1701 City Plaza Drive
Spring, TX 77389
USA

bbenson@absconsulting.com
rmontgomery@absconsulting.com
ssandler@absconsulting.com

Keywords: Chemical Facility Anti-terrorism Standards (CFATS), Cyberattack, Cybersecurity, Information Technology (IT), IT Cybersecurity, Layer of Protection Analysis (LOPA), Operational Technology (OT), OT Cybersecurity, Process Hazard Analysis (PHA), Safety Instrumented System (SIS), Safety Instrumented Function (SIF),

Abstract

It seems like almost every day we hear about some cybersecurity breach that potentially impacts some aspect of a business's operations. In most cases (that we hear about after the fact), these breaches relate to information technology (IT) systems (e.g., emails, business financial records, employee records). However, IT systems are not the only systems being attacked by cybercriminals. Companies operating in the process-industry market space, such as refineries, pipeline operations, chemical plants, paper mills, etc., have operational technology (OT) systems (e.g., supervisory control and data acquisition [SCADA] systems, process controls, safety instrumented systems). These systems can be - and have been - breached by cybercriminals. Some OT cybersecurity breaches have been publicized, but many have not. Nonetheless, these OT cybersecurity breaches have occurred and have often resulted in significant impacts on the affected companies.

The paper provides an overview of OT cybersecurity and common vulnerabilities in OT systems. In addition, the potential impacts of OT cybersecurity breaches on plant operations, plant safety, environmental compliance, and business reputation are discussed. Because of the potential for impacts on the country's critical infrastructure, federal regulations addressing OT cybersecurity are currently being developed. This paper also provides an overview of cybersecurity-related requirements of chemical facility anti-terrorism standards (CFATS) (6 CFR 27) and other potentially applicable regulations. The final aspect discussed is a practical approach that a company may be able to utilize to help it start and progress on protecting itself from an OT cybersecurity breach.

1 Introduction

Cyberattacks are an ever-increasing threat in today's world, and this threat can impact (and may already have impacted) our personal lives and the companies for which we work. Almost every day, cyberattacks impacting the financial, medical, and/or retail sectors are reported. These cyberattacks are typically directed at organizations' information technology (IT) systems and involve some type of data breach. While often not noted as much in the public news, cybercriminals are also targeting operational technology (OT) systems in the manufacturing sector (including process industries), and these attacks are occurring at an increasing frequency. Some recent process industry cyberattacks include:

- ALPHV BlackCat Ransomware attack in Europe that impacted the automated processes associated with the loading and unloading of the tank farm materials at ARA and Germany Oil Terminals [1]

- Clorox ransomware attack that impacted IT and OT systems, resulting in product outages and supply disruptions that generated multimillion-dollar losses [2]

- TRISIS Malware Safety Instrumented System [SIS] attack, which compromised logic in final control elements, thereby resulting in plant shutdowns. [3]

While IT cyberattack impacts are often related to obtaining personal and other sensitive data, OT cyberattack impacts can go beyond data breaches and can result in safety, environmental, and operational impacts. For example, OT cybercriminals target systems such as process control and safety instrumented systems which help to ensure the safe operation of hazardous processes. These attacks can potentially have the same impact as physical security breaches. In addition, these attacks can shut down operations or incoming/outgoing supply chains, often resulting in significant economic and potentially even reputational impacts. Another reality of OT cybersecurity attacks is that the diagnosis and correction of these events are often very challenging and require expertise outside of the companies' current organization to resolve.

OT cybersecurity threat is a risk to facilities operating in the process industries, and these attacks can result in process safety events, environmental exceedances, operational interruptions, and other significant business impacts. The remainder of this paper provides an overview of OT cybersecurity and potential vulnerabilities, a more detailed discussion of process safety and operational impacts, an overview of current OT cybersecurity regulations, and a practical approach for proactively addressing OT cybersecurity threats.

2 OT Cybersecurity

2.1 Overview and Common Vulnerabilities

In operational environments, operations managers and executives are becoming increasingly concerned with cyber threats, and most have made major investments in their cybersecurity programs. These programs are typically governed by the corporate IT departments, and their focus is on protecting the confidentiality of proprietary or personal data and ensuring the availability of critical business systems by implementing a robust strategy that includes:

- Executive-level oversight by a chief information security officer (CISO)
- Governance policies and procedures
- Cybersecurity training
- Cyber vulnerability assessments
- Design of a secure architecture
- Configuration and maintenance of hardware and software
- Incident response plans

While most companies have implemented an IT cybersecurity program, far fewer are extending their cybersecurity programs to address their OT. OT systems detect or cause changes through the direct monitoring and control of physical devices, processes, and events. In short, many companies are investing more resources and energy to protect against threats that may or may not impact operations than in protecting OT systems that are more likely to directly impact operations. It is critical to assess what functions and systems in an operational environment have cyber dependencies and to understand that the prioritization of the security of those critical systems is paramount. If an IT system is compromised, it could result in a loss of confidentiality of process operational data (e.g., access to data historians, process controls to obtain process set points, process recipes); however, if an OT system is compromised, it could most certainly result in downtime for operations or worse (e.g. safety or environmental impact).

Since IT and OT have different purposes, they have nearly opposite system priorities. OT systems emphasize integrity, availability, and confidentiality, in that order, whereas IT systems prioritize confidentiality, availability, and then integrity. Traditionally, OT systems have been isolated, whether virtually or physically, from IT networks. This is not as common in new builds today but still represents a significant percentage of companies we've assessed. OT systems are often managed by companies' engineering or operations departments, which are primarily concerned with verifying that the systems are "up and running" and providing effective control over operations. Since the availability of OT system functionality is paramount, OT systems are designed to be simpler (relative number of users, etc.), and more reliable (e.g., less downtime) than most business/IT systems In addition OT systems typically are expected to have a much longer lifespan (e.g., 30 years) than IT business systems (e.g., 6-10 years). The isolation of OT systems has traditionally been viewed by engineering/operations groups as the ultimate safeguard against outside threats, but the days of physical isolation of OT systems and manual process control are no more. One of the biggest drivers for OT cyber risk to operations observed by our assessors is the forced convergence of IT and OT systems for business optimization, such as when process data is transferred into the plant management system such as inventory control, and maintenance.

Based on our experience, a major OT cybersecurity vulnerability is a belief that OT systems are air-gapped but often are easily defeated by the following types of activities:

- Outside vendor connecting their computer-based equipment (e.g., computers) for maintenance, monitoring, diagnosis, etc. for systems and equipment

- Employees using available computer connections for unauthorized equipment (e.g., phones, iPads)

- On-line vendor access for equipment maintenance and monitoring

- External online connection for plant and engineering personnel to have remote access to systems for troubleshooting, etc.

- Unidentified/unknown hardwired connections of OT and IT system components

Our experience is these types of activities have resulted in OT cyberattacks, and they are often not adequately safeguarded against in IT cybersecurity programs. OT cybersecurity programs address these types of issues and other unique OT system vulnerabilities.

2.2 Potential Process Safety and Operational Impacts

Over the last three to four decades, the process industry advanced from pneumatic control systems and hardwired safety systems to electronic, smart control, and safety systems. These advancements generated many significant operational enhancements, (e.g., advanced process control schemes, process optimization from increased data). They have also generated many significant safety improvements, such as higher reliability safety instrumented systems, additional alarms, and enhanced operational monitoring. With these advancements, there has been a massive increase in the number of instruments and computer-based control and safety systems. This increase in electronic- and computer-based devices and these systems have resulted in new and different types of potential failures, including many hidden failure modes. Unfortunately, cybercriminals target these systems and, in some instances, have created significant impacts on plant operations. Specifically, the process industry has seen instances in which OT cyberattacks shut down company operations, causing significant economic impact and even reputational impacts. In addition to operational impacts, OT cyberattacks have the potential to cause the malfunction or total disruption of critical safety systems (e.g., critical alarms, control systems, SISs, safety instrumented functions [SIFs]). A potential outcome of safety system cyberattacks is that they can result in hidden system failures which may not be discovered until scheduled function tests are performed or an actual demand occurs. The following paragraphs describe approaches for considering these impacts during process hazard analysis (PHA) and layer of protection analysis (LOPA) studies.

During a traditional PHA, Basic Process Control System [BPCS] and SIS that are independent are typically assessed as separate contributions to the risk of a hazard scenario. For example, a process controller may malfunction, forcing the final element (e.g., control valve) into an extreme position such as wide open. The resulting scenario may exceed one or more design parameters, such as pressure or temperature limits. If the components of a SIF within the SIS are all independent from the BPCS components (including the logic solvers), the team will consider the anticipated effectiveness of the SIF towards reducing the mitigated likelihood of the scenario.

When assessing hazards caused by cyber threats, the fact that the BPCS and SIS may be independent does not necessarily preclude a single initiating event (or unrevealed initial event followed by a later cause or initiating event) from leading to a control malfunction and concurrent disablement of the SIF. As such, the hazard potential of cyber threats goes beyond the scope of common causes evaluated during a typical PHA/LOPA workshop. Therefore, the risk profile obtained from a cyber-based analysis will be different than that of a traditional process-based study.

For example, a typical LOPA scenario (e.g., overpressure caused by a BPCS initiating event) may assign credit to an applicable SIF safeguard as an independent protection layer (IPL). However, a cyber-based analysis would evaluate the potential common cause event of a cyberattack, assuming both initiating the event and inhibiting the SIF IPL(s), which will likely result in a higher risk.

When assessing cyber-related hazards, the traditional PHA or LOPA worksheets may be used as the starting point. It is recommended to use software that allows easy modification of causes and safeguards (and IPLs) as well as the ability to sort the scenarios based on risk. To begin, only scenarios that could be caused by a cyber threat should be analyzed, such as the example described above in which a cyberattack potentially initiates the event and/or disables IPLs.

For each applicable hazard scenario, the safeguards and IPLs are reviewed. Those that could be compromised by a cyberattack are assumed to be unavailable. Only safeguards that would be unaffected by a cyberattack will remain in the analysis, such as pressure relief valves, flame arrestors, and heat-activated mitigation systems. Instrumented safeguards such as alarms, safety controls/interlocks, and SIFs would be considered ineffective. The adjusted scenarios would then be sorted by residual risk.

It becomes evident through this analysis that the greatest risks are hazards that can be caused by controller malfunctions and those that rely disproportionately on instrumented systems (e.g., SIL 3 SIF). Overpressure events that are adequately protected by pressure relief valves or rupture disks would have a lower residual risk than similarly-sized hazardous releases resulting from temperature excursions, which rely more on instrumented safeguards (e.g., low-temperature embrittlement of carbon steel).

2.3 OT Cybersecurity Regulations

US government agencies have developed (and are developing) regulations addressing the OT cybersecurity threat, analogous to the regulations developed to address physical security (i.e., CFATS [4]). The CFATS requirements were developed in the post-9/11 world to help prevent dangerous chemicals from falling into the wrong hands. CFATS applies to any "[e]stablishment or individual that possesses or plans to possess any of the more than 300 chemicals of interest (COI) in Appendix A at or above the listed screening threshold quantity (STQ) and concentration." Any organization that meets this criterion must complete a screening survey from the Cybersecurity & Infrastructure Security Agency (CISA). Organizations that CISA designates as high-risk will be notified and required to comply with CFATS (if not already required to comply). This obviously applies to many chemical facilities and could also apply to facilities such as oil and gas operations handling COIs.

Specifically, CFATS includes four subparts with Subpart B, Chemical Facility Security Program, addressing both physical and cybersecurity requirements. Section §27.230 contains the risk-based performance standards for facility security programs, including paragraph §27.230 (a)(8) which addresses cybersecurity requirements [4]. Specifically, this section addresses the following:

- Scope – applicable to SCADA systems, distributed control systems (DCS), process control systems (PCS), industrial control systems (ICS), critical business systems, and other sensitive computerized systems

- Requirements – deter cyber sabotage, including preventing unauthorized onsite or remote access to critical process controls

To assist in CFATS risk-based performance requirements, CISA issued a guidance document entitled "Chemical Facility Anti-Terrorism Standards (CFATS) Risk Based Performance Standards Guidance" [5]. This guidance document outlines security measures and considerations for cybersecurity that include a combination of the policies and practices addressing these topics:

- Security Policy
- Access Control
- Personnel Security
- Awareness and Training
- Monitoring and Incident Response
- Disaster Recovery and Business Continuity
- System Development and Acquisition
- Configuration Management
- Audits

This guidance document also includes risk-based performance standard metrics for cybersecurity. Based on our experience, two of the metrics where organizations often fall short are:

- IT/OT Network Segmentation: Many organizations are beginning to integrate their IT systems with their OT systems. This can be very helpful for efficiency and optimization but may have cybersecurity implications as this can expose the OT systems to untrusted networks. The RBPS specifically requires that the organizations define system boundaries and document external connections.

- Network Visibility: Organizations often have robust IT network monitoring capabilities, but little to nothing in the OT space. Network monitoring is explicitly required under this RBPS.

Another example of federal legislation that is starting to incorporate more cybersecurity guidance that can protect against anti-terrorism threats is the Maritime Transportation Security Act of 2002 (MTSA) [6]. MTSA is a wide-reaching piece of legislation affecting the maritime industry and may also apply to larger chemical facilities and many oil and gas operations involving marine operations. It applies to the following facilities and vessels:

Applicability - Facilities:

- Facilities subject to 33 CFR 126, 127, or 154
- Facilities receiving commercial vessels certified to carry more than 150 passengers
- Facilities receiving commercial vessels subject to the International Convention for Safety of Life at Sea [SOLAS]
- Facilities receiving cargo vessels greater than 100 gross register tons
- Barge fleeting facilities receiving barges carrying cargoes in bulk are regulated in 46 CFR Chapter I, subchapters D or O, or certain dangerous cargoes

Applicability -— Vessels:

- Cargo Vessels, domestic and foreign, weighing more than 100 gross register tons
- Passenger vessels certified to carry more than 150 passengers
- Passenger vessels weighing more than 100 gross register tons
- Vessels subject to SOLAS
- Barges carrying certain dangerous cargo
- Mobile offshore drilling units

The MTSA regulation could be relevant to portside facilities such as gasoline storage petrochemical plants, Liquified Natural Gas (LNG) facilities, or refineries. Broadly, MTSA requires the development of facility and vessel security plans, vulnerability assessments, and access control requirements. This legislation was primarily considered to affect physical security, but the Coast Guard released a navigation and vessel inspection circular (NVIC) 01-20 to give guidance on how this legislation applies to cybersecurity [7]. The NVIC does not introduce new requirements under MTSA, but the guidance document clarifies how a facility might integrate cybersecurity into its security assessment plan.

2.4 Practical Approach to OT Cybersecurity

The almost constant stream of marketing materials from cybersecurity providers has overly-promoted social media regarding "hyper-optimized" assessment approaches and offers of cure-all solutions. This portrayal can generate confusion among operations on how to best assess and manage cybersecurity risk. This appears to be very similar to the confusion that many organizations had in the early 1990s when SIS emerged as the best means to address operational risks.

Existing assessments and analyses such as PHAs, LOPAs, operational risk assessments (e.g., quantitative risk studies), and safety case studies are all excellent starting points to begin to contextualize cybersecurity risk to operations. Cyber vulnerability assessments and other cyber assessments such as validated architecture design reviews, when combined with traditional safety studies, can provide a holistic view of areas where additional safeguards may need to be added to prevent catastrophic consequences.

In many cases, cyber vulnerabilities are likely to exist, but the cyber risk does not necessarily translate to safety and operational risks. How is that possible? Cyber vulnerabilities that are present within operational technology that have adequate safeguards (manual or digital) that are not susceptible to cyber exploitation will likely not present a significant risk to safety and operations.

By combining the process safety/risk and cybersecurity approaches, organizations can answer the "OT Cybersecurity: What is this all about?" question with relative ease. Specifically, the process safety efforts have already identified a company's highest risk assets and implemented risk management strategies such as engineered safeguards (e.g., SIFs, overpressure protection) to manage these consequences. In cybersecurity, an analogous approach can be taken. One such approach is consequence-driven cyber-informed engineering, which begins by prioritizing cybersecurity evaluations-based consequences resulting from cyberattacks on critical assets, such as electronic control and safety systems associated with high-risk assets.

For loss scenarios associated with identified during a consequence-driven cyber-information engineering evaluation, an experienced OT cyber expert begins reviewing cybersecurity risks by:

- Reviewing asset inventories of cyber-enabled systems (e.g., controls networked with a facility with direct or indirect internet connectivity) present in the operating environment
- Reviewing the combined IT/OT network architecture (e.g., are these systems truly air-gapped)
- Working with operations experts to begin to understand the current cybersecurity program maturity (e.g., just thinking about cybersecurity, robust cybersecurity management program)

- Identifying component-level OT information associated with the loss scenarios, such as safety set points, trip functions, etc.

This information, when combined with process safety/risk study results and P&IDs, provides a means to identify the engineered safeguards that should be evaluated for potential impacts from a cyberattack. For example, specific cybersecurity vulnerabilities (e.g., vulnerability to an internal cyberattack to an SIS) are reviewed to evaluate the potential operational and process safety risks due to a specific cyberattack scenario. This exercise informs an asset risk profile that will allow for recommendations to be made that help reduce unacceptable scenarios/losses by including both cyber and additional engineered safeguards (as needed) to mitigate multiple sources of risk. It is possible to visualize this cyber "lens" over existing safeguards and equipment using bowtie diagrams or other "layered" protection models to depict assets and their representative protections from both cyber and non-cyber measures.

A simple way to exercise this capability could be a risk workshop with multi-disciplinary personnel—including process engineers, operators, maintenance personnel, instrumentation and controls engineers, and IT (local and corporate) personnel. The assembled team should conduct a systematic evaluation of each potential loss scenario and the associated cyber-initiated event risks by considering these questions/issues:

1. Could this loss scenario be initiated by a cyber-initiated event (e.g., close pressure control valve and overpressure tank)?
2. Could critical engineered safeguards/IPL be compromised/degraded/impaired by a cyber-initiated event (e.g., SIF trip point)?
3. For each safeguard/independent protection layer (IPL), could its performance be degraded (e.g. disable high-pressure alarm) or failed (e.g., disable the high-pressure alarm and automatic shutdown on the tank) by cyber-initiated event?
    a. If so, categorize degradation potential (i.e. degree of difficulty related to executing the cyber-initiated event) and identify potential corruption vectors (e.g., physical hardware ports, digital remote access, access to integrated safety systems).
    b. If not, note that the safeguard/IPL is not susceptible to cyber attack (e.g. manual pressure relief valve, not integrated with SIS or other automated systems).

(Note: These questions can easily be adapted into a checklist with columns to document checklist question responses, associated cyber-related safeguards [e.g., cyber policies/procedures, detection devices], and recommendations).

Once all loss scenarios have been analyzed, the assessment team develops a risk profile that identifies the highest risk scenarios and compares these to the baseline risk to characterize the company's overall cyber risk. Based on the cyber risk of a loss scenario, the team generates recommendations to address both cyber and physical safeguards. This process can be repeated for all sites within a company to develop an enterprise risk profile. Figure 1 provides a basic flowchart of this practical cyber risk evaluation approach.

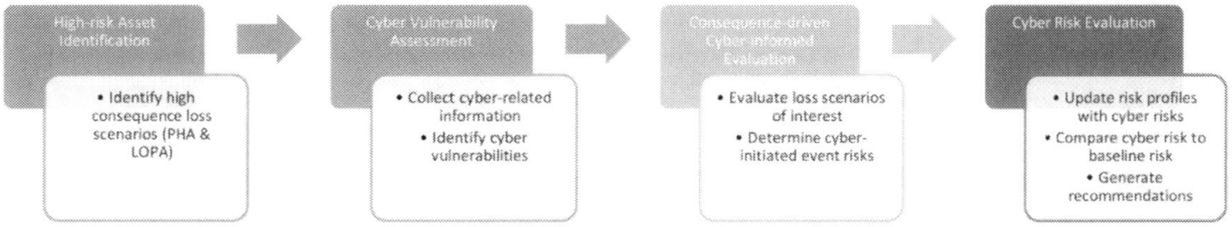

Figure 1: Practical Cyber Risk Evaluation Approach

The results of a cyber vulnerability assessment should contribute to more reliable decision-making, thus enabling more effective management of risk and security across the spectrum of high-performance, data-driven assets found in today's complex facilities.

3 Conclusion

Cybersecurity is an important issue to the chemical industry for several reasons. First, OT cybersecurity attacks present an increasing threat to process industry operations and safety. Second, regulatory requirements are (or likely will be) in place to address this threat, which will require compliance similar to what the industry experienced with process safety regulations and physical security regulations now in place. The information presented in this paper is intended to expose process industry safety professionals to this topic and begin the education process on cybersecurity. Finally, the practical approach presented provides companies with a means to start addressing this risk. We believe now is the time to get ahead of threats because it is no longer a case of "if an OT cyberattack happens" but "when an OT cyberattack happens." Let's work together as an industry to proactively manage this increasing risk.

4 References

[1] S&P Global Commodity Insights. February 3, 2022. https://www.spglobal.com/commodityinsights/en/market-insights/latest-news/oil/020322-cyberattack-causes-chaos-at-key-european-oil-terminals. Accessed February 15, 2024.

[2] Reuters. October 4, 2023. https://www.reuters.com/business/retail-consumer/clorox-expects-quarterly-loss-hit-cyberattack-2023-10-04. Accessed February 15, 2024

[3] Dragos. December 14, 2017. https://www.dragos.com/resource/trisis-analyzing-safety-system-targeting-malware/. Accessed February 15, 2024

[4] Code of Federal Regulations. https://www.ecfr.gov/current/title-6/chapter-I/part-27. Accessed February 15, 2024.

[5] CISA. May 2009. https://www.cisa.gov/resources-tools/programs/chemical-facility-anti-terrorism-standards-cfats/cfats-risk-based-performance-standards. Accessed February 15, 2024.

[6] US Department of Justice. May 2005. https://www.ojp.gov/ncjrs/virtual-library/abstracts/maritime-transportation-security-act. Accessed February 15, 2024.

[7] Federal Register. 3/20/2020. https://www.federalregister.gov/documents/2020/03/20/2020-05823/navigation-and-vessel-inspection-circular-nvic-01-20-guidelines-for-addressing-cyber-risks-at. Accessed February 15, 2024.

Managing the broken SIS through Peak Risk Analysis

Kedar Kottawar
SIS-TECH Solutions, LP
12621 Featherwood Drive, Suite 120, Houston, TX 77034
kkottawar@sis-tech.com

Jyoti Sharma, Angela E. Summers, and Eloise Roche
SIS-TECH Solutions, LP

Keywords: SIS, peak risk, risk management, process safety, compensating measures, bypassing

Abstract

Safety Instrumented Systems (SIS) are used to reduce the frequency of hazardous events in the process industry. When SIS equipment fails or requires online maintenance, the processing facility must decide whether to continue operation or shut down. Continued operation can pose risks beyond what was deemed acceptable per the Company's risk criteria. Understanding the peak risk is important to ensure that all stakeholders are aware of the risk and have compensating measures that ensure continued safe operation until the SIS equipment is returned to service. The paper proposes using Peak Risk Analysis (PRA) to identify unacceptable risks and to document the compensating measures required to support continued safe operation with out-of-service SIS equipment.

1 Introduction

A safety instrumented system (SIS) is an instrumented safeguard that provides an independent and reliable means to reduce the risk of hazardous events involving serious consequences. The risk reduction provided by the SIS can be degraded when SIS equipment is taken out of service while the process continues operation. The impact can be low, such as when redundant SIS equipment is implemented in a way that results in the function still being able to operate as required. The impact can also be significant, such as when only one sensor is used to execute a safety function, and the automatic response to the detected fault results in the safety function being disabled. The peak risk can easily exceed the company's risk criteria if the impact is not managed. Ensuring that the process remains protected during out-of-service periods is the underlying philosophy for the fault detection and response requirements of American National Standards Institute/International Society of Automation (ANSI/ISA) 61511-1 [1].

Compensating measures are temporary safeguards used to control the peak risk caused by SIS equipment being out-of-service during process operation. To be effective, the compensating measure should close the risk gap until the SIS equipment is restored to full operation. Gap closure should be accomplished without exceeding the capabilities of existing systems or operating personnel. Compensating measures should be identified early in the design process so that needed equipment and procedures are in place when the SIS equipment goes out-of-service.

2 Understanding Peak Risk and the Risk Gap

2.1 Why do we care about peak risk?

While it is a useful simplification in risk assessment to evaluate proposed SIS designs in terms of average performance, operators experience risk in real time (Figure 1). When a process deviation occurs in a facility, the resulting outcome is not determined by the average performance of the safety equipment over its mission time. The outcome depends on the availability of the SIS at that specific moment in time. The peak risk is what the operations personnel experience and this risk needs to be managed whenever the safety equipment is degraded or out-of-service. The CCPS book, *Guidelines for Enabling Events and Conditional Modifiers* [2], states "It is not appropriate to use a time-at-risk enabling probability when the initiating cause is activity-based, or opportunity-based. Workers are at risk every time they perform the activity or have an opportunity to make a critical error."

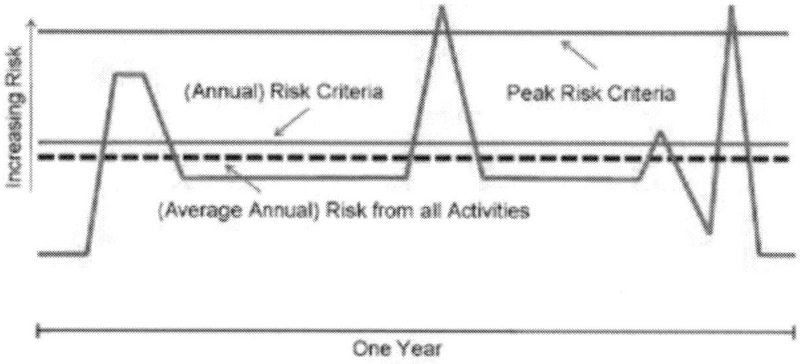

Figure 1. Average Total Risk vs. Annual and Peak Risk Criteria [4]

The need to manage peak risk is reflected in general terms in the US [Occupational Safety and Health Administration Process Safety Management] OSHA PSM regulation [3] that "The employer shall correct deficiencies in equipment that are outside acceptable limits … before further use or in a safe and timely manner when necessary means are taken to assure safe operation." ANSI/ISA 61511-1:2018 more explicitly requires "Compensating measures that ensure continued safe operation shall be provided … when the SIS is in bypass (repair or testing)."

2.2 When might a peak risk event occur?

The peak risk events shown in Figure 1 occur when safety system equipment is out-of-service while the hazardous event could still occur. Two foreseeable events trigger SIS equipment out-of-service periods. The first involves the random failure of SIS equipment, which occurs without warning. The second is through planned online maintenance tasks such as inspection, testing, and preventive maintenance. Planned maintenance lacks an element of surprise, which can create a false impression that the peak risk is still under control even if no compensating measures are applied. Offline testing will not generate a peak risk event since the process is not operational during offline maintenance.

2.3 Is MTTR enough?

The Mean Time to Restoration (MTTR) represents the time it takes to detect a failure in a SIS, repair the failure, and restore the SIS back to full operation. This value appears in the performance calculations that are used to verify that a SIS design can achieve the average performance targets for risk reduction and spurious trip rate. Some might think that including MTTR in the calculation means the risk of the out-of-service period is addressed. This is not true. The calculation considers a single MTTR period in the device's installed life. Since the calculation estimates the performance averaged across the installed life, it does not address the peak risk that operations will be exposed to when process operation is continued with SIS equipment out-of-service.

2.4 How much of a gap is there?

An out-of-service event does not always cause a risk gap. The SIS design, including the fault diagnostic response, bypass philosophy, and maintenance facilities, determines whether a risk gap is created when a SIS device goes out of service. Consider a SIS that detects abnormal operations using a single sensor and acts using a single valve. Either device going out-of-service disables the SIS. In contrast, the SIS could be implemented using a 1oo2 redundant equipment configuration. This would make the SIS fault-tolerant to a single random device failure. However, during planned online maintenance, whether a single device is being serviced or the entire SIS function is offline will depend on the maintenance philosophy for the facility and the bypass design. Bypassing any single device would cause only a partial loss of risk reduction. However, if the maintenance philosophy is to bypass the entire SIS during online testing, the risk gap will be the total risk reduction the SIS is expected to provide, despite the investment in redundant devices.

Because of the potential differences between the bypass philosophy and the automatic response to detected failure, both random and planned activities should be analyzed to determine the risk gap triggered by SIS design choices (Figure 2).

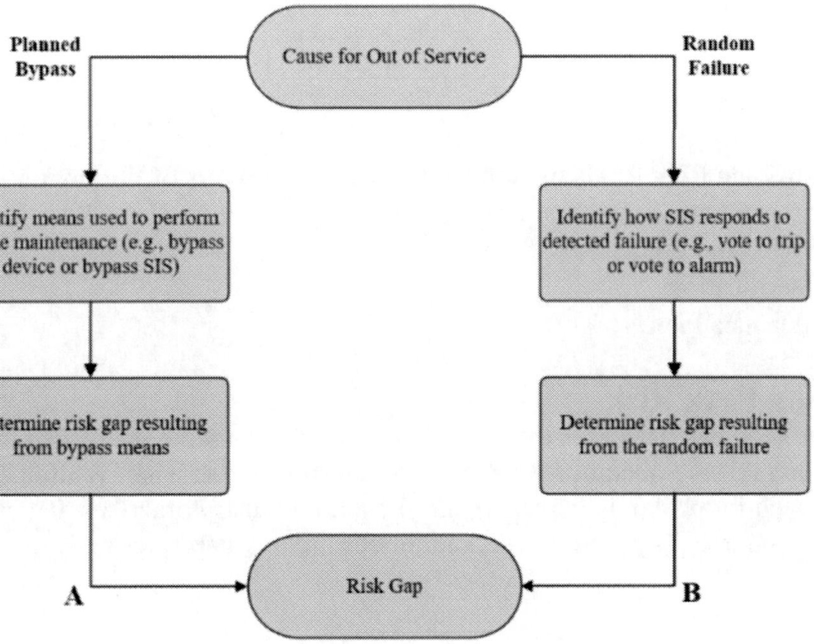

Figure 2. Determining Risk Gap Flowchart

The loss of risk reduction can be quantitatively analyzed by applying techniques similar to the ones used to verify the design of the fully functional SIS. For each SIS device failure, a new average Probability of Failure on Demand (PFD_{avg}) can be calculated. When the PFD_{avg} for the degraded SIS exceeds the target PFD_{avg}, the loss of risk reduction is the "peak risk" that can be expressed as a Risk Reduction Factor gap (RRF_{Gap}) or PFD_{Gap}.

$$PFD_{gap} = \frac{PFD_{Target}}{\left(PFD_{SIS} - PFD_{Sub-SIS} + PFD_{Deg-Sub-SIS}\right)}$$

PFD_{Target} : PFD_{avg} required by the Safety Integrity Level (SIL) assigned to the SIS.
PFD_{SIS} : PFD_{avg} achieved by the SIS when not degraded.
$PFD_{Sub-SIS}$: PFD_{avg} achieved by redundant/subsystem devices when not degraded.
$PFD_{Deg-Sub-SIS}$: Reduced PFD_{avg} achieved by redundant/subsystem devices modeled in the $PFD_{Sub-SIS}$ except one or more devices are degraded and removed from the calculation.

$$RRF_{Gap} = \frac{1}{PFD_{Gap}}$$

2.5 Performing the Peak Risk Analysis (PRA)

Peak Risk Analysis is best done during the design phase so that the equipment needed to manage the peak risk will be designed and implemented along with the rest of the SIS. The maintenance plan assumed during the design should be documented clearly, detailing the requirements for online and offline testing. Maintenance facilities can then be specified and installed to support testing and repair. Likewise, the automation system will need to be designed to respond to detected failures or bypass requests in the intended fashion.

The PRA requires experienced individuals who are well-versed in the site's risk assessment goals, operator procedures, and maintenance practices. These individuals are supported with the following documents:

- Hazards and Risk Analysis Report (e.g., LOPA: Layers of Protection Analysis)
- Piping and Instrumentation Diagrams (P&IDs)
- Safety Requirements Specification
- SIL Verification
- Maintenance Plan

3 Managing Peak Risk

Once the risk gap is understood, a plan for managing the identified risk gap should be developed to assure safe operation until the SIS is restored to service. While no standards define a specific approach for peak risk management, there are industry practices that can be considered in determining whether a proposed solution addresses the risk gap (Figure 3).

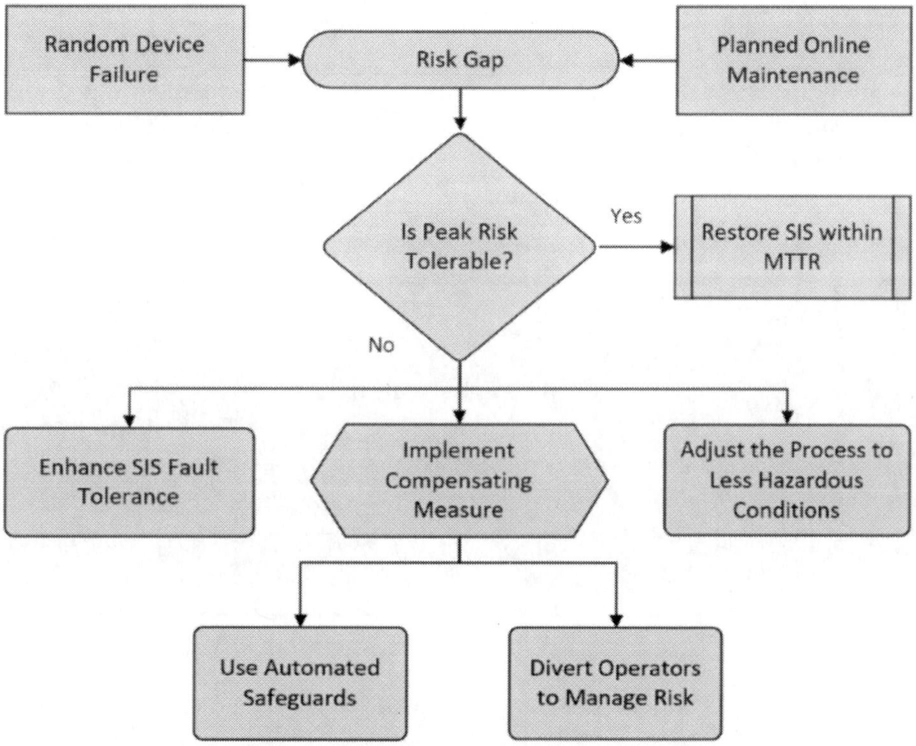

Figure 3. Simplified Peak Risk Analysis Flowchart

3.1 Adjusting the Process to Less Hazardous Conditions

One way to avoid needing a compensating measure is to eliminate the potential hazardous event by temporarily moving the process to a less hazardous state. This could involve limiting the operating modes or changing the process conditions. Perhaps operations can limit production to less hazardous products while on-line maintenance of the SIS is being performed. For tank farms, maybe only one or two raw materials are severely hazardous, so unloading the higher-hazard chemicals could be delayed until the detected failure is repaired. The goal is to lower the hazardous event risk without stopping operations completely. Unfortunately, adjusting process operations to less hazardous conditions while continuing some degree of operation is not always possible, leading to the need for other safeguards.

3.2 Enhance SIS Fault Tolerance

There are some SIS sensor configurations that become inherently safer in response to the first device going out-of-service. If modifying the SIS design would be a reasonable option, choosing one of these architectures may avoid the need to implement a compensating measure for that sensor out-of-service event.

Consider if the sensors are implemented in a 2oo3 voting configuration where faults trigger the trip condition (Figure 4). The risk reduction is actually increased when the first sensor fails since the SIS is degraded to 1oo2. There is no risk gap, so no compensating measure is needed. However, if the out-of-service condition is configured to alarm only (i.e., vote away from trip), the resulting 2oo2 configuration decreases the risk reduction provided by the SIS by two orders of magnitude, creating a significant risk gap.

Similarly, a 2oo2 sensor voting configuration that uses a sensor going out-of-service as a vote to trigger the trip condition degrades to a 1oo1 configuration, which improves the risk reduction by a factor of 2.

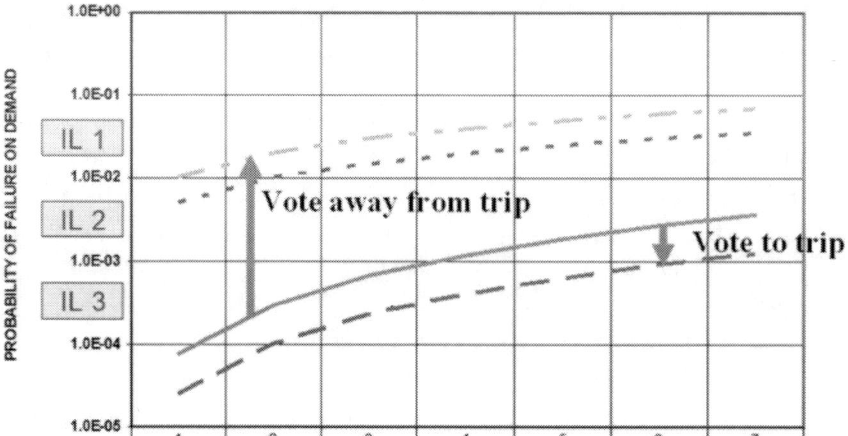

VOTING M-out-of-N	1	2	3	4	5	6	7
1oo1	5.00E-03	1.00E-02	1.50E-02	2.00E-02	2.50E-02	3.00E-02	3.50E-02
1oo2	2.50E-05	1.00E-04	2.25E-04	4.00E-04	6.25E-04	9.00E-04	1.23E-03
2oo2	1.00E-02	2.00E-02	3.00E-02	4.00E-02	5.00E-02	6.00E-02	7.00E-02
2oo3	7.50E-05	3.00E-04	6.75E-04	1.20E-03	1.88E-03	2.70E-03	3.68E-03

TEST INTERVAL (YEARS)

Figure 4. Degradation of Automation Graph

3.3 *Use Automated Safeguards*

Automated safeguards (also referred to as engineered safeguards) can be used as compensating measures, providing a "backup plan" to manage the risk gap caused by the failure of a SIS device or while online maintenance is performed. Ideally, these automated safeguards would be identified either during the risk analysis or early in the design phase of the SIS. For example, a safeguard that was not identified as a protection layer might provide acceptable protection during the out-of-service period.

The key is that the compensating measure must be available and effective when the out-of-service condition occurs until the SIS is restored to full operation [4]. To be available, the automated safeguard must be sufficiently independent of the SIS equipment that is out-of-service and of the initiating causes of the hazardous event the SIS is protecting against. To be effective, the compensating measure must be capable of detecting the abnormal condition and acting to stop its propagation within the process safety time [5]. It should provide sufficient risk reduction to close the risk gap for the time required to restore the SIS. The devices used to implement the compensating measure would be included in the asset integrity program to ensure they are inspected, tested, and maintained [1,6].

3.4 Divert Operators to Manage Risk

Operators are trained to monitor the process and to execute manual actions when abnormal operation occurs. An operator can be part of an effective compensating measure when there is sufficient time to detect and act. Even when well-planned, trained, and practiced [7], industry experience has determined that there should be at least 10 minutes of response time in addition to the time needed to execute the procedure. The recommended cognition time increases if the procedure requires the operator to do extensive troubleshooting to determine the corrective action. For some hazardous events, the process safety time is simply too short for an operator to be an effective substitute for an automated safeguard [8].

There is also a practical limitation on the amount of risk reduction that can realistically be achieved by an operator. Numerous performance-shaping factors affect the likelihood of human action. The stress of the hazardous event propagation can negatively impact many of these factors [9]. As depicted in the hierarchy of controls documented in the CCPS Guidelines for Inherently Safer Chemical Processes [10], fully automated measures are generally more effective than those dependent on operator actions. If the risk gap is over 10, operator response alone will likely be insufficient.

Ensuring sufficient independence of the operator from the initiating causes of the hazardous event and the other protection layers can also be a challenge. If the hazardous event is caused by human error, will the operator responsible for acting as the compensating measure during the out-of-service period be independent of that error? Similarly, if an operator's response to process deviation is already listed as a protection layer against the hazardous event, how independent can the operator acting as the compensating measure be? For the compensating measure to be effective, the operator's attention must be focused on the potential need to act quickly. Managing multiple ongoing operational and even safety priorities could impede the operator's ability to properly execute the specified action in a timely manner.

4 Peak Risk Analysis (PRA) Example

Let's consider a furnace system (Figure 5). One of the primary hazards associated with furnaces is the accumulation of unburned fuel in the firebox, which potentially leads to an uncontrolled deflagration. One of the possible initiating causes is the fuel gas pressure regulator valve malfunctioning, leading to high fuel gas pressure and flame out. Typically, the process safety time from the point of pressure regulator failure to the presence of 100% LEL (Lower Explosive Limit) is on the order of seconds. To prevent the hazardous event, a SIS is installed to detect high fuel gas pressure and automatically shut down the fuel gas block valves.

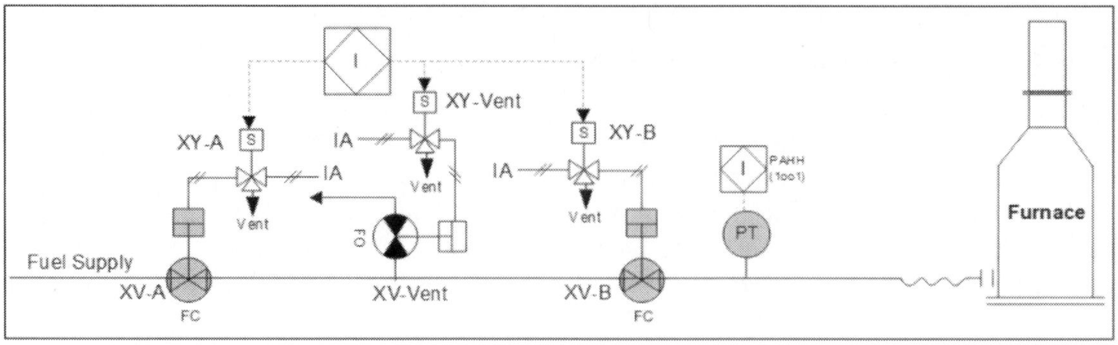

Figure 5. Simplified Process Overview

4.1 *Initial Design*

The initial design shown in Figure 5 relies on a simplex pressure transmitter PT (1oo1 voting). The SIS closes the double block valves, XV-A/B (1oo2 voting), on detection of high pressure. The SIS also actuates XV-Vent, but it is not part of the safeguarding action, so XV-Vent is not included in the SIL calculation. For the purposes of this example, let's assume that the risk reduction requirement is 100 or SIL 2.

4.1.1 *Performance of the initial design without degradation*

The results of the SIL calculation (Figures 6 and 7) show that the SIS as designed meets the target performance, assuming a test interval of 1 year for the pressure transmitter and fuel gas block valves. The calculation assumes 3 year testing for the logic solver given its high availability.

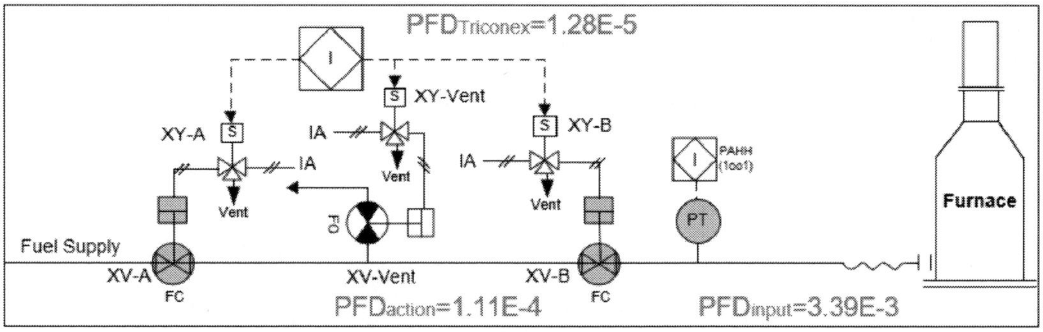

Figure 6. Safety Function Initial Design with PFDavg

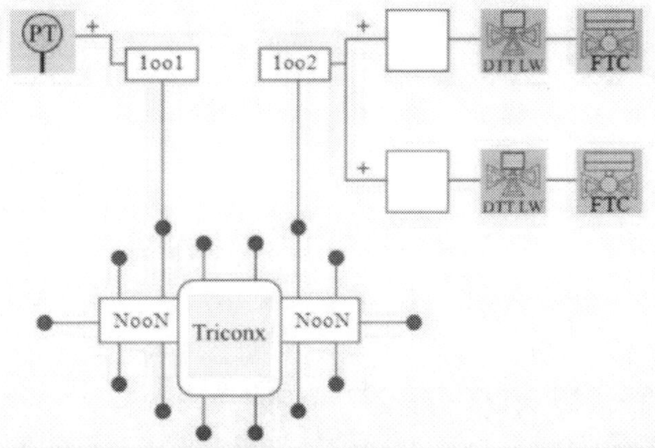

Calc	Targets	Results	Targets Met?
RRF	100	284.7185	
PFDavg	8.00E-03	3.51E-3	
IL	2	2	
STR (1/yr)	1.00E-01	6.05E-2	
MTTFs (yr)	10.00	16.54	
HFT	0	0	

Subsystem	Voting	PFDavg	STR	CCF	HFT	Tag	Description	PFDavg	STR	DI/PI (hr)	TI (yr)	Overhaul Interval (yr)	MTTR (hr)	DC	PTC
Fuel Gas Pressure	1oo1	3.39E-3	1.25E-2	N/A	0										
	1oo1				0	PT	Pressure Transmitter	3.39E-3	1.25E-2	0.5	1	20	72	NO DC	100
Fuel Gas Block Valves	1oo2	1.11E-4	4.67E-2	1	1										
	1oo1				0	XY-A	XV-A Solenoid	2.90E-3	1.67E-2	0	1	20	72	NO DC	100
	1oo1				0	XV-A	Fuel Gas Valve	3.76E-3	6.67E-3	0	1	20	72	NO DC	100
	1oo1				0	XY-B	XV-B Solenoid	2.90E-3	1.67E-2	0	1	20	72	NO DC	100
	1oo1				0	XV-B	Fuel Gas Valve	3.76E-3	6.67E-3	0	1	20	72	NO DC	100
Triconex Logic Solver	2oo3D	1.28E-05	1.30E-03		1										
						TMR	Triconex Logic Solver				3				

Figure 7. Initial Design SIL Calculation [11]

4.1.2 *Performance of the initial design with sensor degradation*

For the initial design, what happens when the pressure transmitter becomes unavailable, either due to random failure or online maintenance? Since the SIS detects a hazard using only one transmitter, the loss of the transmitter results in the SIS being disabled. The risk gap is an RRF of 100. For this process application, it is not feasible to move to an inherently safer process condition while the transmitter is under repair. The risk gap is larger than 10 and the process safety time is very short, so diverting an operator to manage the risk would likely be ineffective. Assuring continued safe operation during this peak risk event will depend on either implementation of an automated safeguard as a compensating measure or enhancing the fault tolerance of the SIS.

4.2 *Proposed Design Change*

As discussed previously, one way to address a risk gap is to enhance the SIS design to include redundancy. In this example, the proposed design changes the pressure transmitter configuration from 1oo1 to 2oo3 (Figure 8). The automation programming treats a detected failure as a vote to trip, which changes the voting logic from 2oo3 to 1oo2. To protect against human errors during the online maintenance (e.g., pulling out the in-service transmitter), the voting logic will change from 2oo3 to 2oo2 on bypass of a sensor instead of treating the bypass condition as a vote to trip. The automation design will limit the maintenance bypass to a single transmitter without affecting the process variable indications or alarms from the other two transmitters. The SIS logic solver will generate an alarm if any transmitter deviates from the mid-selected transmitter by more than 5% for more than 1 minute.

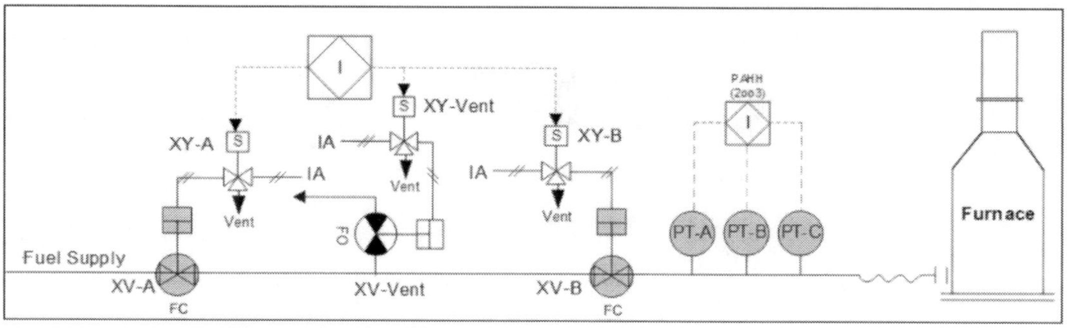

Figure 8. Simplified Process Overview (Proposed Design)

4.2.1 Performance of the proposed design without degradation

The revised SIS meets the target performance at a longer test interval of 3 years for all SIS devices (Figures 9 and 10). This improvement is largely due to the external diagnostics provided by the comparison of signals from the redundant pressure transmitters.

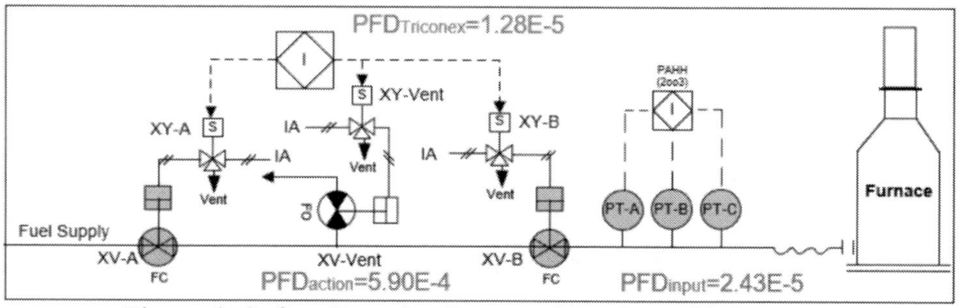

Figure 9. Safety Function Proposed Design with PFDavg

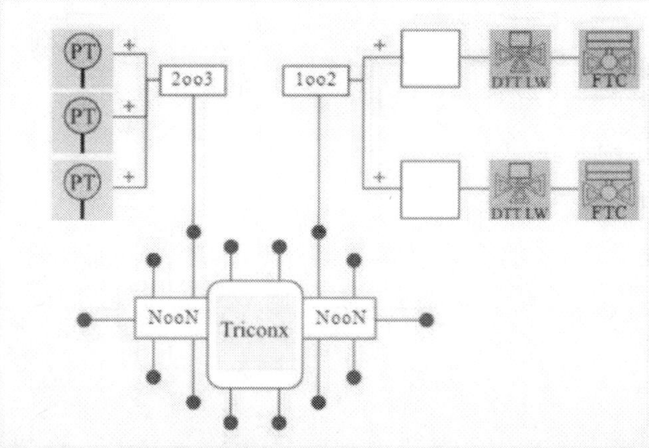

Calc	Targets	Results	Targets Met?
RRF	100	1595.7727	
PFDavg	8.00E-3	6.27E-4	
IL	2	3	
STR (1/yr)	1.00E-01	4.80E-2	
MTTFs (yr)	10.00	20.84	
HFT	0	1	

Subsystem	Voting	PFDavg	STR	CCF	HFT	Tag	Description	PFDavg	STR	DI/PI (hr)	TI (yr)	Overhaul Interval (yr)	MTTR (hr)	DC	PTC
Fuel Gas Pressure	2oo3	2.43E-5	7.73E-6	2	1										
	1oo1				0	PT-A	Pressure Transmitter	1.06E-3	1.25E-2	0.5	3	20	72	DC 3	100
	1oo1				0	PT-B	Pressure Transmitter	1.06E-3	1.25E-2	0.5	3	20	72	DC 3	100
	1oo1				0	PT-C	Pressure Transmitter	1.06E-3	1.25E-2	0.5	3	20	72	DC 3	100
Fuel Gas Block Valves	1oo2	5.90E-4	4.67E-2	1	1										
	1oo1				0	XY-A	XV-A Solenoid	8.62E-3	1.67E-2	0	3	20	72	NO DC	100
	1oo1				0	XV-A	Fuel Gas Valve	1.12E-2	6.67E-3	0	3	20	72	NO DC	100
	1oo1				0	XY-B	XV-B Solenoid	8.62E-3	1.67E-2	0	3	20	72	NO DC	100
	1oo1				0	XV-B	Fuel Gas Valve	1.12E-2	6.67E-3	0	3	20	72	NO DC	100
Triconex Logic Solver	2oo3D	1.28E-05	1.30E-03		1										
						TMR	Triconex Logic Solver				3				

Figure 10. Proposed Design SIL Calculation [11]

4.2.2 Performance of the proposed design with random sensor failure

Now, what happens when the pressure transmitter becomes unavailable due to a random failure? The SIS is configured to vote the first fault towards the trip, which changes the effective transmitter voting from 2oo3 to 1oo2 (Figures 11 and 12). The degraded SIS still meets target performance. No compensating measure would be required.

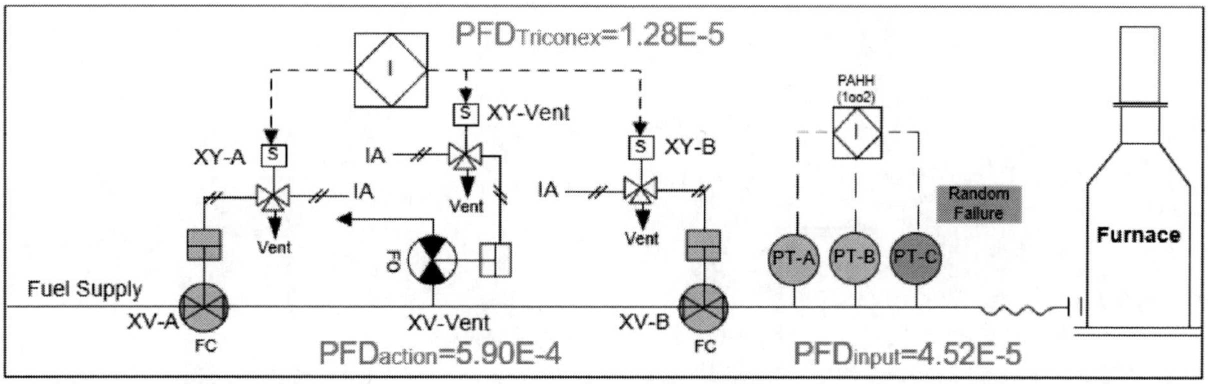

Figure 11. Proposed Design Process Overview with PFDavg (Random Hardware Failure)

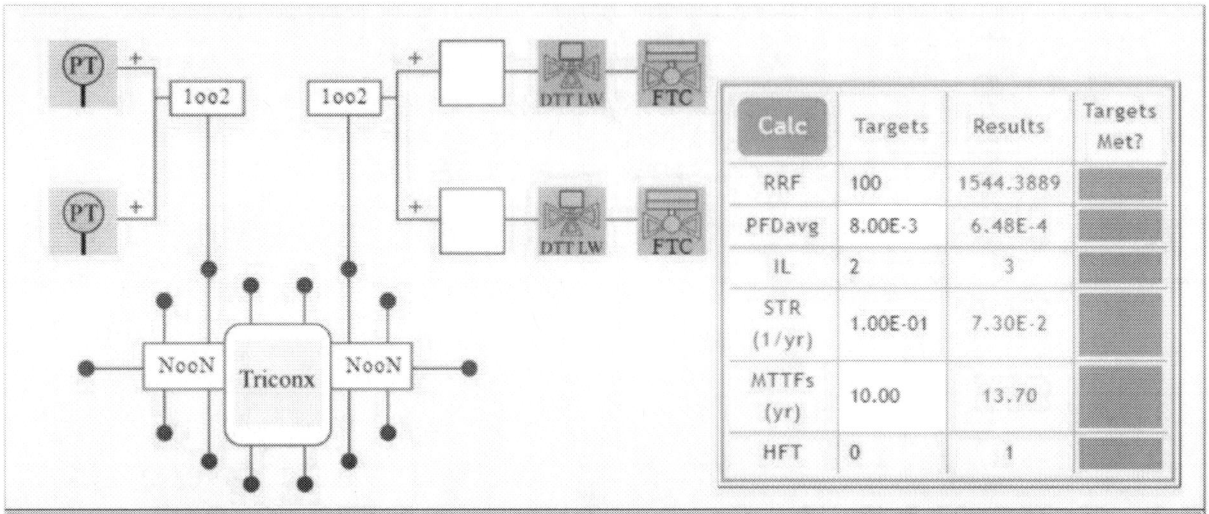

	Targets	Results	Targets Met?
RRF	100	1544.3889	
PFDavg	8.00E-3	6.48E-4	
IL	2	3	
STR (1/yr)	1.00E-01	7.30E-2	
MTTFs (yr)	10.00	13.70	
HFT	0	1	

Subsystem	Voting	PFDavg	STR	CCF	HFT	Tag	Description	PFDavg	STR	DI/PI (hr)	TI (yr)	Overhaul Interval (yr)	MTTR (hr)	DC	PTC
Fuel Gas Pressure	1oo2	4.52E-5	2.50E-2	2	1										
	1oo1				0	PT-A	Pressure Transmitter	2.06E-3	1.25E-2	0.5	3	20	72	DC 2	100
	1oo1				0	PT-B	Pressure Transmitter	2.06E-3	1.25E-2	0.5	3	20	72	DC 2	100
Fuel Gas Block Valves	1oo2	5.90E-4	4.67E-2	1	1										
	1oo1				0	XY-A	XV-A Solenoid	8.62E-3	1.67E-2	0	3	20	72	NO DC	100
	1oo1				0	XV-A	Fuel Gas Valve	1.12E-2	6.67E-3	0	3	20	72	NO DC	100
	1oo1				0	XY-B	XV-B Solenoid	8.62E-3	1.67E-2	0	3	20	72	NO DC	100
	1oo1				0	XV-B	Fuel Gas Valve	1.12E-2	6.67E-3	0	3	20	72	NO DC	100
Triconex Logic Solver	2oo3D	1.28E-05	1.30E-03		1										
						TMR	Triconex Logic Solver				3				

Figure 12. Proposed Design SIL Calculation for Degraded 1oo2 Voting [11]

4.2.3 Performance of the proposed design with sensor in maintenance bypass

Per the proposed design, the SIS is configured to exclude a bypassed sensor from the vote. This changes the transmitter voting from 2oo3 to 2oo2. The degraded SIS still meets the performance target (Figures 13 and 14). Therefore, maintenance bypass of a sensor in this design would also not require any compensating measures.

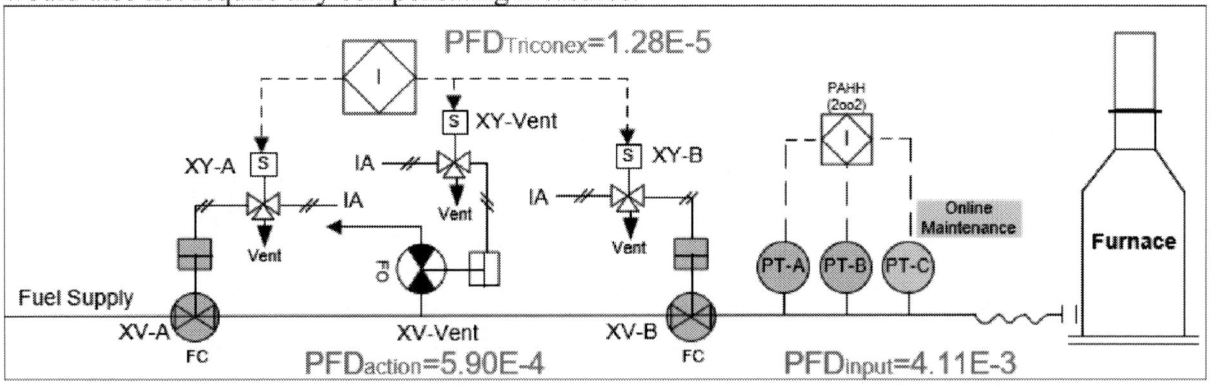

Figure 13. Proposed Design Process Overview with PFDavg (Online Maintenance)

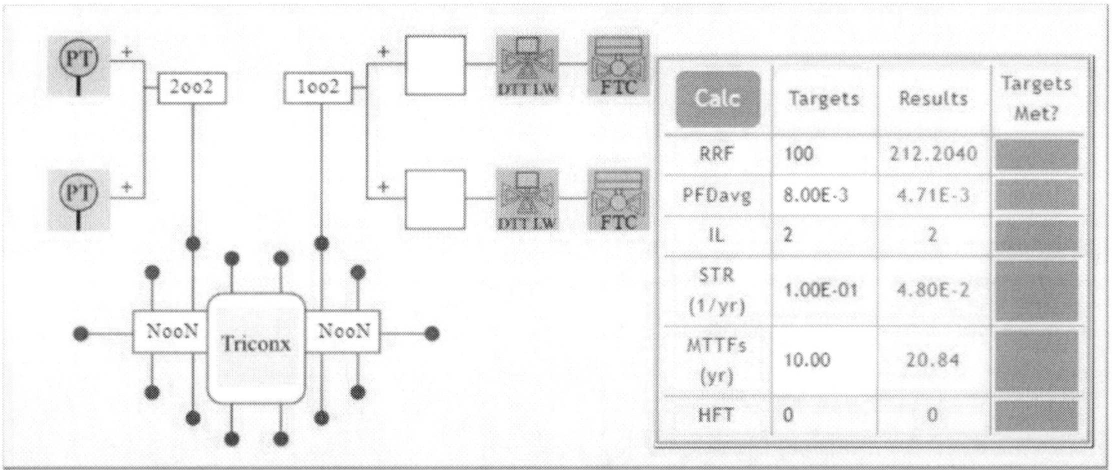

Subsystem	Voting	PFDavg	STR	CCF	HFT	Tag	Description	PFDavg	STR	DI/PI (hr)	TI (yr)	Overhaul Interval (yr)	MTTR (hr)	DC	PTC
Fuel Gas Pressure	2oo2	4.11E-3	2.58E-6	N/A	0										
	1oo1				0	PT-A	Pressure Transmitter	2.06E-3	1.25E-2	0.5	3	20	72	DC 2	100
	1oo1				0	PT-B	Pressure Transmitter	2.06E-3	1.25E-2	0.5	3	20	72	DC 2	100
Fuel Gas Block Valves	1oo2	5.90E-4	4.67E-2	1	1										
	1oo1				0	XY-A	XV-A Solenoid	8.62E-3	1.67E-2	0	3	20	72	NO DC	100
	1oo1				0	XV-A	Fuel Gas Valve	1.12E-2	6.67E-3	0	3	20	72	NO DC	100
	1oo1				0	XY-B	XV-B Solenoid	8.62E-3	1.67E-2	0	3	20	72	NO DC	100
	1oo1				0	XV-B	Fuel Gas Valve	1.12E-2	6.67E-3	0	3	20	72	NO DC	100
Triconex Logic Solver	2oo3D	1.28E-05	1.30E-03		1										
						TMR	Triconex Logic Solver				3				

Figure 14. Proposed Design SIL Calculation for Degraded 2oo2 Voting [11]

5 Conclusion

Continuing process operation during either the repair of a failed SIS device or execution of planned online maintenance of the SIS can create a peak risk event to the operating personnel. Managing this risk, and thus assuring safe operation during this period of time, can be done through a variety of means. These include adjusting the process to a less hazardous condition, enhancing the SIS design to be more fault tolerant of a device being out-of-service, or deploying a compensating measure to address the risk gap until the SIS can be restored to full operating condition.

Understanding which of these approaches will be effective requires consideration of the magnitude of the peak risk, the initiating causes of the hazard, the nature of available safeguards, and the process safety time of the hazardous event. Identifying this "backup plan" during SIS design ensures any necessary hardware is included up-front and the associated procedures are in place. Having well-designed and well-managed compensating measures in place goes a long way in ensuring that the choice to remain online with a known SIS fault is safe. Though this paper focuses on SIS, PRA can be applied to any automated safeguard.

6 References

[1] ISA, ANSI/ISA-61511-1:2018, Functional Safety: Safety Instrumented Systems for the Process Industry Sector - Part 1: Framework, Definitions, System, Hardware and Software Requirements, Research Triangle Park, NC (2018).

[2] CCPS/AIChE, Guidelines for Enabling Conditions and Conditional Modifiers in Layer of Protection Analysis, Wiley-Interscience, New York (2020).

[3] U.S. OSHA. 1992-2016. Occupational Safety and Health Standards: Process safety management of highly hazardous chemicals, 29 CFR 1910.119. Washington D.C.: OSHA.

[4] D Lewis and F Henselwood, "Safety System Impairment and the Need to Manage Peak Risk" Presented at AIChE 8th Global Congress on Process Safety, Houston, TX, April 1-4, 2012.

[5] E Roche and A Summers, "Are your safeguards fast enough to be effective?" Presented at AIChE 13th Global Congress on Process Safety, San Antonio, TX, March 27, 2017.

[6] ISA, ANSI/ISA-84.91.01-2021. Identification and Mechanical Integrity of Process Safety Controls, Alarms and Interlocks in the Process Industry Sector, Research Triangle Park, NC (2021).

[7] CCPS/AIChE, Guidelines for Preventing Human Error in Process Safety, Center for Chemical Process Safety, New York (2004).

[8] CCPS/AIChE, Guidelines for Initiating Events and Independent Protection Layers in Layer of Protection Analysis, Wiley-Interscience, New York (2014).

[9] CCPS/AIChE, Guidelines for Safe and Reliable Instrumented Protective Systems, Wiley-Interscience, New York (2007).

[10] CCPS/AIChE, Guidelines for Inherently Safer Chemical Processes, Wiley-Interscience, New York (2020).

[11] SIS-TECH: SIL calculation software. "SIL Solver Enterprise v2.0", by SIS-TECH Solutions, LP, Houston, Texas (2023)

Optimization of QRA Tools & PowerBI Interrogation to Support Agile Risk-Informed Decision Making

Ganesh Mohan
Senior Quantitative Risk Engineer
Chevron Technical Center, Houston, TX, USA
Email: ganesh.mohan@chevron.com

Jim Salter
Chevron Fellow & Senior Principal Quantitative Risk Engineer
Chevron Technical Center, San Ramon, CA, USA
Email: jsalter@chevron.com

Rod Travis
Technical Safety Engineering Manager
Chevron Technical Center, San Ramon, CA, USA
Email: Rod.Travis@chevron.com

Keywords: Consequence Modeling, Quantitative Risk Assessment, QRA, Detailed Design and Manage Risks

Abstract

Quantitative Risk Assessments (QRAs) have helped assure the oil & gas industry and stakeholders (neighbors and regulators) that our facilities can be operated safely and have identified areas where the risk of major accident events can be better managed. QRA tools available in the industry necessitate a premium of cost & time in the quest for required quality. This creates a challenge for projects & operating assets in making risk informed decisions in a timely manner.

QRAs are typically executed by third-party contractors, taking months to complete at a significant cost. The reliance on third parties has contributed to a lack of understanding of the results of these analyses by the end users. This also inhibits the ability to use QRA to make informed decisions related to management of change or evaluate risk reduction alternatives once an asset is operating, given the third-party interface and long lead time required to make changes.

Chevron in partnership with a commercial software vendor, optimized both onshore and offshore QRA applications, reducing cycle time & cost by over 70%. These applications support agile project development, provide the data needed to make day to day operations decisions, analyze various potential risk reduction strategies in real time, and reduce reliance on third parties.

Chevron also developed in-house Power BI tools to interrogate QRA results such that the top risk contributors can instantly be traced back all the way down to a single release event. The published QRA results in Power BI can be accessed and interrogated by any user with no background or skillset to operate a QRA tool. This enables the customers to easily understand the QRA results and its top risk contributors.

Demonstration at this conference will include the fidelity of these tools, cost savings and interrogation of big data using Power BI to arrive at meaningful recommendations that have greater impacts to risk reduction. This presentation will also illustrate how these tools have been transformative in supporting agile project development and helping our operations better understand and manage process safety.

1 Introduction

In 2006, Chevron established an enterprise-wide requirement to quantitatively determine the risk of harm to people as the result of its operations (both Individual and Societal Risk). This was largely in response to the drive from within the industry and regulators (particularly outside US) that were prompted by numerous industrial incidents over the preceding years, notably those at Flixborough, Seveso, Bhopal, Piper Alpha and Buncefield [1].

Detailed guidance on the application of Quantitative Risk Assessments (QRA) in the chemical process industry was provided in 1989 via CCPS's first edition of "Guidelines for Chemical Process Quantitative Risk Analysis" [2]. Early applications of QRA methodologies were employed in Europe, notably at Canvey Island east of London and Rijnmond, in the Netherlands and gained traction following the Seveso directive, its UK equivalent, the Control of Industrial Major Accident Hazards Regulations and the North Sea Safety Cases which became mandatory throughout the UK North Sea offshore industry following the Cullen report on the Piper Alpha disaster [3].

Existing facilities demonstrate proof of safety levels through modeling of QRAs. QRAs are technically challenging and time-consuming activities that require high levels of expertise from the analyst as they need to identify main risk drivers both onsite and offsite. QRAs can also provide insights to support Management of Change (MOC), Turn Around Reviews (TAR), Building Siting Assessments (BSA), Egress, Evacuation & Rescue Assessments (EERA), etc.

For greenfield projects, most of the inherent risk profile is defined in the earliest stages of project design due to process selection, layout, and location. However, performing QRAs early in a project is problematic due to lack of definition of the design. Then, as projects evolve changes to the design can occur rapidly, particularly during early Front-End Engineering Design (FEED) such that traditional QRA approaches are unable to keep pace. However, the efforts associated with design changes increase exponentially as the project progresses such that it can be difficult to resolve risk issues that are identified by QRAs later in the project lifecycle (per Figure 1).

Because QRAs are time consuming and complex to execute, most QRA work is typically performed by specialist third party consultants using proprietary tools. However, the use of consultants can produce inconsistent results due to differing assumptions and modeling tools. Care must be exercised in the type of models used and the personnel who run them. For example, over-prediction of blast overpressures may lead to higher costs and unnecessary mitigation. On the contrary, under-predictions can have serious consequences in the event of an incident. Critical to the analysis is using the appropriate model and having qualified personnel conduct the modeling.

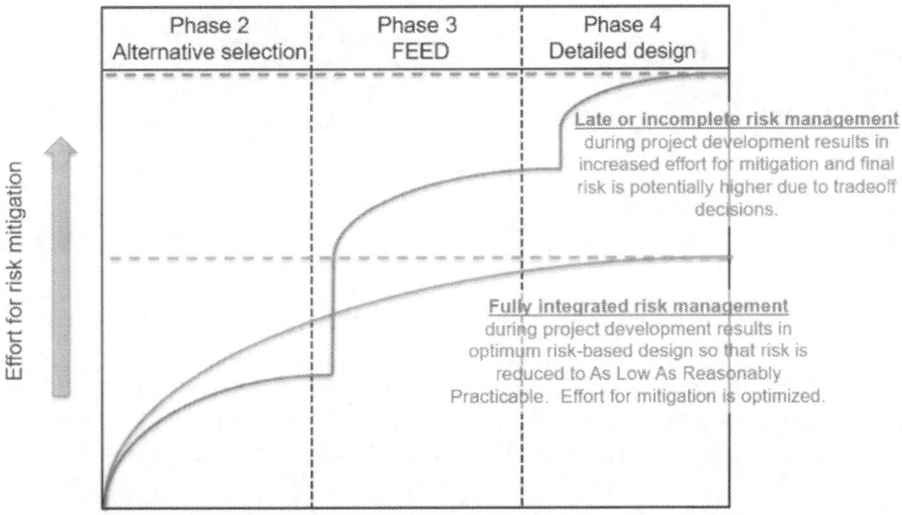

Figure 1: Effort for Risk Mitigation on Projects

As a result, Chevron chose to develop tools to produce QRA-lite products called Chevron Onshore Risk Quantification App (CORQA) and Chevron Offshore Risk Screening App (CORSA) which allow relatively non-specialist personnel to build a screening level QRA model for all types of facilities that Chevron operates. These QRA-lite products are founded on standard QRA techniques that have the following main advantages:

- Uses pre-populated process design blocks and/or equipment packages.
    - Both frequency and consequence source terms are defined but scalable for the specific application.
    - Can be combined to represent most potential projects in Chevron's operations.
- Usable with limited training of non-specialist staff.
- Use of commercial calculation engines reduces support time and includes the latest modeling.

2 Chevron Onshore Risk Quantification App (CORQA)

CORQA is a simplified front end that provides units operations with pre-defined scenarios and equipment blocks. Under the hood, it uses a commercial QRA software for the consequence and risk calculations. The front-end user interface is simple and easy to use whilst providing flexibility in the definitions of equipment counts and process conditions such that an entire onshore facility can be built quickly on a scaled map as shown in Figure 2.

Figure 2: Example of onshore facility built in CORQA

An example of the pre-populated process units is shown in Figure 3. The release scenarios are identified by the asterisks and the part counts are pre-defined for the major assemblies within the process unit, but the number of assemblies is defined by the user.

Figure 4 shows typical release point information for each of the scenarios defined for the unit operation. The compositions are pre-defined, but the scenario is modifiable to allow variations in temperature, pressure, and inventory as well as the concentration of toxics within the stream (in this case H2S) to be defined.

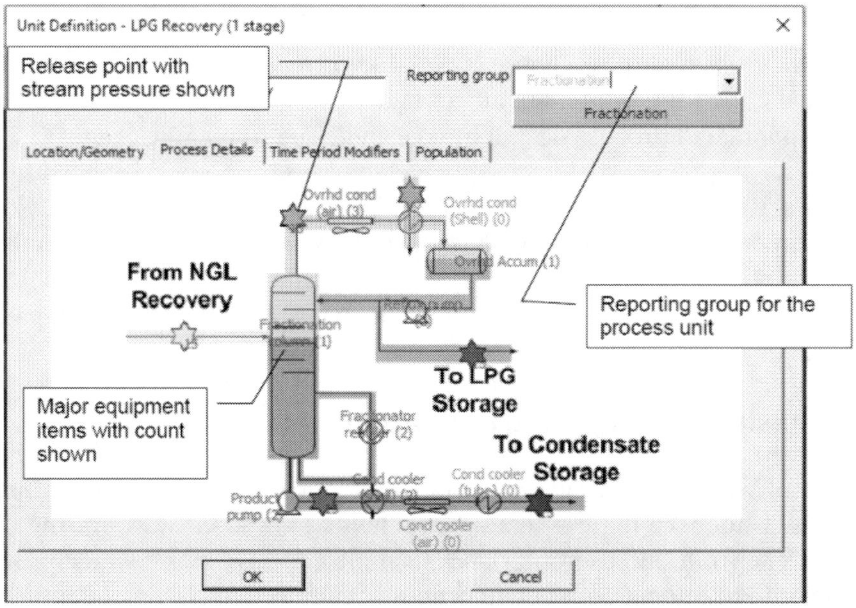

Figure 3: Typical Process Unit Setup (LPG Recovery) in CORQA

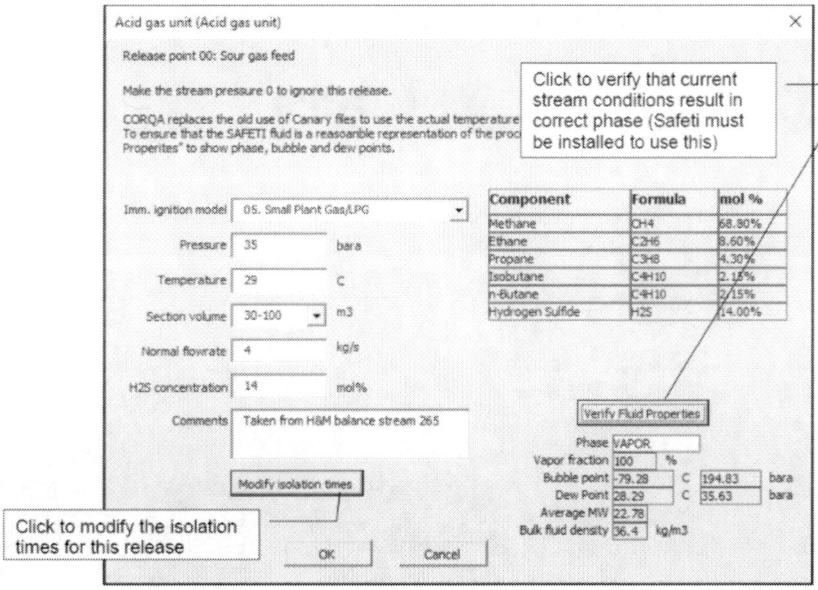

Figure 4: Typical Release Point Inputs (Sour Gas Feed) in CORQA

3 Chevron Offshore Risk Screening App (CORSA)

CORSA is a customized Chevron user interface to a commercial cloud based QRA software. Conceptually like CORQA, an offshore facility can be built as a simple Process Flow Diagram (PFD) model using pre-populated equipment packages. The packages have pre-defined release scenarios included with them as well as parts counts for each major assembly. As in CORQA, process conditions can be varied but compositions are fixed. See Figure 5 for an example PFD built in CORSA and Figure 6 for typical separator assembly inputs. CORSA also provides 3D visualization for easy verification while building an offshore geometry.

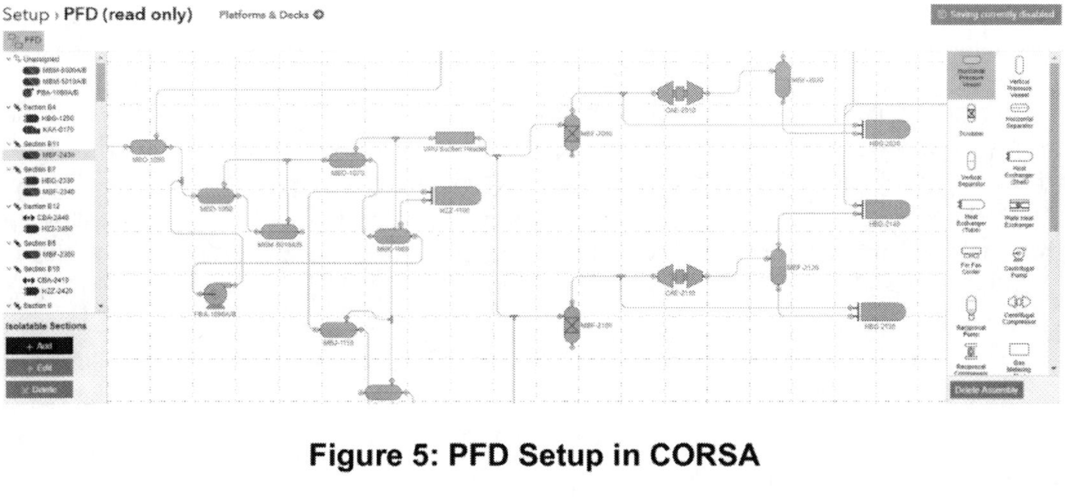

Figure 5: PFD Setup in CORSA

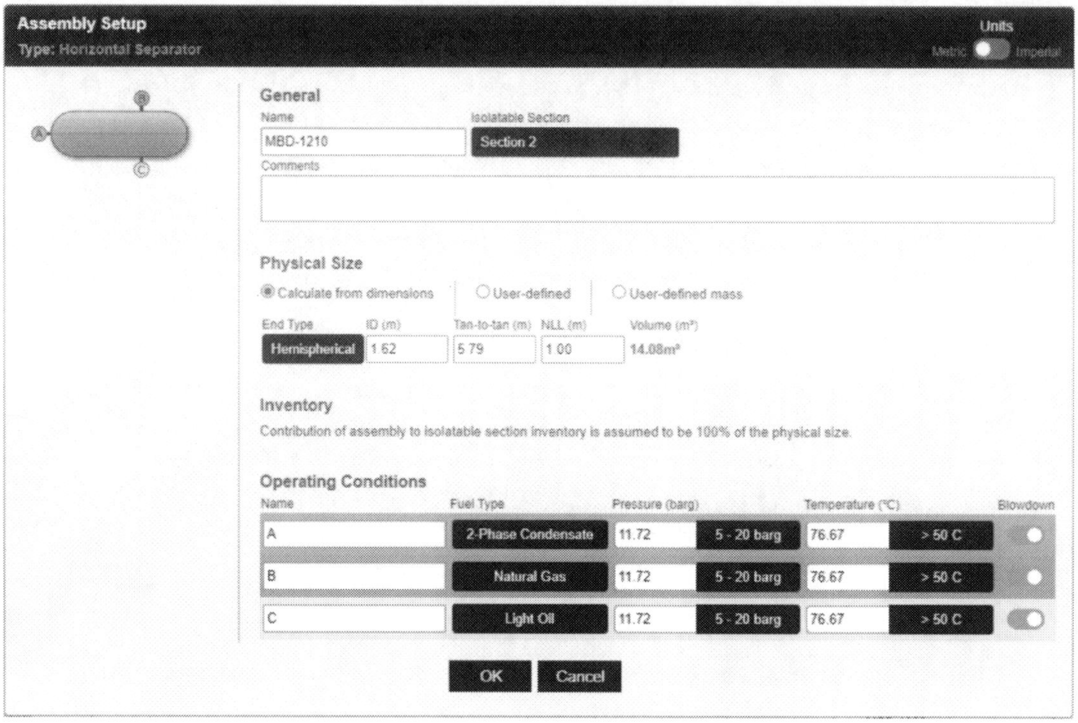

Figure 6: Assembly Setup in CORSA

4 Comparison of QRA tools

Both CORQA and CORSA require only PFD level information as inputs due to the presence of inbuilt package count frequencies and volume calculations at equipment level (separators, columns, tanks, pumps, heat exchangers, etc.) that are customized for Chevron facilities worldwide. QRA analysts appreciate excessive time saved in avoiding detailed Process & Instrumentation Diagram (P&ID) markup, parts count, inventory, and frequency calculations. However, they provide detailed results like any other QRA tool that uses P&ID level information for its calculations.

Table 1 provides a comparison of steps to model using a standard QRA tool vs the Chevron tools. Depending on the size of a facility, developing assumptions register (step 0), marking up P&IDs (step 1) and parts count / inventory calculation (step 2) for a regular QRA tool alone can take several weeks to months more than CORQA & CORSA. Although all other steps remain the same, the Chevron tools are also more user friendly and helps saving more time and cost to the end user.

Table 1. Comparison of Steps to Model in a typical QRA tool vs CORQA & CORSA

Steps	Typical QRA	CORQA	CORSA
0	Develop Assumptions Register	CORQA tool is prepopulated to Chevron's standard & hence no additional effort required.	CORSA tool is prepopulated to Chevron's standard & hence no additional effort required.
1	Identify & markup isolatable sections on Process & Instrumentation Diagrams (P&IDs)	Identify & markup isolatable sections on PFDs & match with available CORQA process units	Identify & markup isolatable sections on PFDs
2	Detailed Parts Count & Inventory Calculation for Isolatable Sections	Count the number of major equipment items (e.g., vessels, heat exchangers, pumps, tanks, etc.) within a typical CORQA Process Unit. Calculate inventory at equipment level with an assumed multiplier to accommodate for associated piping, etc.	Build PFDs within CORSA using equipment level data (e.g., vessels, heat exchangers, pumps, tanks, etc.) and group them by isolatable sections. Calculate inventory at equipment level and enter CORSA. The tool automatically adds additional inventory associated with associated piping, etc.
3	Gather release stream properties for each isolatable section and add them to the QRA model.		
4	Determination of leak frequency for representative release sizes (typically 4) for each isolatable section using historical database		
5	Determination of typical windspeeds, directions and stability		
6	Specific consequence modelling (to establish extent of potential harm) based on fluid characteristics, operating pressure / temperature, volume and flowrate at representative hole sizes and weather conditions. Modelling by Consultant provided Tool.		
7	Location of process equipment on plot plan		
8	Location / Type of Occupied Buildings on plot plan		
9	Location of personnel on plot plan		
10	Population density surrounding the facility		
11	Identification of Potential Explosion Sites on plot plan		
12	Identification of Ignition sources on plot plan		
13	Consequence modeling & risk calculation		

Validation exercises performed on the QRA-lite tools show generally good agreement with more detailed QRA approaches in several cases studied within Chevron. Although somewhat conservative (by design), the QRA-lite approach provides similar risk drivers to more detailed approaches when the same consequence engines are used. The approach has been successfully utilized in countries where the use of QRA is regulated.

The value of using these tools is multi-faceted:

- Chevron maintains ownership of the models, allowing them to be used for day-to-day risk management decisions rather than a one-time exercise.
- The models produced are known to be aligned with the requirements of Chevron's Risk Management process without a need for discussion / alignment on a set of assumptions.
- Models are significantly more consistent between facilities allowing better allocation of resources to hazard mitigation.
- Consistency of result presentation makes it easier for our business units to maintain their competency in result interpretation.
- The simplicity of the interface allows QRA analysts to keep up to date with the rapid rate of design changes often found in pre-FEED and FEED activities.

The short-term financial value of the approach was highlighted recently when we requested quote for a third party to carry out a full QRA for a major refinery. The cost was estimated to be more than $1,000,000. Using CORQA, we performed a similar analysis for less than $300,000. Similarly for an integrated offshore platform, the third party QRA work can range from $300,000 to $1,000,000 depending on its complexity (multiple bridge connected facility). Using CORSA, the analysis typically costs less than $100,000.

5 Power BI Interrogation of QRA Results

Both tools produce detailed QRA results such as Individual Risk per Annum (IRPA), FN Curves (Frequency of N or more Fatalities per year), Overpressure, Thermal and Toxic Exceedance curves. These results are presented in a Power BI format that enables interrogation through drilldown to individual release cases and hole size. Such interrogation allows mitigations to have the potential to affect the top contributors to risk.

As an example, Figures 7 to 11 shows Power BI drilldown of an overpressure exceedance curve that identifies the domination of rupture and large hole size releases from a handful of gas/gas heat exchangers. This enables the QRA analyst or end user to review the QRA input data for this scenario – verify frequency calculations and release properties such as pressure, temperature, flow rate, inventory, etc. Following a similar drilldown process, the user can also investigate the next highest contributing process stream and so on. Once it has been verified and confirmed that the scenario contributions are modeled correctly, then the analyst can develop potential mitigations such as relocation of a building (greenfield projects or addition of a new building), increase blast rating of the building, reduced inventories, etc.

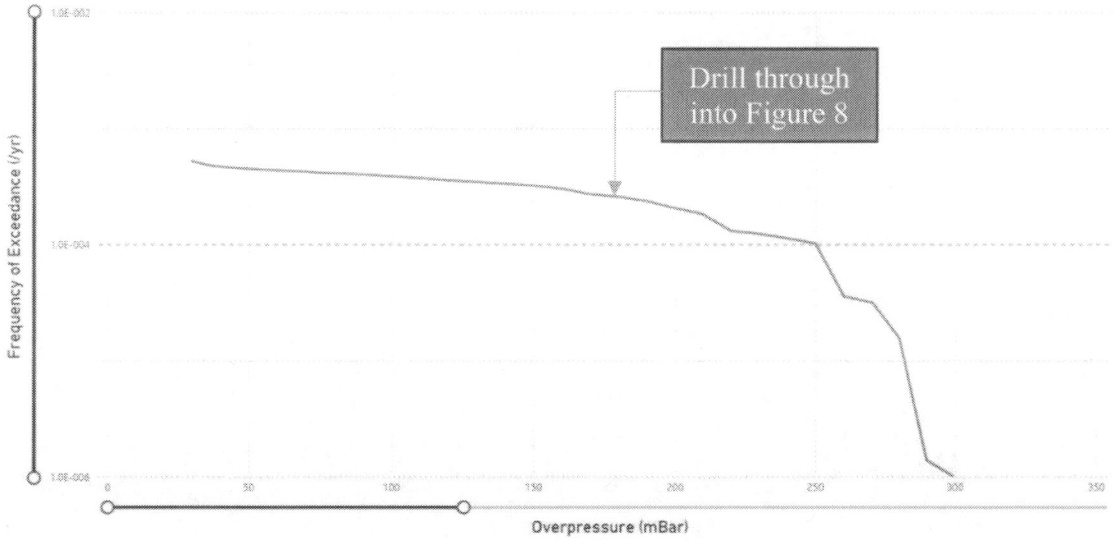

Figure 7: Overpressure Exceedance Curve Drilldown 1 of 5

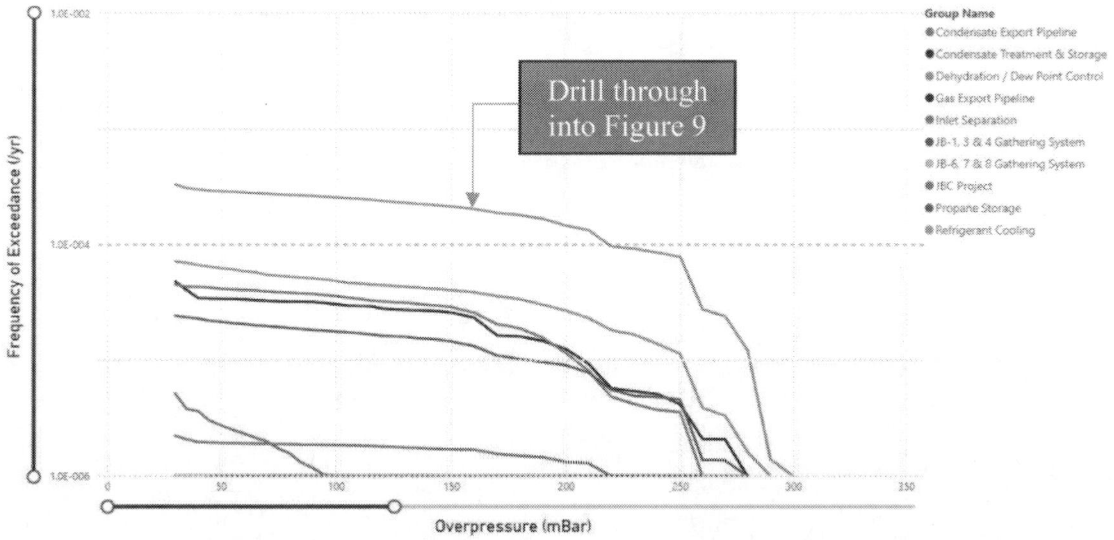

Figure 8: Overpressure Exceedance Curve Drilldown 2 of 5

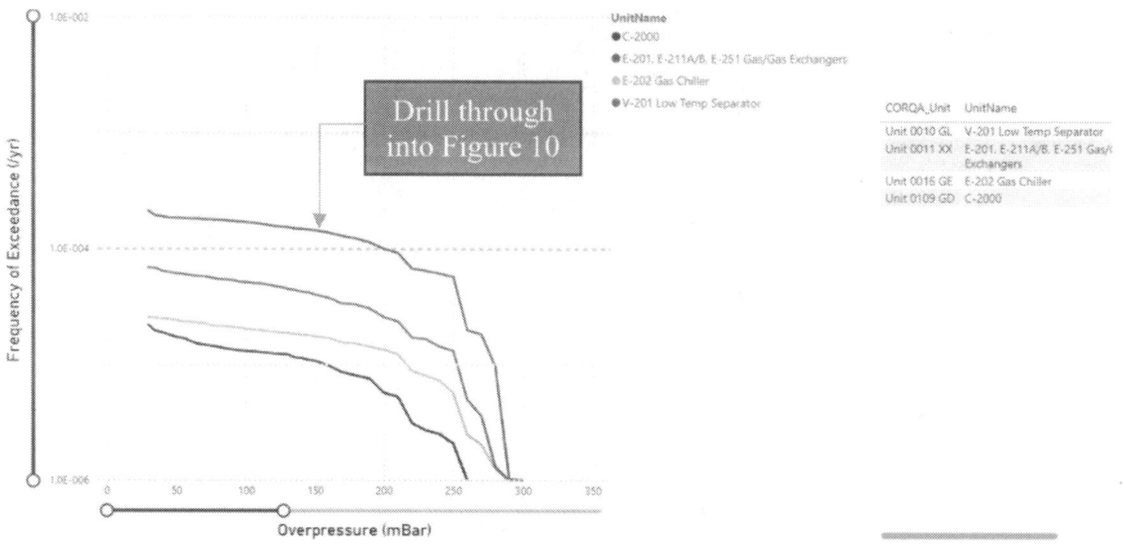

Figure 9: Overpressure Exceedance Curve Drilldown 3 of 5

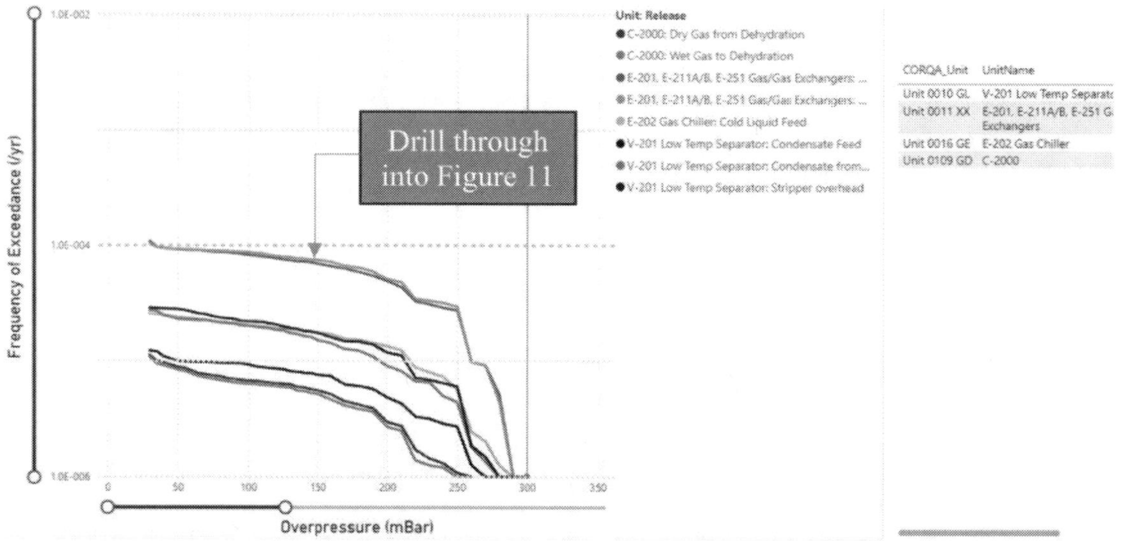

Figure 10: Overpressure Exceedance Curve Drilldown 4 of 5

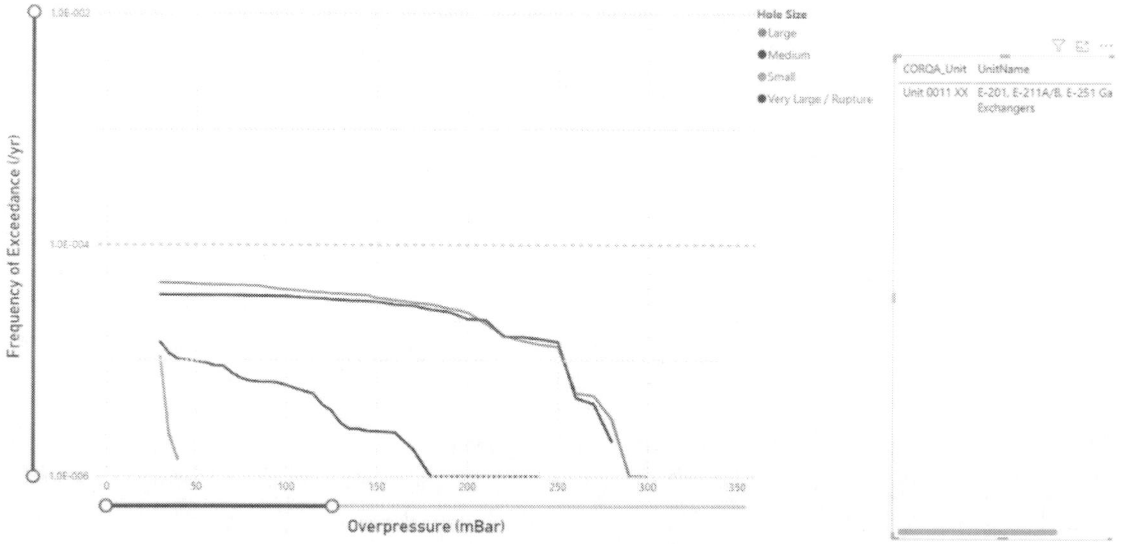

Figure 11: Overpressure Exceedance Curve Drilldown 5 of 5

The published QRA results in Power BI can be accessed and interrogated by any user with no background or skillset to operate a QRA tool. This enables the customers to easily understand the QRA results and its top risk contributors, enabling risk-informed decision making by process engineers and operations without requiring specialist skills or tools.

6 Conclusion

QRA techniques need to be streamlined and communicated to the design team to provide the most effective design information. Risk optimized designs require early and frequent feedback. QRA tools need to provide consistent means of assessing a project's potential risk profile that can be applied before design details are known & also easily modified to reflect changes in the design when needed.

Both CORQA and CORSA have been successfully deployed enterprise-wide at Chevron. In summary, these QRA-lite products demonstrate the following:

- Satisfies the requirements of Chevron's Risk Management process through assessment of individual & societal risks, vulnerability of egress routes & muster areas and the survivability of essential systems.
- Generates design accident loads for buildings & structures, consequence analysis for PHAs and supports layout optimization.
- Provides information in a matter of weeks or even days that previously took months, and therefore supports Chevron's vision of agile project development and helps us better understand and manage process safety.
- Through Power BI integration, it provides the ability to interrogate to a level of detail not previously available, allowing greater understanding of the risk drivers and day to day decision making.
- Provides an enhanced capability to support the development of meaningful and effective risk reduction plans.

- Using commercial software under the hood for both CORQA & CORSA enables the tools to remain up to date with the latest modeling without expensive software updates.
- Chevron independently customizes & updates the front end (QRA Input) and back end (Power BI Output) of the tools based on its internal users / customer requirements.

The key benefit in having a QRA model inhouse enables Chevron to maintain an evergreen model that can be updated whenever the facility is modified throughout its lifetime. This improves the management of risk of a facility as well as providing considerable savings in cost, time, and resources in maintaining the models. The ability to interrogate results using Power BI, empowers the end customers to arrive at meaningful recommendations that optimize risk reduction. At Chevron, these tools have been transformative in supporting agile project development and helping our operations better understand and manage process safety risk.

With the advancement of Artificial Intelligence (AI) and the ever-expanding global computational resource, our vision of the future of risk-informed decision making is that the QRA tools will continue to grow in sophistication. We see opportunities to perform dynamic risk modeling, where the facility risk can be calculated in real time based on understanding the current condition of process control devices and alarms. We see a future where a digital twin of a facility can be used to develop QRA models using AI and simulated within a CAD environment via Application Programming Interface (API). As computational power continues to grow, these QRA simulations can be performed almost instantly, and AI can interpret the results to provide practical recommendations to the operators. With the pace of evolution of AI and computational power, such a vision may not be far off.

7 References

[1] Mannan M.S., *Lees' Loss Prevention in the Process Industries*. 4th ed. Oxford, UK: Butterworth-Heinemann; 2012.

[2] CCPS. *Guidelines for Chemical Process Quantitative Risk Analysis*. 1st ed. New York, NY: American Institute of Chemical Engineers; 1989.

[3] Pasman, H., & Reniers, G. Past, present, and future of Quantitative Risk Assessment (QRA) and the incentive it obtained from Land-Use Planning (LUP). *Journal of Loss Prevention in the Process Industries* 2014, **28**: 2-9.

Using Time-Varying QRA to Evaluate Hazards During a Turnaround or Temporary Operations

Colin D Armstrong
AcuTech Group, Inc.
1919 Gallows Rd, Suite 900
Vienna VA, 22182
CArmstrong@acutech-consulting.com

Dr. Andrés Ruiz-Tagle and Richard MacNguyen
AcuTech Group, Inc.
1919 Gallows Rd, Suite 900
Vienna VA, 22182
ARuizTagle@acutech-consulting.com
RmacNguyen@acutech-consulting.com

Keywords: QRA, SIMOPS, Turnaround, Dynamic Risk, Safeti

Abstract

During periods of increased risk, such as turnarounds (T/A) or other simultaneous operations (SIMOPS), Process Hazard Analysis (PHA) and Hazard and Operability Studies (HAZOP) are often used to qualitatively assess risk to onsite populations. However, these tools lack the ability to quantitatively assess process risks as they change throughout operational phases, and PHA can easily become overburdened due to the complexity of the facility and number of ongoing operations.

This paper presents a methodology for performing a time-varying quantitative risk assessment (QRA) over a temporary period of increased risk, such as a turnaround. This paper will present options within DNV's Safeti 8.72 to perform the QRA and export results that clearly illustrate dynamic individual risk. Lastly, this paper will demonstrate the benefits of this methodology using a practical case study.

1 Introduction

During periods of increased risk, such as turnarounds (T/A) or other simultaneous operations (SIMOPS), Process Hazard Analysis (PHA) and Hazard and Operability Studies (HAZOP) are often used to qualitatively assess risk to onsite populations. However, these tools lack the ability to quantitatively assess process risks as they change throughout operational phases, and PHA can easily become overburdened due to the complexity of the facility and number of ongoing operations.

Quantitative Risk Assessment (QRA) and facility siting studies are often used to assess the risks of fires, explosions, and toxic hazards at refining and petrochemical facilities around the globe. These studies give powerful insights into the risk profiles at a facility, the nature of the hazards to people, and the processes that contribute to risk across the site. A QRA is generally used to assess the risks of normal operations to people onsite and offsite and people located near the process in onsite occupied buildings.

However, accidents during normal operations are only a fraction of the catastrophic events that occur in the process industries. In fact, some of the most catastrophic incidents have occurred outside of normal operating periods, such as turnarounds (T/As), startups (S/U), shutdowns (S/D), and simultaneous operations (SIMOPS). According to CCPS, process safety incidents occur five times more often during "other than normal" operations, including during shutdown and startup[1]. There are several factors that contribute to these statistics and the historic events that underly them:

1. Increased risk of process upsets during non-routine operations, such as startup and shutdown, due to temporary modifications to equipment, process conditions, and procedures.
2. Increased risk of misalignment or human error as part of shutdown, isolation, de-inventory, maintenance, and startup activities due to the complexity of work scope.
3. Increased risks to process from external factors (e.g., adjacent work activities, vehicle traffic, lifting devices).
4. Increased intensity and stress on workers.
5. Increased numbers of workers present in and around hazardous process areas.
6. Increased use of temporary occupied structures.

For facilities that have completed a QRA or consequence-based siting study of the normal plant operation, non-routine operations may be assessed with a relatively straightforward approach that provides valuable information for planning. A time-varying, dynamic QRA may be used to assess a complete facility T/A, including SIMOPS. An example study is presented here, using the DNV Phast and Safeti software for performing a QRA to evaluate hazards during a T/A. The example illustrates the methodology, presentation of results, and development of practical risk-based recommendations to protect workers. These methods are applicable for a range of facility siting studies, the assessment of other non-routine operations, and a variety of approaches/software tools as well.

2 Defining the Scope

The first step is to define the time periods (TPs) that will be assessed in the dynamic QRA. This is dependent on the operation being planned. During a large multi-unit T/A, or other SIMOPS case, start-up, shutdown, and maintenance/construction activities are staged over a period of time. The first phase of the dynamic QRA is to define the distinct operating/risk time periods for the operation. Coordinate with the planning team to review the S/D, maintenance, and S/U operations that will be executed, and the order of operations. The following items may be considered to define the time periods for the QRA:

1. Commencement of S/D operations- increases risk of upsets.
2. Completion of S/D and de-inventory- reduces risk of upsets.

3. Commencement of S/U operations- increased risk of upsets.
4. Return to normal operation, S/U complete- Risk returns to baseline level.

For a multi-unit T/A, S/D and S/U operations for multiple units must be considered through overlapping time periods to define operating and risk profile periods for the overall operation. The time periods do not refer to specific dates, but rather milestones related to the four criteria listed above. It is critical that these periods be communicated as T/A milestones, not specific dates. S/U and S/D milestones need to be achieved for the risk profiles to shift to the next period, and delays or unexpected issues are common during non-routine operations, so schedules often shift from their original plan.

As the T/A progresses, the time period is defined based on the progress of S/D and S/U activities. Figure 1 shows an illustration of T/A time periods and the associated operating profile of units for a dynamic QRA. In this example, 18 units at a facility are evaluated through a T/A involving 24 distinct time periods. In the example below, TP1 represents baseline normal operations before any S/D activities start, and TP2 starts when all units begin S/D operations. TP3 is reached when Units 1-27 and 6-43 complete shutdown activities, thereby eliminating loss of containment scenarios in the unit.

Figure 1 - Turnaround Time Periods

	Time-period																							
Process Unit #	1	2	3	4	5	6	7	8	9	10	11	12	13	14	15	16	17	18	19	20	21	22	23	24
1-30						X							O											
1-80				X										O										
1-27			X																			O		
1-47						X						O												
1-71					X												O							
1-18							X									O								
3-63									X								O							
3-80										X									O					
3-83						X								O										
3-65						X													O					
6-94				X											O									
6-43			X																			O		
6-73										X				O										
6-64								X						O										
6-97				X										O										
7-90											X						O							
8-97									X			O												
9-96						X														O				
Number of Non-Operating units	0	0	2	5	6	11	12	13	15	17	18	16	15	10	9	8	5	5	3	2	2	0	0	0
Number of Normal Ops units	18	0	0	0	0	0	0	0	0	0	0	0	1	1	4	7	7	9	10	12	13	13	17	18

Color coding		Description
Normal ops		Unit operates normally
Online starts	O	Start-up operations start on unit
Online		Start-up operation are being carrierd on the unit
Shutdown		Shut-down operations are being carried on the unit
Offline starts	X	Unit stops to operate
Offline		Unit is not operating

Whereas the risk increases from TP1 to TP2 as units begin shutdown activities, the risk is reduced as units complete S/D activities, with TP11 being the lowest risk period during the T/A. During TP11, all units involved in the T/A are completely S/D and de-inventoried. As units begin startup operations, the risk increases until TP24 is reached, where all units have completed S/U and reached normal operations.

As units complete S/D and are de-inventoried for maintenance activities, the potential loss of containment (LOC) scenarios associated with the unit are removed or reduced, consistent with the T/A plans for S/D and de-inventory. As each time period is assessed, the LOC scenarios to be considered are defined based on the process equipment that is operating during that time period. As illustrated in Figure 2, the number of scenarios evaluated reduces as units are S/D, and rebounds to the original value as all units are started up. Note that even when all operating units are S/D, a number of LOC scenarios are still present. These may be potential LOCs from storage tanks holding products for S/U, process units that were not S/D, or utilities that remain in operation.

Figure 2 - Loss of Containment Scenarios in the Study

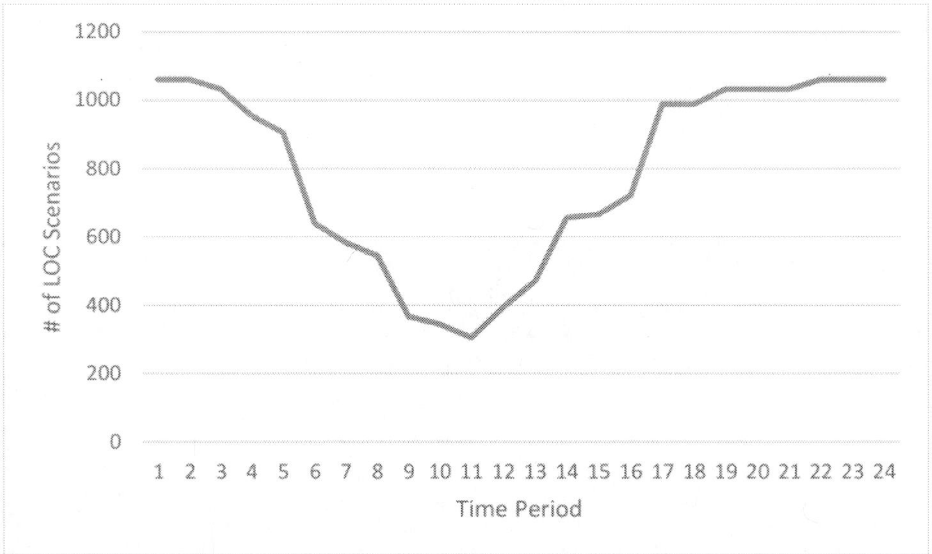

While a process area may be S/D and de-inventoried, eliminating the potential for a flammable or toxic release event, the unit will still present potential areas of obstruction for vapor cloud explosion hazards if flammable vapors can disperse into the unit. In fact, some major work efforts may erect temporary structures that increase or expand areas of obstruction.

A non-routine operation will frequently be accompanied by new populations of workers onsite. To define the scope of the dynamic QRA, it is crucial to identify areas where workers will be mobilized and define the types of work operations and temporary buildings that will be considered for use. These will generally be areas within and around units that will be S/D for construction or T/A activities.

It should also be noted that temporary workers are often exposed to hazards and risks that are not associated with the risks of catastrophic releases of highly hazardous chemicals that are the focus of a QRA. These risks include working at height, confined spaces, dropped objects, and vehicle impacts among others. Such risks are managed through job safety planning and pre-job hazard analyses (PJHAs) among other work safety activities but are not addressed by a dynamic QRA.

While accounting for these factors in the QRA may seem daunting, such studies can be executed with relative ease by implementing the right approach, starting from the initial site risk model.

3 Dynamic Assessment Approach

In the Safeti risk model, risk results are developed by combining results from separate model runs called run rows into a single combination. Run rows are generally used to divide models into toxic- and flammable-specific hazards/models, which may use different model parameters, different weather conditions, and evaluate risk to different populations, among others. Accordingly, a QRA of normal operations will generally have several run rows, representing, for instance, combinations of daytime and nighttime conditions, toxic and flammable hazards, and other hazard parameters. Additionally, Safeti allows the creation of a combination matrix to evaluate the risk of multiple run-row combinations.

The use of run rows and the combinations matrix may be exploited in Safeti for modifying a baseline static QRA model into a dynamic QRA representing time-varying conditions. The first step to do so is to divide the risk model into run rows that align with the operating profiles through the time periods identified for the non-routine operation. In our T/A example, each process unit is separated into its own unique run row. Separating the model into run rows for the dynamic QRA will mean dividing each run row into new rows for each unit/area. The result will be that the number of run rows in the study will be multiplied, with the total number determined by the complexity of the site and the study. In our example, the study expanded from 28 run rows in the static QRA to 78 run rows in the dynamic analysis. It is most effective to create a combination matrix to determine the number of combinations that will be assessed in the study, which will represent the combination of operating process units for each time-period considered in the study. Run rows can be created and imported using Safeti's Excel import function, which can be most helpful in handling the large number of cases that need to setup.

Once the study is set up with the run rows required for the study, a set of combination factors needs to be set up for each time period. For each time period, the combination factors are adjusted to account for the increase or decrease in risk contribution from each process area during the time period.

Figure 3 - Example Subset of Combination Factors Over Time

Unit	Day Night	Hazard Type	Confined Unconfined	TP1	TP2	TP3	TP4	TP5	TP6	TP7	TP8	TP9	TP10	TP11	TP12	TP13	TP14	TP15	TP16	TP17	TP18	TP19
Unit 13	Day	FLAM	Open	0.4	4	4	4	4	4	4	4	4	4	4	4	4	4	4	4	4	4	0.4
Unit 13	Night	FLAM	Open	0.6	6	6	6	6	6	6	6	6	6	6	6	6	6	6	6	6	6	0.6
Unit 13	Night	FLAM	Open	0.6	6	6	6	6	6	6	6	6	6	6	6	6	6	6	6	6	6	0.6
Unit 13	Day	FLAM	Open	0.4	4	4	4	4	4	4	4	4	4	4	4	4	4	4	4	4	4	0.4
Unit 7	Day	FLAM	Open	0.4	4	4	4	4	0	0	0	0	0	0	0	4	4	4	4	4	4	0.4
Unit 7	Night	FLAM	Open	0.6	6	6	6	6	0	0	0	0	0	0	0	6	6	6	6	6	6	0.6
Unit 7	Day	TOXIC and FLAM	Open	0.4	4	4	4	4	0	0	0	0	0	0	0	4	4	4	4	4	4	0.4
Unit 7	Night	TOXIC and FLAM	Open	0.6	6	6	6	6	0	0	0	0	0	0	0	6	6	6	6	6	6	0.6
Unit 3	Day	FLAM	Open	0.4	4	4	4	4	0	0	0	0	0	0	0	0	0	0	0	0	6	0.6
Unit 3	Night	FLAM	Open	0.6	6	6	6	6	0	0	0	0	0	0	0	0	0	0	0	0	6	0.6
Unit 6	Day	FLAM	Open	0.4	4	4	4	4	4	4	4	4	0	0	0	0	0	0	0	0	3	0.3
Unit 6	Day	FLAM	Open	0.4	4	4	4	4	4	4	4	4	0	0	0	0	0	0	0	0	3	0.3
Unit 2	Day	FLAM	Open	0.4	4	4	4	4	4	4	4	4	4	0	0	0	0	0	0	3	3	0.3
Unit 2	Day	FLAM	Confined	0.4	4	4	4	4	4	4	4	4	4	0	0	0	0	0	0	3	3	0.3
Unit 6	Day	FLAM	Open	0.4	4	4	4	4	4	4	4	4	4	0	0	0	0	0	0	3	3	0.3
Unit 6	Night	FLAM	Open	0.6	6	6	6	6	6	6	6	6	0	0	0	0	0	0	0	0	6	0.6
Unit 6	Night	FLAM	Open	0.6	6	6	6	6	6	6	6	6	0	0	0	0	0	0	0	0	6	0.6
Unit 2	Night	FLAM	Open	0.6	6	6	6	6	6	6	6	6	6	0	0	0	0	0	0	6	6	0.6
Unit 2	Night	FLAM	Confined	0.6	6	6	6	6	6	6	6	6	6	0	0	0	0	0	0	6	6	0.6
Unit 6	Night	FLAM	Open	0.6	6	6	6	6	6	6	6	6	6	0	0	0	0	0	0	6	6	0.6
Unit 4	Day	FLAM & TOX	Open	0.4	4	4	4	4	4	4	0	0	0	0	0	0	4	4	4	4	4	0.4
Unit 4	Day	FLAM & TOX	Open	0.4	4	4	0	0	0	0	0	0	0	0	0	0	4	4	4	4	4	0.4
Unit 4	Night	FLAM & TOX	Open	0.6	6	6	6	6	6	6	0	0	0	0	0	0	6	6	6	6	6	0.6

As units enter the S/D phase, combination factors for each run row in that unit, and interconnected or adjacent units (as appropriate), are increased to reflect the elevated risk level. Increasing the combination factors increases the outcome frequencies for each unique outcome in the risk analysis by the selected factor. Each outcome is a unique combination of even tree branches in the analysis, which may include release scenario, orientation, weather condition, wind direction, and ignition time. In a highly interconnected facility, like an integrated oil refinery, risk may increase across multiple units during S/D of a single unit, whereas a chemical plant may operate independent units that are not affected by operational changes at other units. Consideration should be given to the magnitude and scope of the risk factors to be applied for the operation. While the operators take additional measures to control risk during non-routine operations, historical incident trends indicate that the risks increase during these periods. In the T/A example used in this study, a highly integrated processing facility is considered, and the risk factors for all 18 units are increased by one order of magnitude during the T/A to account for the range of risk factors that elevate the risk of incidents during the T/A. The factors used in each study, and the application of factors across the site's operations should be reviewed and selected with input from a team that is knowledgeable and experienced in the site's operations and the process risks.

As units complete S/D, the risk factors (combination factors) for that unit's hazard scenarios (run rows) may be reduced to 0. If parts of the unit remain in operation or contain flammable inventory, the LOC scenarios may be kept in a separate run row with the elevated risk factor used during S/D, as appropriate.

The result will be a matrix of run row combination factors which reflect the increased and decreased risk of operations associated with each unit/area through the non-routine operation's time periods. A subset of the combination factors for the example case study are shown in Figure 3. Figure 4 illustrates the sum total of the combination factors for all run rows in the example study over the T/A period, which provides a simple corollary to the site's overall risk profile. As we see, risk profile at the site increases significantly as non-routine operations begin with the units across the facility in S/D mode. As units are taken offline, the risk level reduces, but does not reach 0, as some hazardous processes remain in operation throughout the T/A.

Figure 4 - Example Sum of Combination Factors Through T/A

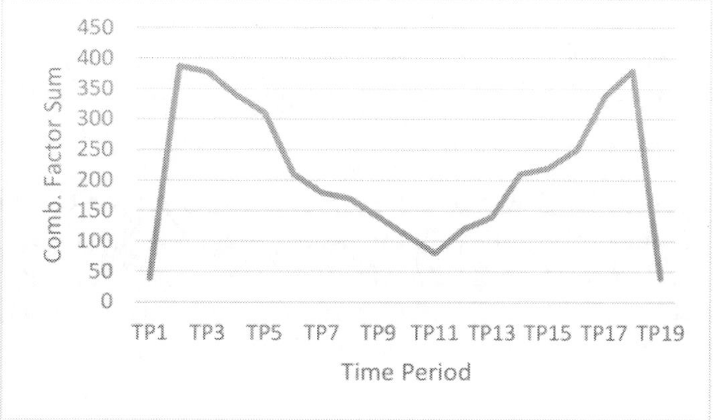

Depending on the initial scope of the study and the proposed operations, additional vulnerability definitions will likely need to be added for the new populations that will be located at the site in new temporary structures (e.g. tents/trailers), or outdoor work locations. During non-routine operations, workers will frequently be sited in trailers and tents of varying construction types.

If temporary buildings are sited, and the study follows the requirements of API RP 753[1], light wood trailers must be considered irrespective of the risk level, and sited with adequate standoff, outside of Zone 1, and no less than 330 feet from a process area congested region. In addition, non-essential personnel are not allowed to occupy portable buildings within Zone 1. Risk based calculations cannot be used to site light wood trailers or non-essential personnel within Zone 1, or within 330 feet of congested regions.

4 Risk Criteria

Risk evaluations of temporary operations can lead operators to consider whether the transitory risks should be more acceptable than baseline risks that remain present over years of operations. Some people may reason that higher risk levels over a short period are acceptable, and in general, humans are willing to bear increased risks for short periods. However, additional factors must be considered when applying risk criteria to non-routine operations.

Risk levels in a QRA are generally given as values of individual risk and societal risk, and both risk measures are generally presented as risk per annum, or annualized risk. Accordingly, it may be considered whether the risk during a non-routine operation, lasting less than one year, may be annualized to account for the short period of increased risk levels. API 753 addresses this in

[1] API. RP 753: Management of Hazards Associated with Location of Process Plant Portable Buildings. 2024; 2nd edition.

Section 5.3.1.2, and states that when evaluating individual risk, "No credit shall be taken for cases where the building will not be used or located at the site for less than a year"[2]. For societal (aggregate) risk calculations, operators may account for the limited duration of the non-routine operation, which contributes only a portion to the site's annualized aggregate risk.

Considering location specific individual risk (LSIR) further, it is important to recall the basis of LSIR, and the measure that is being expressed. While LSIR is expressed as a predicted likelihood of impacts by a fatal hazard on an annualized basis, as the risk to a person present at a location for a year, the risk measure is also expressing the instantaneous risk exposure to a person in a location at a moment in time. Accordingly, the LSIR for a worker is not affected by the time they are present; permanent and temporary workers are both exposed to the same risk levels at any moment in the same location.

Furthermore, it should also be considered that temporary contract workers present at the site during non-routine operations may work at multiple sites throughout the year. Turnaround crews and construction workers often rotate from one facility to another, working temporarily in several facilities throughout the year. Accordingly, the overall risk to these workers may be divided between multiple sites, but the industry, as a whole, presents a combined risk to the temporary worker. If operators relaxed risk criteria for temporary workers, the result would be a significant increase in the overall risk to the temporary worker over the year.

Accordingly, the prudent decision is to not relax risk criteria during non-routine operations, but rather maintain the same risk tolerance as during normal operations. Understandably, these criteria may be harder to meet during periods of increased risk, when units are executing S/U or S/D.

5 Developing Results

The results of a dynamic QRA may be illustrated using tables, contours, or plots. Individual risk contours may be generated in sets or combined into dynamic graphics which illustrate the change in risk over the period of operation. F-N Curves which account for the period of increased population and risk can be viewed dynamically to understand the risk profile during the non-routine operation or developed to illustrate the annualized risk profile considering the fraction of time in the non-routine operation.

Figure 5 and Figure 6 illustrate an example set of individual risk contours, illustrating the extent of intolerable risk areas over multiple time periods. These figures can be used in the planning phase to understand when personnel may be safely sited in areas around the facility during non-routine operations. These contours can be combined and integrated into a dynamic animation representation of the risk profile to improve visualization and understanding. An example of such a graphic can be found here: (https://youtu.be/PHaPyub1EBA).

Figure 5 - Example Risk Contours for Defined Time Periods

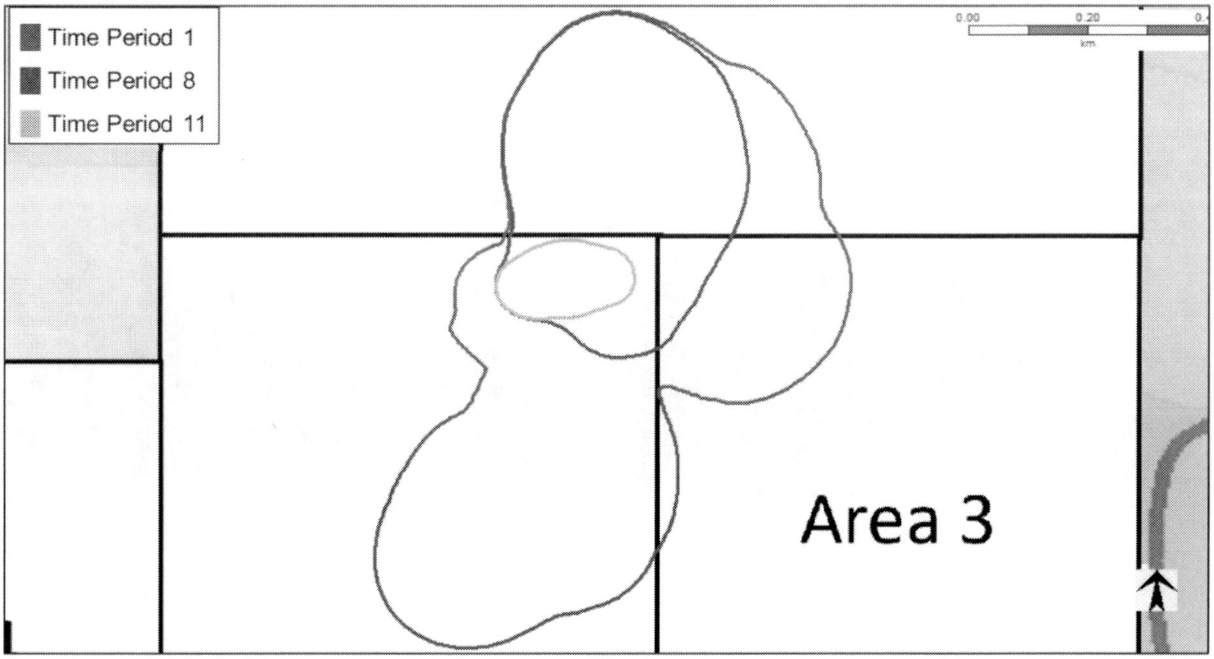

Figure 6 - Example Risk Contours for Various Vulnerabilities Over Time

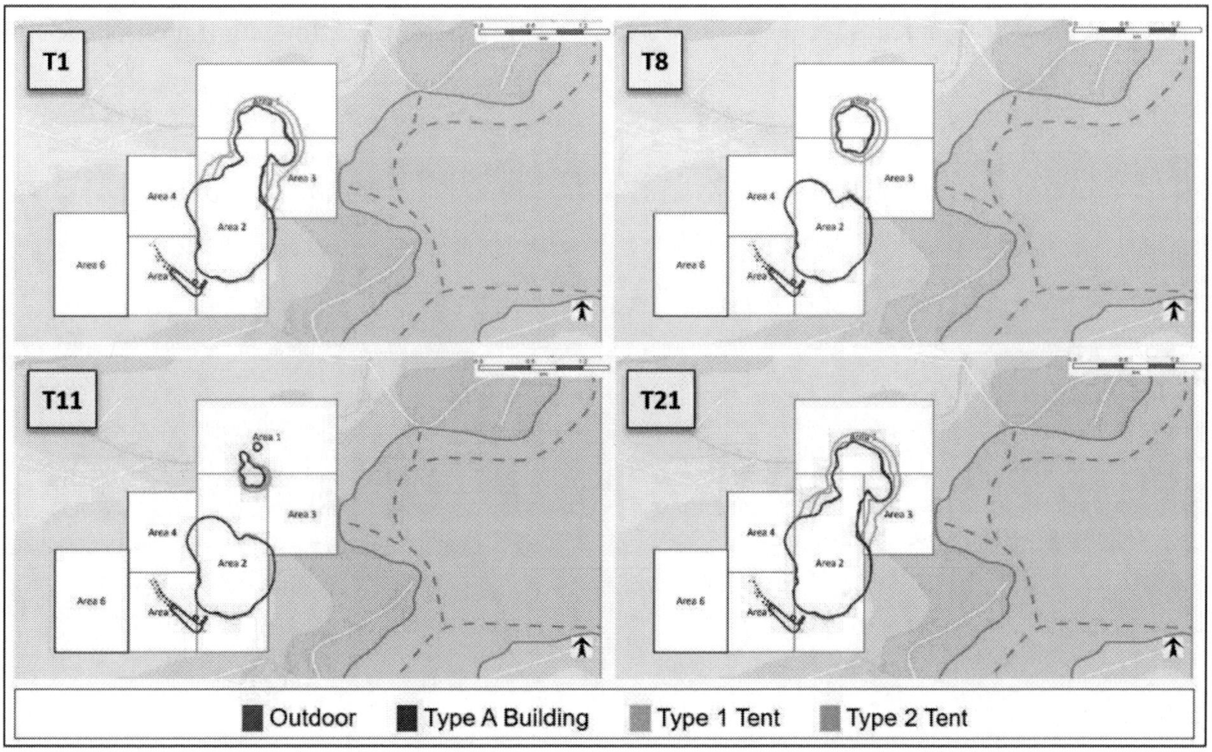

Figure 7 illustrates an example of tabular results for a specific building type across multiple proposed work areas at the facility. In the example, the risk to occupants of temporary tents is considered. As illustrated, we can see that tents can be sited at Location D prior to the T/A, but the risk of occupancy during S/D of adjacent units (TP2-TP3) is too high to allow occupancy. The tent may be occupied again once S/D in that area is complete, until S/U operations begin and risk levels increase again (TP4-TP-13).

Figure 7 - Example Tabular Results - Acceptable Risk Periods for Temporary Tents by Area

Areas	1	2	3	4	5	6	7	8	9	10	11	12
A	2.9E-5	2.9E-5	2.9E-5	2.8E-5	2.8E-5	6.5E-6	4.3E-6	3.4E-6	3.4E-6	1.0E-6	1.0E-6	1.0E-6
B	6.6E-5	6.6E-5	6.6E-5	6.6E-5	6.6E-5	2.4E-5	4.3E-6	2.8E-6	2.8E-6	6.1E-7	6.1E-7	9.1E-7
C	1.4E-3	1.4E-3	1.4E-3	7.7E-4	7.7E-4	7.7E-4	9.1E-5	9.1E-5	8.9E-5	7.1E-5	7.0E-5	7.0E-5
D	1.1E-4	1.1E-4	1.1E-4	8.1E-5	8.1E-5	7.0E-6	7.0E-6	6.6E-6	6.6E-6	4.8E-6	1.9E-6	1.9E-6
E	2.1E-4	2.1E-4	2.1E-4	1.9E-4	1.9E-4	1.1E-5	1.1E-5	1.1E-5	1.0E-5	1.3E-7	1.3E-8	1.3E-8
F	1.3E-4	1.3E-4	1.3E-4	1.3E-4	1.3E-4	6.0E-7	6.0E-7	4.6E-7	4.6E-7	4.7E-9	4.7E-9	4.7E-9
I	2.8E-4	2.8E-4	2.8E-4	2.8E-4	2.8E-4	6.6E-5	5.9E-5	6.6E-5	6.6E-5	5.3E-5	5.3E-5	2.0E-4
J	3.4E-4	3.4E-4	3.4E-4	3.4E-4	3.4E-4	4.8E-5	4.8E-5	4.7E-5	4.7E-5	3.5E-5	3.5E-5	2.1E-4
K	4.9E-4	4.9E-4	4.9E-4	4.1E-4	2.6E-4	2.5E-4	6.3E-6	5.8E-6	5.8E-6	2.1E-6	2.1E-6	2.1E-6
L	7.2E-4	7.2E-4	7.2E-4	1.7E-4	4.4E-5	3.9E-5	5.0E-6	4.4E-6	4.4E-6	2.0E-6	2.0E-6	2.0E-6
M	3.6E-4	3.6E-4	3.6E-4	3.6E-4	3.6E-4	3.6E-4	3.5E-4	3.5E-4	2.9E-4	2.3E-4	2.3E-4	2.3E-4
N	3.5E-4	3.5E-4	3.5E-4	3.5E-4	3.5E-4	3.5E-4	3.5E-4	3.5E-4	8.1E-5	2.9E-5	2.9E-5	2.9E-5
O	4.7E-4	4.7E-4	4.7E-4	4.7E-4	4.7E-4	1.5E-4	1.5E-4	1.5E-4	1.0E-4	2.2E-6	2.2E-6	2.2E-6

Areas	13	14	15	16	17	18	19	20	21	22	23	24
A	1.0E-6	2.8E-5	2.8E-5	2.9E-5	2.9E-5	2.9E-5	2.9E-5	2.9E-5	2.9E-5	2.9E-5	2.9E-5	2.9E-5
B	9.1E-7	4.7E-5	4.7E-5	6.6E-5	6.6E-5	6.6E-5	6.6E-5	6.6E-5	6.6E-5	6.6E-5	6.6E-5	6.6E-5
C	7.0E-5	7.3E-4	7.3E-4	1.4E-3	1.4E-3	1.4E-3	1.4E-3	1.4E-3	1.4E-3	1.4E-3	1.4E-3	1.4E-3
D	7.4E-5	1.0E-4	1.0E-4	1.0E-4	1.0E-4	1.0E-4	1.1E-4	1.1E-4	1.1E-4	1.1E-4	1.1E-4	1.1E-4
E	1.8E-4	2.0E-4	2.0E-4	2.0E-4	2.0E-4	2.0E-4	2.1E-4	2.1E-4	2.1E-4	2.1E-4	2.1E-4	2.1E-4
F	1.3E-4	1.3E-4	1.3E-4	1.3E-4	1.3E-4	1.3E-4	1.3E-4	1.3E-4	1.3E-4	1.3E-4	1.3E-4	1.3E-4
I	2.0E-4	2.8E-4	2.8E-4	2.8E-4	2.8E-4	2.8E-4	2.8E-4	2.8E-4	2.8E-4	2.8E-4	2.8E-4	2.8E-4
J	2.1E-4	3.4E-4	3.4E-4	3.4E-4	3.4E-4	3.4E-4	3.4E-4	3.4E-4	3.4E-4	3.4E-4	3.4E-4	3.4E-4
K	2.1E-6	9.4E-5	9.4E-5	3.4E-4	4.9E-4	4.9E-4	4.9E-4	4.9E-4	4.9E-4	4.9E-4	4.9E-4	4.9E-4
L	2.3E-6	5.6E-4	5.6E-4	5.9E-4	7.2E-4	7.2E-4	7.2E-4	7.2E-4	7.2E-4	7.2E-4	7.2E-4	7.2E-4
M	2.3E-4	2.3E-4	2.3E-4	2.5E-4	3.0E-4	3.0E-4	3.6E-4	3.6E-4	3.6E-4	3.6E-4	3.6E-4	3.6E-4
N	2.9E-5	2.9E-5	2.9E-5	3.0E-5	3.0E-4	3.0E-4	3.5E-4	3.5E-4	3.5E-4	3.5E-4	3.5E-4	3.5E-4
O	1.8E-5	1.9E-5	1.9E-5	1.9E-5	6.3E-5	6.3E-5	5.4E-4	5.4E-4	5.4E-4	5.4E-4	5.4E-4	5.4E-4

Figure 8 illustrates an example set of F-N curves through the time periods of a T/A. These curves are generated by defining assumed population profiles throughout the T/A. In the example, we can see several useful pieces of information. First, we see that the maximum number of fatalities increases during the T/A, as curves for TP2-TP16 extend further to the right. We can also see that TP16 will exceed the maximum tolerable risk criteria. A large population is being considered to be located in an area of elevated risk, resulting in a high risk of a large catastrophic hazard. Relocation of the planned workforce, delay of the adjacent unit S/U, or other risk mitigation measures to protect the workers can be considered to meet the risk criteria. These risk figures can assist in locating temporary workers. Exporting F-N Details report from the Safeti model study into a dynamic risk spreadsheet can allow planners to test various staffing configurations to meet the criteria.

Figure 8 - Example F-N Curves by Time Period

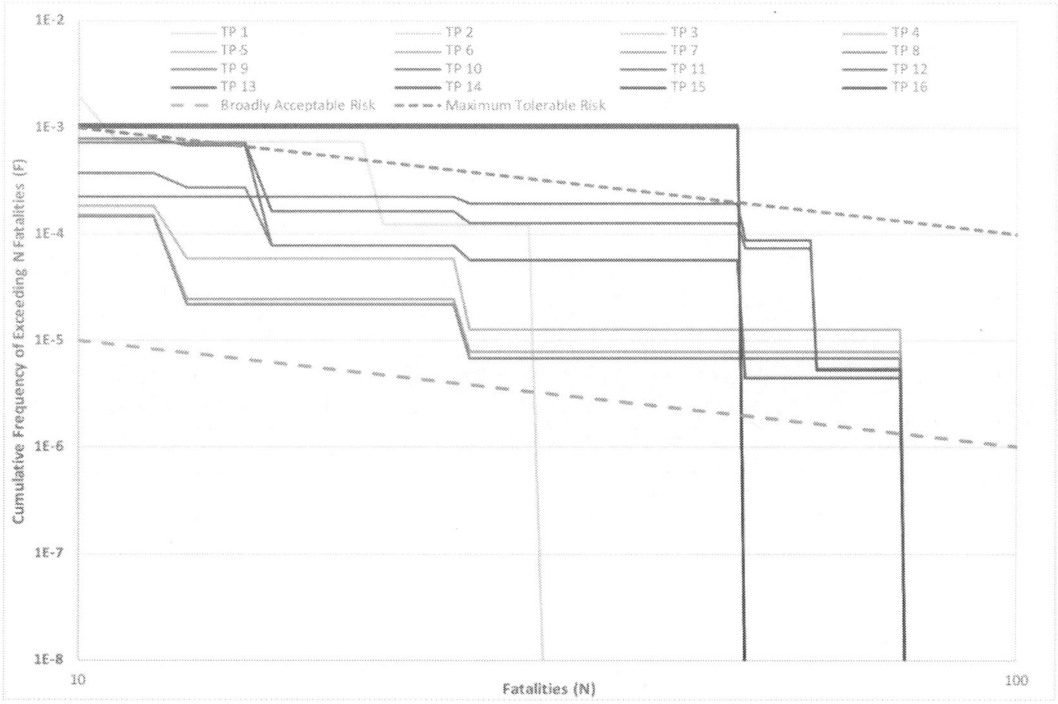

6 Reducing Risk During the Non-Routine Operations

The results of a T/A QRA may present useful findings to mitigate risk to personnel during a planned T/A, or to plan a T/A in a way that reduces risk to personnel. Common themes and findings from a T/A QRA may include the following considerations:

1. Co-locate equipment that will be S/D together in adjacent areas, or time process unit shutdowns and downtimes by area. Equipment that continues to present risk in the area will limit the ability to locate personnel during T/A, even after units are shutdown. Area shutdowns increase workers' distance from operating units during the S/D.
2. Consider emptying tanks located near process units during T/A to reduce the risk to T/A personnel and temporary structures. Operating tank farms in proximity to temporary workers presents risks that limit the ability to safely site personnel near units during S/D. Consider shifting storage operations to eliminate or limit the use of tanks near work areas.
3. Consider constructing/locating temporary buildings, tents, and other structures prior to the T/A to minimize risk to personnel from S/D operations to avoid workers assembling/siting buildings during periods of increased risk.
4. A system to manage and control occupancy must be in place once buildings are sited to ensure the structures are not occupied until a safe time period is reached. Access control

and signage, as well as periodic reviews can ensure buildings are not occupied during periods of elevated risk.

5. Consider the potential for adjacent or interconnected units' risk profiles to be affected during non-routine operations. Periods of S/U, S/D, and maintenance present increased risks to adjacent units, whose operations may be impacted by changes in feed/product routing, reduced staffing levels, extended shifts, operating outside of normal capacity ranges, or other risk factors.

7 Conclusions

Completing a QRA to assess the risk throughout a T/A is just one step of the process for safely siting personnel through changing operations. By implementing a robust safety management program throughout the facility's operational phases, organizations can mitigate potential hazards and ensure the safety of personnel. This collaborative effort involves proactive planning, clear communication, adherence to safety protocols and procedures, and continuous monitoring and supervision of activities. By fostering a culture of safety, organizations can navigate the complexities of non-routine operations while minimizing risks and maximizing operational efficiency.

Adapting a baseline static QRA to a dynamic QRA that assess a T/A or other non-routine operation is a relatively straightforward process, and the results generated from the study can clearly illustrate acceptable siting locations and acceptable periods of occupancy for temporary buildings and work areas throughout the facility. These results can give planners critical information that is needed to ensure people are safely mobilized before, during, and after a T/A, or other SIMOPS case.

8 References

[1] CCPS. Guidelines for Safe Process Operations and Maintenance. 1995; p 113, citing Large Property Damage Losses in the Hydrocarbon-Chemical Industries. Marsh McLennan, 14th edition, New York, NY, 1992.

[2] API. RP 753: Management of Hazards Associated with Location of Process Plant Portable Buildings. 2024; 2nd edition.

A Novel Approach to Strengthening Conventional Light Wood Trailer Portable Buildings

Gabriel A. Shelton, P.E.
Baker Engineering and Risk Consultants, Inc.
3330 Oakwell Court, Suite 100, San Antonio, Texas 78218
GShelton@BakerRisk.com

Thomas H. Anderson, P.E., and Daniel A. Vogel, E.I.T.
Baker Engineering and Risk Consultants, Inc.
3330 Oakwell Court, Suite 100, San Antonio, Texas 78218
TAnderson@BakerRisk.com, DVogel@BakerRisk.com

Keywords: Facility Siting, APR RP 753, Light Wood Trailer

Abstract

The stated purpose of the American Petroleum Institute's (API) Recommended Practice (RP) 753 document [1], Management of Hazards Associated with Location of Process Plant Portable Buildings, Second Edition (2024), is to "provide guidance for reducing the risk to personnel located in portable buildings from potential explosion, fire and toxic release hazards." Process plant owners and/or operators routinely house personnel within conventional portable building complexes (e.g., multiple light wood trailers) located within their facilities in accordance with the guidance provided by API RP 753. However, they often find that changes to their facilities (construction of new process areas, expansion of existing process areas, etc.) can mean that the location of their existing conventional portable building complexes no longer conform to the minimum requirements of API RP 753.

Conventional light wood trailer portable buildings not conforming to the minimum requirements of API RP 753 must be relocated, which may involve unexpected capital expenses associated with site preparation and installation of utilities for the relocation effort. This can leave process plant owners/operators looking for alternative solutions such as in-place strengthening of the conventional portable buildings to keep them in place, rather than relocating them. However, intrusive structural strengthening processes, such as adding new interior wall studs or roof beams, are often not a practical option, as conventional light wood trailers are usually leased by the process plant owner/operator from a third-party vendor and structural modifications can, therefore, not be made.

To address this need for alternative solutions but still meet the requirements of API RP 753 to protect occupants of conventional portable buildings from blast hazards, Baker Engineering and Risk Consultants, Inc. (BakerRisk®) has developed a novel approach to strengthening conventional light wood trailer portable buildings: structural upgrades that are designed to be non-intrusive, installed from the building exterior, and fully removable once the lease period has ended.

Full scale vapor cloud explosion testing is presented for both an upgraded and a non-upgraded conventional light wood trailer. Free-field blast overpressures applied during testing were similar to the predicted blast loads in the 2005 Texas City accident that resulted in numerous fatalities in light wood trailers. The efficacy of the upgrade is demonstrated in this testing, with the upgraded trailer exhibiting Low building damage, and the non-upgraded conventional trailer failing catastrophically.

1 Introduction

Developed as a direct result of the BP Texas City explosion, the American Petroleum Institute (API) Recommended Practice (RP) 753 was developed to provide guidance for reducing the risk to personnel located in portable buildings from potential explosion, fire, and toxic release hazards. Light wood trailers are a convenient and cost-effective method to temporarily house personnel in various locations that are ever-changing due to construction and turn-arounds. However, changes to processes and the same construction expansions may result in the same light wood trailers falling out of compliance with the requirements of siting portable buildings as set forth in API RP 753.

Relocating light wood trailers can be expensive and undesirable due to the necessity of relocating utility infrastructure and site preparation, resulting in facility owners/operators looking for alternative options such as structural retrofits and upgrades. On the other hand, developing structural upgrades for conventional light wood trailers can be problematic from a practical standpoint, since these types of trailers are often leased from third-party vendors and therefore intrusive structural strengthening approaches such as adding new interior wall studs and roof beams cannot be implemented without purchasing the trailers, which is likely cost prohibitive.

To meet the requirements of API RP 753 to protect occupants of conventional portable buildings from blast hazards while also addressing this need for alternative solutions, BakerRisk® has developed a novel approach to strengthening conventional light wood trailer portable buildings: Structural upgrades that are designed to be non-intrusive, installed from the building exterior, and fully removable once the lease period has ended. Full scale vapor cloud explosion testing is presented for both an upgraded and a non-upgraded conventional light wood trailer. Free-field blast overpressures applied during testing were similar to the predicted blast loads in the 2005 Texas City accident that resulted in numerous fatalities in light wood trailers. The efficacy of the upgrade is demonstrated in this testing, with the upgraded trailer exhibiting Low building damage, and the non-upgraded conventional trailer failing catastrophically.

2 Overview of API RP 753

One of the direct results of the BP Texas City event, which led to the deaths of 15 workers, was a major review and development of API RP 753 with a focus on managing hazards associated with the placement of portable buildings. These fatalities stemmed from the close proximity of occupied portable trailers to nearby process units during a unit turnaround. Due to all the fatalities occurring to personnel that were either inside or around the portable trailers during the event, the Chemical Safety Board (CSB) issued a recommendation to the American Petroleum Institute to include the siting of portable buildings to the API RP 752 or to issue a new recommended practice to ensure the safe placement of portable buildings [2]. In response to the recommendation from the CSB, API issued the first edition of API RP 753 in 2007.

As stated in a previous report from BakerRisk® [3], and with reference to, and quoting from, the first edition of API RP 753 [4]:

> API RP 753 was developed under the assumption that light wood trailers, defined as consisting of nominal 2"×4" dimensional wood-framed construction with a thin outer skin, were “generally representative of the weakest constructed portable building used in the

processing industries." Based on this premise the API developed a set of curves or zones to be used for the siting of portable buildings.

The Zone 1-2 Boundary was defined as being the maximum of the distance to the 0.9 psi overpressure contour calculated by assuming a TNO Blast Severity Level 5 for a given congested volume of 330 feet. The 0.9 psi overpressure corresponds to the expected cracking of wall studs on the side-on surfaces of a light wood trailer for a long duration blast load (duration ≥ 200 ms) and the 330 foot minimum stand-off distance is "intended to provide time for building occupants to react to flammable gas hazards."

The Zone 2-3 Boundary was defined as being the distance to the 0.6 psi overpressure contour calculated by assuming a TNO Blast Severity Level 7 for a given congested volume. The 0.6 psi overpressure corresponds to the expected cracking of wall studs on the reflected surface (i.e., surface that faces the explosion) of a light wood trailer for a long duration blast load (duration ≥ 200 ms); however, it is expected that the wall studs would remain in place.

According to the recently published second edition of API RP 753 [1], portable buildings that could be impacted by external vapor clouds can be assessed by either Zoning Method A or Zoning Method B. Zoning Method A utilizes the chart provided in Figure 4 of API RP 753 while Zoning Method B allows the owners/operators to conduct a blast load analysis on specific congested volumes to develop the Zone 1-2 and Zone 2-3 boundary contours based on 0.9 psi and 0.6 psi overpressure contours, respectively. Both zoning methods can be utilized in siting of portable buildings, but Zoning Method B provides a more detailed method and can be used to identify zone for explosion hazards other than external vapor clouds (e.g., bursting pressure vessels, etc.).

As shown in Table 1 of API RP 753, no occupied light wood trailer can be sited in Zone 1. Other types of occupied portable buildings (i.e., other than light wood trailers) can be sited in Zone 1 but require a siting evaluation to be conducted and can only house essential personnel. Within Zone 1, risk-based approaches are not permitted to site light wood trailers or non-essential personnel. For Zone 2, occupied light wood trailers and other types of occupied portable buildings can be sited in this zone but require a siting evaluation to be conducted. Within Zone 2, since it would be further than 330 feet away from process area congestion, the facility owner/operator would have to consider the Guiding Principles of API RP 753 and ensure that personnel restrictions are followed, as shown in the evaluation flowchart shown in Figure 2 of API RP 753. Light wood trailers and other portable buildings can be sited within Zone 3 with no restrictions but, similar to siting in Zone 2, the facility owner/operator would have to consider if the siting is in accordance with Guiding Principles 1, 2, 3, and 4 of API RP 753 and ensure that personnel restrictions are followed. A detailed flowchart for siting portable buildings for external vapor cloud explosions (VCEs) can be seen in Figure 3a and 3b of API RP 753.

If the portable building in question requires a siting evaluation for explosion hazards such as light wood trailers in Zone 2 or other types of portable buildings within Zone 1 or Zone 2, it must be evaluated in the same manner as a permanent building in accordance with Section 7 of the recently published fourth edition of API RP 752 [4]. For siting other process hazards such as fire or toxic material releases, if the portable building could be impacted by either of these hazards a siting evaluation is required in accordance with Section 8 and 9 of API RP 752, respectively. As with siting for explosion hazards, it would have to be considered if the portable building in question is in accordance with API RP 753's Guiding Principles 1, 2, 3, and 4 for toxic material releases and fire hazards.

3 Description of Light Wood Trailers

Light Wood Trailer Description

API RP 753 defines a light wood trailer as a "portable building with a wall design consisting of [notched] 2"×4" studs with a thin outer skin." It is noted in API RP 753 that nominal "2"×4" studs refers to timber framing members that have cross-sectional actual measurements of 1.5 in. by 3.5 in. (38 mm by 89 mm)." API RP 753 also states that the wall studs "were assumed to be stud grade southern pine, spaced 16 inches [(406 mm)] apart… [with a] thin (0.030 inch) aluminum skin… [spanning] over the studs." The stud span was assumed to be 8 feet (2438 mm) and the allowable static bending capacity of the wood was assumed to be 875 psi (6.03 MPa) [5]. Furthermore, it was assumed that the overall strength of the light wood trailer "was not limited by the strength of the connections at the ends of the studs or by the performance of the roof" [5].

Trailer Construction Variety

It is important to note that the light wood trailer construction described above was initially selected, as described in API RP 753, to represent "the weakest type of wood construction likely to be encountered in processing facilities in the United States." However, API RP 753 acknowledges that "there may be other portable building construction, such as self-framing panel construction, that may have inferior performance under blast loading" and states that other construction types such as "metal trailers" and "wood trailers with larger studs (e.g., 2"×6" studs)… may be stronger" than the initially selected light wood trailer construction.

It should also be acknowledged that the wood species used, the strengths of the wood member connections, the individual wood member spans, and the size/type of roof framing, among other items, will affect the blast capacity of any given light wood trailer. Therefore, if the construction of a light wood trailer being assessed differs from the light wood trailer construction described in API RP 753, further analyses to quantify the blast capacity are warranted. As an example of a change that will potentially affect the blast capacity of light wood trailers: the latest American Wood Council Design Standard [6] states that the allowable static bending capacity of stud grade southern pine is currently 650 psi (4.48 MPa). This represents a reduction in the material strength of over 25% for stud grade southern pine being used in 2"×4" studs since the initial publication of API RP 753.

Constraints to Trailer Modifications

If an assessment of an existing light wood trailer finds that the blast capacity is deficient, it is conceivable that structural modifications (i.e., structural upgrades) can be designed and implemented in place to ensure that the existing light wood trailer is no longer deficient. However, there may be constraints to modifying an existing light wood trailer. For example, the process plant owner/operator may have leased the light wood trailer from a third-party vendor, and because of the lease, structural modifications (such as adding new interior wall studs or roof beams) cannot be made. An alternative in-place strengthening solution that is designed to be non-intrusive, installed from the building exterior, and fully removable once the lease period has ended is, therefore, potentially advantageous to process plant owners/operators.

4 Case Study

Consider the fictitious site layout shown in Figure 1. The site layout shown will illustrate siting of portable buildings for explosion hazards in accordance with the current API RP 753. For this example, two congested volumes are present: 1.) a 250 ft. × 250 ft. × 12 ft. congested volume (shown as the "Process Area" with a total congested volume of 750,000 cubic feet) and 2.) a 180 ft. × 180 ft. × 20 ft. congested volume (shown as the "Trees" with a total congested volume of 648,000 cubic feet). Both congested volumes are in close proximity to existing occupied portable light wood trailers. In this example, it has already been determined that these portable buildings are intended for occupancy and that they can be impacted by credible external vapor cloud explosions. Note that API RP 753 provides two zoning methods for use in the siting of portable buildings and states that "[p]ortable buildings intended for occupancy… should be located using either the simple Zoning Method A or more detailed Zoning Method B." Both zoning methods are illustrated herein.

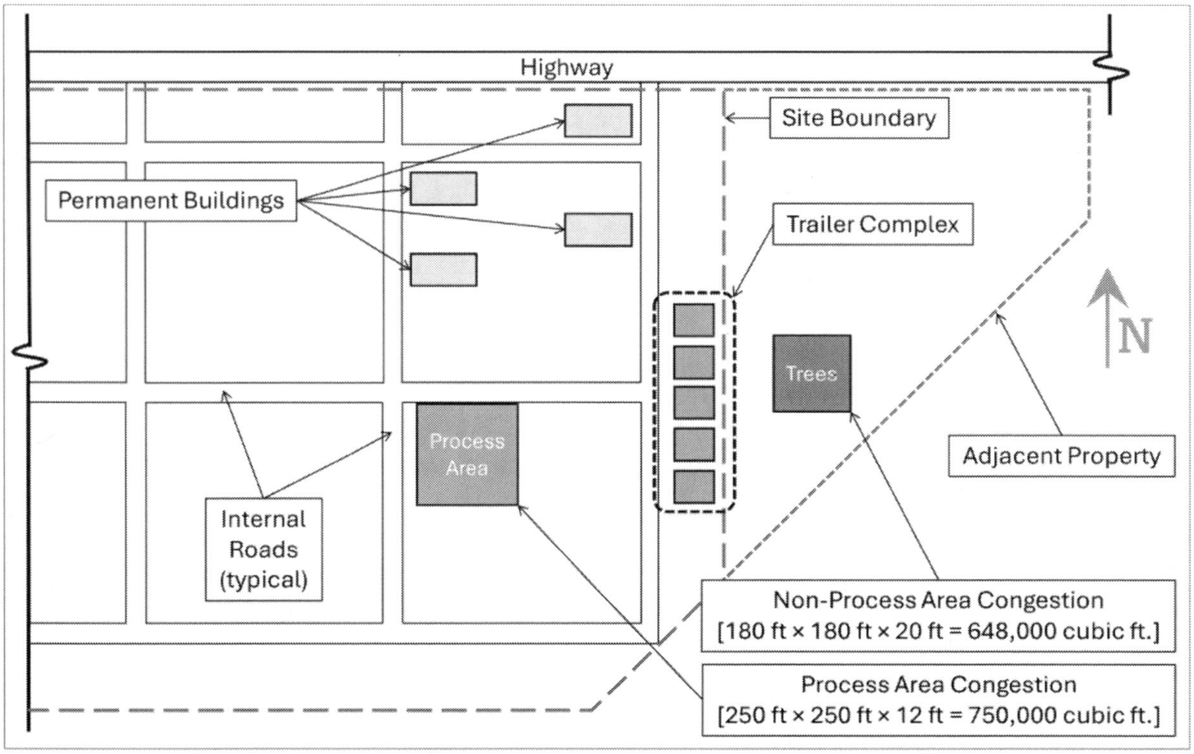

Figure 1. Fictitious Site Layout for API RP 753 Analyses

API RP 753 Zoning Method A

Section 7.2.2.1 of API RP 753 presents Zoning Method A. This method utilizes separation distances, as "measured from the edge of the congested volume," based on the congested volume(s) present to define the Zone 1-2 and Zone 2-3 boundaries. The Zone 1-2 and Zone 2-3 separation distances can be determined by using either Figure 4 or Figure C.1 of API RP 753. In this example, the separation distances from the two congested volumes are summarized in Table 1 and illustrated on the fictitious site layout shown in Figure 2. Note that the existing occupied portable light wood trailers fall within Zone 1; it is important to note that light wood trailers intended for occupancy "shall not be located within Zone 1" as stated in API RP 753.

Table 1. API RP 753 Zoning Method A – Separation Distances

Congestion	Congested Volume (cubic feet)	Zone 1-2 Boundary (feet)	Zone 2-3 Boundary (feet)
Process Area	750,000	512	1,750
Trees	650,000[1]	485	1,665

[1] The calculated total congested volume of 648,000 cubic feet has been rounded up to 650,000 cubic feet to allow the use of the *Separation Distances* shown in API RP 753 Figure C.1 without interpolating between values.

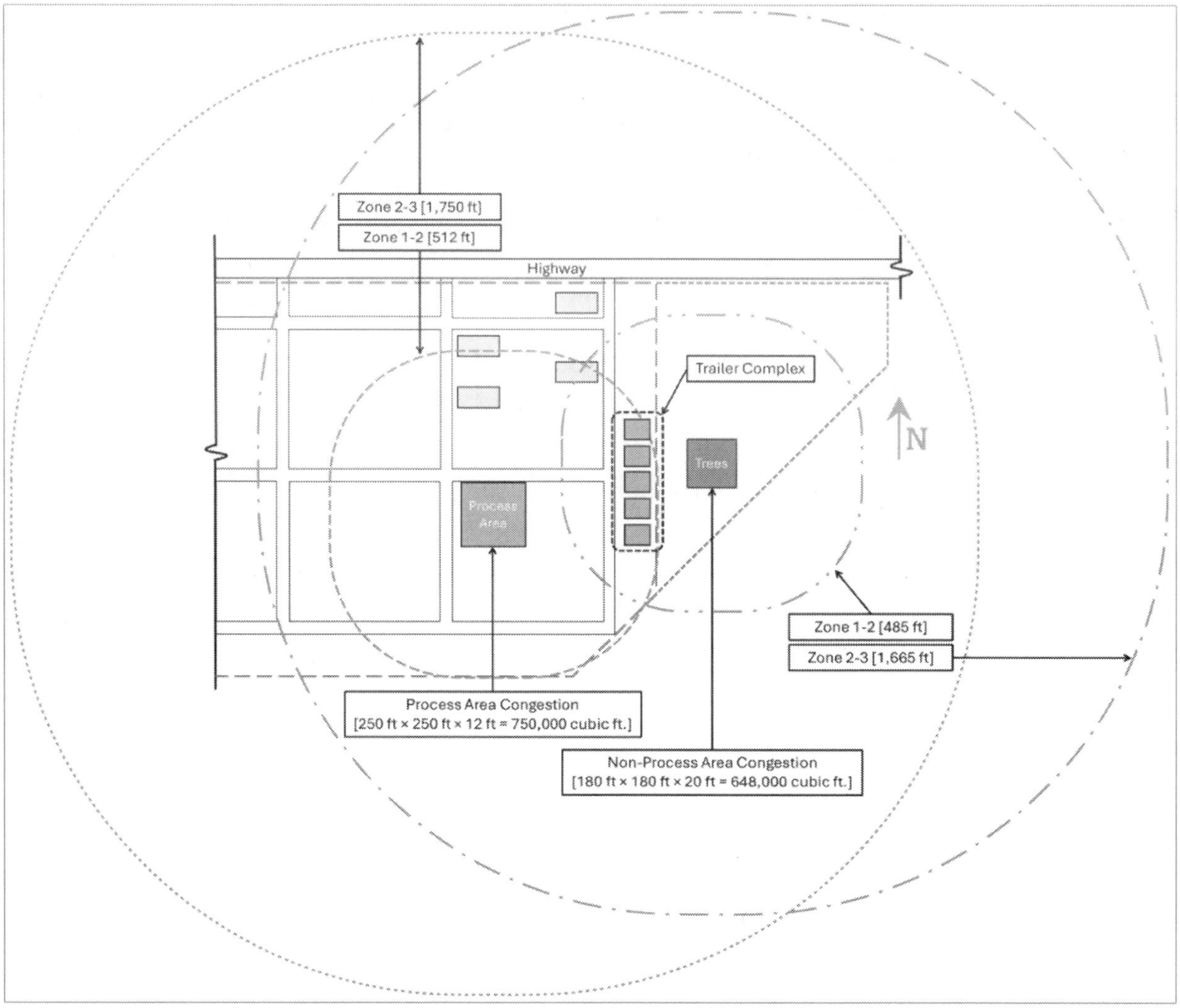

Figure 2. API RP 753 Zoning Method A – Zone Boundary Contours

API RP 753 Zoning Method B

Section 7.2.2.2 of API RP 753 presents Zoning Method B. According to API RP 753, this method allows the facility owner/operator to utilize "a blast load analysis approach… to establish the Zone 1-2 and Zone 2-3 boundary contours based on the calculated 0.9 psi and 0.6 psi free-field overpressure contours." For the zone boundary contours related to process area congested volumes, API RP 753 states that they "shall not be less than 330 feet from the edge of process area congested volumes greater than 7,500 cubic feet." Furthermore, for the zone boundary contours related to non-process area congested volumes, API RP 753 states that the offset of the boundary contours "from the edge of non-process area congested volumes may be based on the [0.9 psi and 0.6 psi] overpressure contours." This implies that the 330 feet minimum separation distance applies only to process area congested volumes.

In this example, the Baker-Strehlow-Tang (BST) method [7,8] was utilized to perform a simplified blast load analysis assuming both congested volumes were filled with a flammable vapor cloud. The resulting free-field blast overpressure contours (and associated zone boundaries) are shown in Figure 3. Note that the free-field blast overpressure at the location of the light wood trailers is in excess of 0.9 psi (the light wood trailers are subject to free-field blast overpressures of between 0.9 psi and 1.5 psi). Therefore, the existing occupied portable light wood trailers fall within Zone 1. Again, as stated in API RP 753, light wood trailers intended for occupancy "shall not be located within Zone 1". As a point of interest, note that the 330 feet minimum separation distance from the edge of the process area congested volume does not extend beyond the 0.9 psi free-field blast overpressure contour, and, therefore, the 0.9 psi free-field blast overpressure contour solely defines the Zone 1-2 boundary.

Mitigation Options

Given the fact that, in the example discussed above, the light wood trailers intended for occupancy are located within Zone 1 (which is specifically prohibited by API RP 753), the facility owner/operator has the following options for these light wood trailers (in accordance with the guiding principles of API RP 753):

1. Remove the occupants from the light wood trailers;
2. Relocate the light wood trailers and manage their occupancy in accordance with the requirements of API RP 753;
3. Remove the light wood trailers and replace them with "portable buildings other than light wood trailers" that are designed to protect the occupants from explosion hazards (in conformance with a siting evaluation for explosions as presented in Section 7.4 of API RP 753) and manage the occupancy in accordance with API RP 753;
4. Modify the light wood trailers to protect the occupants from the explosion hazards (in conformance with a siting evaluation for explosions as presented in Section 7.4 of API RP 753) and manage the occupancy in accordance with API RP 753. The modified light wood trailers would fall into the "portable buildings other than light wood trailers" category and would, therefore, be able to be located within Zone 1.

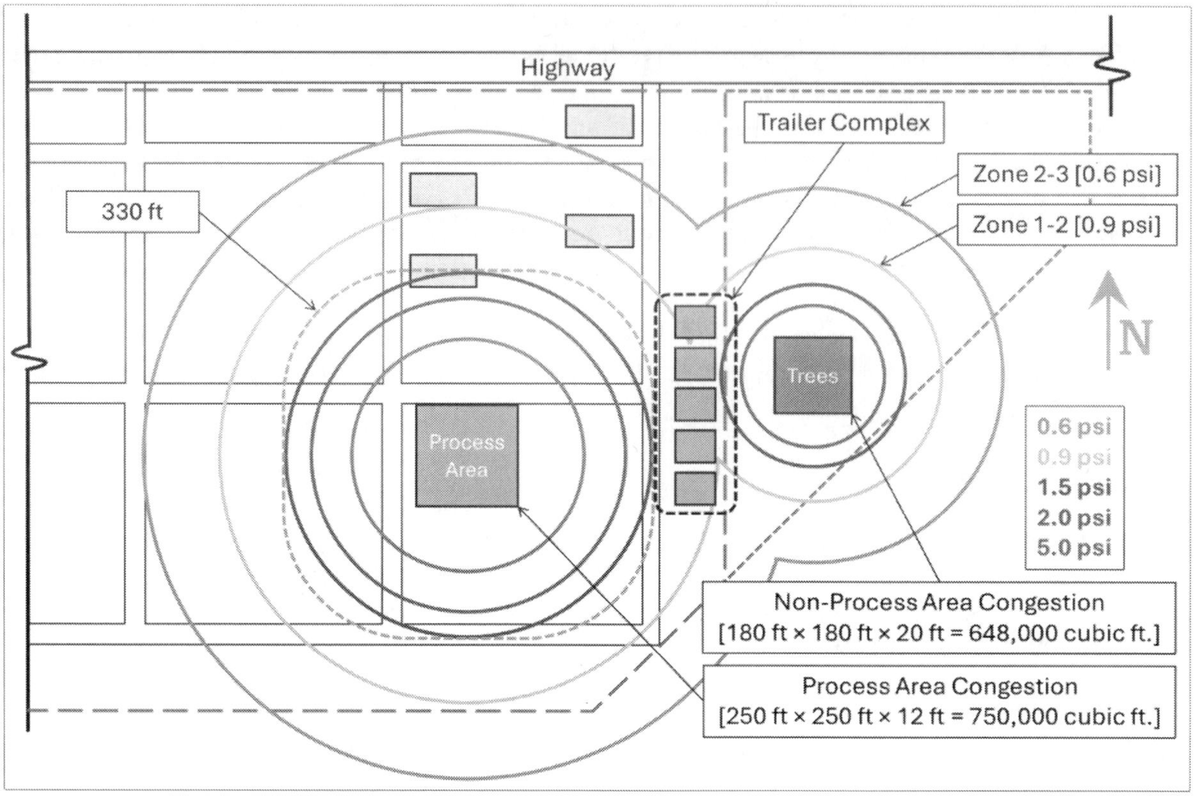

Figure 3. API RP 753 Zoning Method B – Free-Field Overpressure and Zone Boundary Contours

The results of the testing performed by BakerRisk on a modified light wood trailer (as presented in the sections to follow) demonstrate that the occupants would be protected from the explosion hazards predicted for the fictitious site layout presented above.

5 Test Description

Test Specimen Description

The test specimen was a light wood trailer with nominal dimensions of 40 feet long × 10 feet wide. Actual length is 36'-0" due to the length of the tow-hitch tongue. The chassis of the trailer consists of two longitudinal, 10-inch-deep steel I-beams with Z-shaped, bent plate crossmembers between the beams and tapered bent plate outriggers outside of the beams. Crossmembers and outriggers are spaced at 4-feet on-center. The chassis is typically supported below the I-beams during transport via by two sets of axles rated at 7000 lbs. each and the tow-hitch tongue, and during installation via eight concrete masonry unit (CMU) pier footings and strapped down to the ground with six metal frame tie-down straps.

The trailer floor is a tongue and groove floor deck supported by 2×6 floor joists at 16-inch spacing and rim-rails. The trailer walls were constructed of 2×4 wood studs that are notched for furring strips at 24-inches. The walls are clad with 26-gauge corrugated aluminum metal panels. The roof consists of rounded trusses that are spaced at 24-inches on center and clad with a roofing membrane over OSB decking. The walls, floor, and roof are insulated with R-19 (R-13 in walls) batting type insulation. Figure 4 shows the trailer installed at the BakerRisk Box Canyon Test Facility. The trailer has a heating, ventilation, and air conditioning (HVAC) system that is mounted to the exterior of the front wall. There are two personnel-sized doors on the starboard side of the trailer and seven windows located on the back (1 window) and sides (3 windows each) of the trailer. Figure 4 shows the trailer installed at the BakerRisk Box Canyon Test Facility.

Figure 4. Photos Showing the Front/Port side and Rear/Starboard Side of the Test Specimen Trailer (Vendor Name Redacted)

Figure 5 shows the general layout plan of the test specimen trailer.

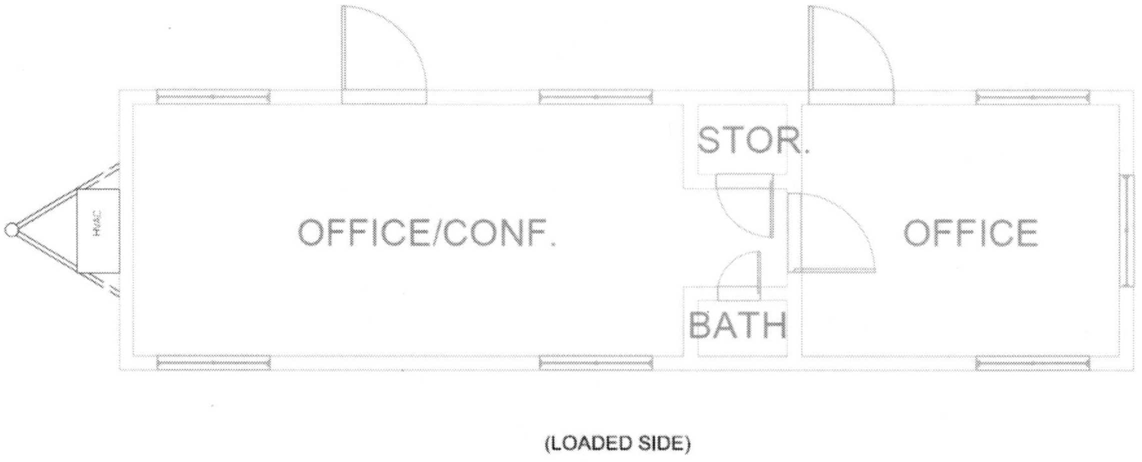

Figure 5. General Layout Plan of Test Specimen Trailer

The test specimen trailer was selected based on the balance of a variety of factors, including ease of transport and installation, the aspect ratios of the length/width/height, and cost.

Design of Upgrade

The test specimen was evaluated under two conditions: In the as-is condition and in an upgraded condition. The upgrade was designed to accommodate certain restraints. Since light wood trailers are typically procured from vendors in a leasing arrangement, the intent of the upgraded design is to minimize direct connection to the trailer.

The upgrade was designed to remain elastic at the target load and consists of a wood shell consisting of removable wooden panels constructed from readily available two-by (2×) dimensional lumber clad with exterior grade plywood. All the materials are available at common lumber yards and "big box" hardware stores. The panels are designed to be modular roof and wall panels that can be constructed off-site and then transported to site via truck and readily implemented on-site with a minimal crew and a fork-lift and man-lift. In addition to the wood shell the upgrade also has connections designed to connect the panels together in a semi rigid frame and ballast connected to the frame to resist sliding and overturning.

The trailer doors were removed and replaced with standard off-the-shelf doors attached directly to the new wood shell. The windows were covered by ¾-inch plywood on the interior and exterior, fastened into the wall studs with deck screws.

Test Set-Up

The testing was conducted at the BakerRisk Box Canyon Test Facility in Kinney County, Texas. Blast loads were imparted on the test specimen trailer by the deflagration load generator (DLG) Test Rig, which can produce blast waves that are representative of full-scale vapor cloud explosions (VCEs) in industrial settings. During a test, the DLG test rig is configured with idealized congestion and filled with a near-stoichiometric fuel-air mixture that is ignited near the center of the confined fuel cloud. By controlling the congestion level and standoff distance between the test rig and the test article, the desired blast load on the target structure can be achieved. A plan view of the trailer test set-up is shown in Figure 6

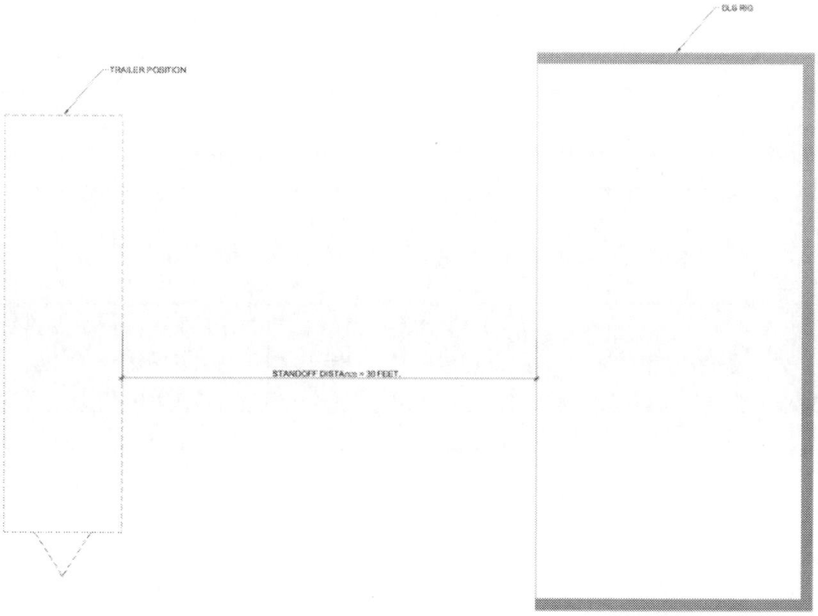

Figure 6. Plan View of Trailer Test Set-up

The trailer was positioned so that the centroid of the roof was 35 feet from the face of the DLG Test Rig along the DLG centerline. The trailer was oriented with portside facing the DLG Rig resulting in the doors facing the leeward, or non-loaded direction.

Due to cost constraints, an approach of testing a single trailer with two different upgrade conditions was selected over conducting a single test with two different trailers with different upgrade conditions. This approach required that some minor repairs be implemented to the trailer prior to the second test.

Table 2 summarizes the upgrade conditions for each test in the test matrix.

Table 2. Test Matrix

Test Number	Upgrade Condition
Test 1	Upgraded and Tethered
Test 2	Non-upgraded

Test Results

Blast Loads

The DLG Test Rig was configured with congestion simulated by an array of 17 × 26 vertically oriented pipes that are fastened to the rig via heavy chain. The deflagration was caused by the ignition of a propane cloud that was mixed with air to have a concentration of 4.33%. The ignition was initiated at a location roughly 30 inches back from the center of the Rig. Table 3 summarizes the approximate free-field blast loads for both tests. Pressure gauges to measure the free-field blast loads were placed on the ground on either side of the trailer, along the trailer centerline, as shown in Figure 7. The idealized wave shape for the free-field blast loads is shown in Figure 8.

Table 3. Approximate Blast Loads for Both Tests

Test #	Description	Average Free Field Pressure (psig)	Free Field Impulse (psi-msec)	Positive Phase Duration (msec)
Test 1	Upgraded & Tethered	1.8	45	50
Test 2	Non-upgraded	2.0	50	50

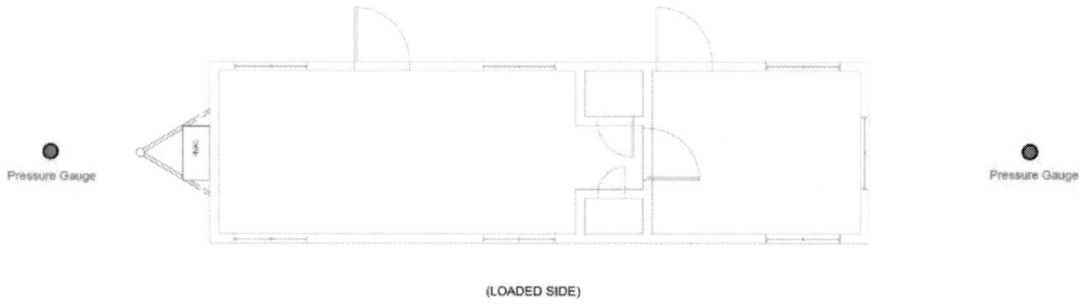

Figure 7. Approximate Locations of Pressure Gauges

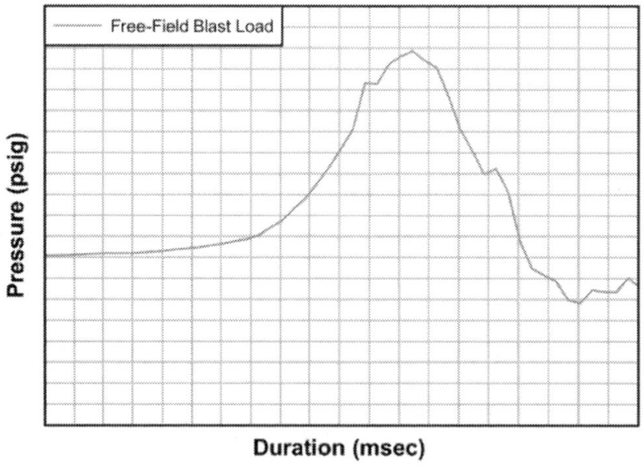

Figure 8. Idealized Blast Pressure Wave for Free-Field Gauges

Description of Damage – Test 1

Photographs showing the upgraded and tethered trailer before and after Test 1 are shown in Figure 9 and Figure 10, respectively.

Figure 9. Test 1 - Before Photo

Figure 10. Test 1 - After Photo

Minor damage to the upgrade panels and the trailer was observed after Test 1. One plywood sheet on the back panel of the trailer was partially loose from the 2× wall frame as shown in Figure 11.

Figure 11. Photo Showing Panel Damage after Test 1

The rear upgraded door was flung open but remained attached to the hinges. The door hardware was damaged yet operable, and the door was able to close and open. Photos showing the rear door damage is shown in Figure 12.

Figure 12. Photos of the Rear Door Open (Left) and Damage to Hardware and Edge-Seam (Right) After Test 1

The bulkhead box around the trailer tongue was damaged as shown in Figure 13.

Figure 13. Damage to Bulkhead Box Around Trailer Tongue After Test 1

Batten boards that covered the joint between the adjacent wall panels and the joint between the ends of the adjacent roof panels were pulled out slightly, indicating that the wall panels translated inward from the original position. A photo of an example of this is shown in Figure 14.

Figure 14. Batten Board at Adjacent Wall/Roof Joints After Test 1

On the trailer interior, the interior wall paneling on the transverse (short) walls at the front of the trailer, and the transverse wall of the bathroom were detached from the wood studs and deflected out of plane. Photos showing these deflections are shown in Figure 15. This deflection indicates that the transverse walls were exercised in resisting the lateral loading applied by the reaction of the upgraded walls during the test.

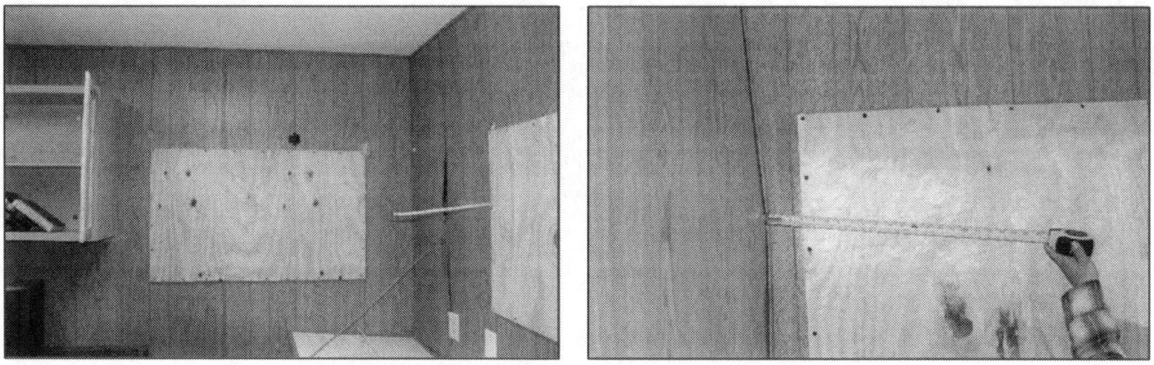

Figure 15. Interior Wall Panel Damage After Test 1

There was no measurable translation of the base of the trailer, upgrade panels, or ballast relative to markings on the concrete test slab after Test 1. Video footage indicates that the top of the trailer translated several inches in the direction of the loading but returned close to its original position after Test 1.

Between Test Repairs

Several minor repairs were made to the test specimen trailer between Test 1 and Test 2. The end walls at the front and back of the trailer, where some interior paneling had deflected, were repaired by removing the paneling, replacing any studs that were cracked with new studs of equal grade (Southern Yellow Pine, grade 2), and reinstalling the interior paneling. Exterior corrugated aluminum paneling was also re-fastened where screws had either pulled through the paneling or pulled out of studs.

Description of Damage – Test 2

Photographs showing the upgraded and tethered trailer before and after Test 2 are shown in Figure 16 and Figure 17, respectively.

Figure 16. Test 2 - Before Photo (Vendor Name Redacted)

Figure 17. Test 2 - After Photo

Significant widespread damage was observed for the non-upgraded trailer in Test 2. The roof and walls collapsed and were obliterated into many small pieces of debris that were thrown distances of up to 200-feet. The blast load failed the reflected wall at the top and bottom connections. This allowed blast loads to push into the interior space of the trailer and push the roof upwards and toward the leeward direction. The blast also loaded the underside of the trailer chassis and floor, causing the chassis to be flipped upside down. The CMU piers collapsed. The trailer tie-down straps yielded. Generally, all the walls and roof collapsed. The windows all had high hazard debris conditions. The doors were thrown as part of the wall assembly but remained connected to their respective frames. Internal non-structural debris was thrown as well.

Building Damage Levels

Building Damage Levels are qualitative amounts of damage to a building based on the ERASDAC computer program. [9]

Table 4. Building Damage Level Descriptions

Building Damage Level	Damage Description
1	Onset of visible damage to reflected wall of building.
2.0 (2A)	Reflected wall components sustain permanent damage requiring replacement, other walls and roof have visible damage that is generally repairable.
2.5 (2B)	Reflected wall components are collapsed or very severely damaged. Other walls and roof have permanent damage requiring replacement.
3	Reflected wall has collapsed. Other walls and roof have substantial plastic deformation that may be approaching incipient collapse.
4	Complete failure of the building roof and a substantial area of walls.

Based on the descriptions in Table 5 the upgraded, tethered trailer was a BDL-1 after Test 1, while the non-upgraded trailer was a BDL-4 after Test 2.

Expected Personnel Occupant Vulnerability [9]

Occupant vulnerability (OV) is a model that predicts the percentage of building occupants who sustain serious injury, meaning that, as a minimum, there is an immediate danger to life unless immediate medical attention is provided. OV is a function of the fraction of the population that is in an interior space or a perimeter space in the building, the fraction of the width of walls with glass, the wall cladding and roof cladding types, and the respective wall and roof damage vulnerability numbers. Calculating the OV of the building using this methodology, the OV for the trailer would be 0.0% for Test 1 and at least 77.8% for Test 2. The calculated OV for each test, assuming that the perimeter space population fraction is 100% for the non-upgraded trailer and 0% for the upgraded trailer and that the Glass Vulnerability was ~0.05 (based on an average impulse of 55 psi-ms) is shown in Table 6.

Table 5. Building Damage Level and Occupant Vulnerability of Both Tests

Test	BDL	OV
Test 1	BDL 1	0.0%
Test 2	BDL 4	77.8%

Other Considerations

Something that the occupant vulnerability calculation does not explicitly account for is non-structural debris. This is debris that is not part of the structural support of the building (trailer). Examples of non-structural debris include items such as wall mounted equipment, bookshelves, furniture, etc. Wall mounted equipment can include fire extinguishers, lock-out-tag-out boxes, monitors, chalkboards, and dry erase boards. These non-structural debris can become accelerated by the sudden movement of wall components impacted by blast loads and be thrown away from the walls due to failure of the connection hardware that was holding the debris to the wall component. These debris objects can cause significant injury to personnel occupying the trailers that are impacted by the debris.

Non-structural debris, if not connected properly, can exacerbate the occupant vulnerability of the trailer. This additional vulnerability or risk can be mitigated if connections are designed properly, or other measures are put into place, such as appropriate standoff distances of furniture and equipment from the walls of the trailer.

Similarly, while windows are accounted for in the OV calculations, additional vulnerability to the perimeter population caused by windows can be mitigated by simple upgrades to the windows such as encasing them in plywood to prevent them from breaking and the glass shards from injuring building occupants. Also, consideration of doors must be accounted for. Doors can fail and become debris hazards, significantly increasing the vulnerability of occupant to the blast loads. As such, doors must remain elastic, or at a minimum, the hardware must remain operable.

6 Conclusions

API RP 753 was developed to provide guidance for reducing the risk to personnel located in portable buildings from potential explosion, fire, and toxic release hazards. Light wood trailers are a convenient and cost-effective method to house temporary personnel in various locations that are ever-changing due to construction and turn-arounds. However, changes and expansions to processes can result in light wood trailers falling out of compliance with API RP 753.

Relocating light wood trailers can be expensive and challenging due to cost and the need to relocate utilities and prepare the site pad, resulting in facility owners/operators looking for alternative solutions such as structural retrofits and upgrades. On the other hand, developing structural upgrades for conventional light wood trailers can be problematic from a practical standpoint, as stated previously, since these types of trailers are often leased from third-party vendors and intrusive structural strengthening approaches such as adding new interior wall studs and roof beams cannot be implemented without purchasing the trailers, which is likely cost prohibitive.

BakerRisk has developed a novel approach to strengthening conventional light wood trailer portable buildings to meet the requirements of API RP 753 to protect occupants of conventional portable buildings from blast hazards while also addressing this need for alternative solutions. The structural upgrades were designed to be non-intrusive, installed from the building exterior, and fully removable once the lease period has ended.

Testing was conducted on a single test specimen in an upgraded and non-upgraded condition with free-field blast overpressures similar to the blast loads calculated in the 2005 Texas City accident that resulted in numerous fatalities in light wood trailers. The efficacy of the upgrade is demonstrated in this testing, with the upgraded trailer exhibiting Low building damage (BDL-1), and the non-upgraded conventional trailer failing catastrophically (BDL-4).

7 References.

[1] API RP 753: Management of Hazards Associated with Location of Process Plant Portable Buildings, Second Edition Reference, American Petroleum Institute, January 2024

[2] U.S. Chemical Safety and Hazard Investigation Board (2005) Recommendation 2005-4-I-TX-2

[3] Anderson, T., Malik, D., Real World Applications of API RP 753, Proceedings of the Global Conference on Process Safety 2017

[4] API RP 752: Management of Hazards Associated with Location of Process Plant Permanent Buildings, Fourth Edition Reference, American Petroleum Institute, January 2024

[5] Polcyn, M.A., Bennett, R.H., and, Baker, Q.A., "Determination of Pressure Levels for Siting Light Wood Trailers at Chemical Plants and Refineries," presented at the 41st Annual Loss Prevention Symposium, April 22-26, 2007.

[6] *Design Values for Wood Construction – NDS Supplement.* (2024). https://awc.org/publications/2024-nds-supplement/ . Accessed 6 February 2024.

[7] Baker, Q.A. et al., (2010) Guidelines for Vapor Cloud Explosion, Pressure Vessel Burst, BLEVE and Flash Fire Hazards, Second Edition, Center for Chemical Process Safety (CCPS) & John Wiley and Sons, New York, NY.

[8] Pierorazio, A.J., Thomas, J.K., Baker, Q.A., and Ketchum, D.E., (2005) "An Update to the Baker-Strehlow-Tang Vapor Cloud Explosion Prediction Methodology Flame Speed Table," Process Safety Progress, 24(1): 59-65.

[9] Oswald, C. J., Baker, Q. A., "Vulnerability Model for Occupants of Blast Damaged Buildings," Proceedings of the Ninth International Symposium on Interaction of the Effects of Munitions with Structures, May 1999.

A Practicable Approach to Reactive Pressure-relief System Design for Multi-purpose Vessels

Casey Houston, P.E., CPPM
Chemitics, LLC
4200 Vernon Ave S
Minneapolis, MN 55416
casey@chemitics.com

Surendra Singh, PhD
Belmont Scientific
210 Stedman St, Unit 3
Lowell, MA 01851
singh@belmontscientific.com

Keywords: Hazard Identification and Risk Analysis, Process Safety and Reactors

Abstract

A practicable approach to pressure-relief system design for reactive hazards is a multistep procedure. First, the process information, including the potential chemical inventories, equipment design conditions, and operating conditions, is gathered and structured.

From the identified chemical inventories, the potential reactive chemistry is screened starting with a review of the safety data sheets (SDS) and reference literature including Sax's Dangerous Properties of Industrial Materials, Bretherick's Handbook of Reactive Chemical Hazards, the "*Process Safety Progress*" archives, and the AIChE Design Institute for Emergency Relief Systems (DIERS) online knowledgebase. Chemical incompatibility databases are referenced, and reactivity matrices are constructed from readily available resources (U.S. EPA, NOAA, and Coast Guard, for example). With the potential reactions identified, a review of relevant process safety incidents specific to the chemical inventories is catalogued.

The potential reactivity is then classified according to characteristics such as the calculated adiabatic reaction temperature (CART), the Melhem Index, and the Vent Area Pressure Index. In addition to the severity of the reaction, equilibrium estimates are performed to identify potentially hazardous reaction products, further potential side reactions, and toxic or flammable compounds that require special effluent handling. Based on the identified worst-case reactive inventories, a minimum set of calorimetry tests can be performed to assess pressure-relief system requirements.

1 Introduction

Chemical process industry (CPI) companies, particularly toll or contract manufacturing and blending facilities, often utilize vessels and reactors in multiple services that may exhibit runaway reactions. Performing calorimetry testing and sizing relief systems for every potential mode of operation is often unrealistic and costly.

This paper presents a practicable multistep approach to pressure-relief system design for reactive hazards in multi-purpose vessels as shown in Figure 1. This approach ensures safely designed pressure-relieving systems while minimizing the necessary calorimetry tests by screening for reactivity with literature reviews and analytical techniques.

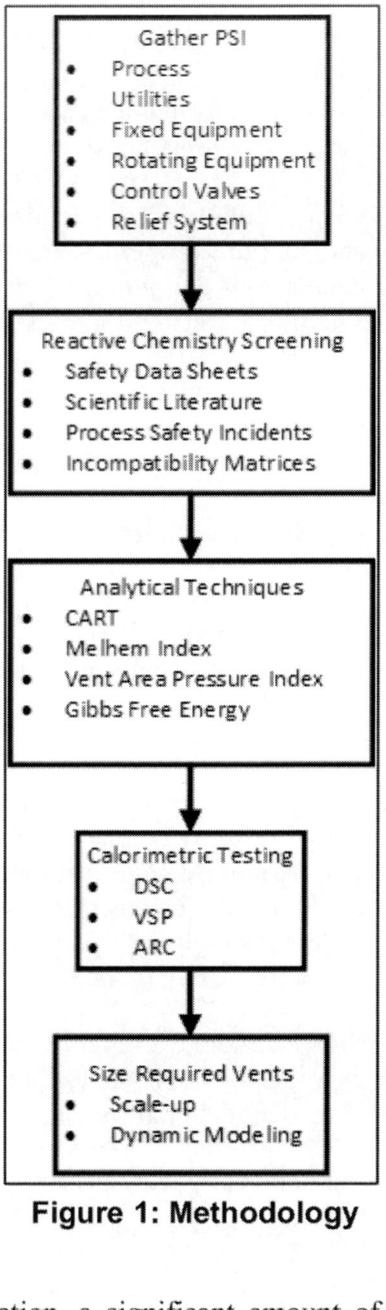

Figure 1: Methodology

2 Multi-purpose Vessels

Multi-purpose vessels are often used in batch or semibatch processing to manufacture specialty chemicals. Batch processing may be performed in-house by a major chemical producer or outsourced to a contract or toll manufacturer because low demand for a specialty chemical does not justify a continuous process or to avoid the risk associated with the production. [1]

Contract and toll manufacturers are relied upon as a fast and efficient way of producing a variety of specialty chemicals for multiple clients [2]; therefore, to quickly shift between products, tollers often use a single vessel to produce multiple products from multiple raw materials.

Due to the range of raw materials and products, the pressure-relief system design challenges associated with multi-purpose vessels requires particular care to ensure compliance with OSHA 29 CFR 1910.119, ASME BPVC Section VIII, API Standards 521 and 2000, and RAGAGEP. Following the proposed approach will ensure adequately sized vents for runaway reactions and identify potentially hazardous unintended reaction products for which safe discharge location can be engineered.

3 Gather Process Safety Information

Consistent with any pressure-relief system design evaluation, a significant amount of process safety information is required. Elements of process safety information required for relief system design documentation include [3]:

- Process Design and Description
  - Intended Chemistry
- Utility and Piping Design
- Fluid and Mixture Properties
- Pressure Relief Devices
- Fixed Process Equipment
  - Vessels
  - Heat Exchangers
  - Heaters / Steam Boilers
- Rotating Equipment
  - Pumps and Compressors
- Control Valves

Of particular importance for reactive-relief multi-purpose vessels, knowledge of the fluid and mixture properties are critical to ensure adequate relief system design. Creating a database of the potential chemical inventories is paramount to managing the reactivity. Thermo-physical properties for typical compounds can be found in databases like AIChE DIPPR, the NIST Chemistry WebBook, or those included with process simulators. For specialty compounds, which are often the raw materials or products in batch production in multi-purpose vessels, thermo-physical properties may need to be based on company generated data or estimated with UNIFAC [4] or Joback [5] structural functional groups.

4 Reactive Chemistry Screening

With a database of the chemical inventories established, the potential reactive chemistry can be screened with public domain software and open literature. The goal of the screening is to identify hazards associated with potential chemical inventories.

4.1 Safety Data Sheets

Safety Data Sheets (SDS) are a useful starting point for evaluating potential reactivity. However, only simple reactive hazards and chemical interactions are typically covered. The reactivity and incompatible materials are often simplified to reactive groups. Additionally, inconsistencies can be found between SDS for the same material from different suppliers. This is evident in the SDS for di-tert butyl peroxide (DTBP) as shown in Figure 2 and Figure 3.

SDS example B in Figure 3 clearly lists hazardous decomposition products whereas SDS example A simply states that under normal conditions hazardous decomposition products should not be produced.

For inconsistencies between supplier SDS, the CAMEO Chemicals database [6] can be useful.

Section 10. Stability and reactivity	
Reactivity	: No specific test data related to reactivity available for this product or its ingredients.
Chemical stability	: The product is stable.
Possibility of hazardous reactions	: Under normal conditions of storage and use, hazardous reactions will not occur.
Conditions to avoid	: Avoid all possible sources of ignition (spark or flame). Do not pressurize, cut, weld, braze, solder, drill, grind or expose containers to heat or sources of ignition.
Incompatible materials	: Oxidizers
Hazardous decomposition products	: Under normal conditions of storage and use, hazardous decomposition products should not be produced.
Hazardous polymerization	: Under normal conditions of storage and use, hazardous polymerization will not occur.

Figure 2: DTBP SDS §10 Example A

10. Stability and reactivity	
Reactive Hazard	Yes
Stability	Stable under normal conditions. Oxidizer: Contact with combustible/organic material may cause fire.
Conditions to Avoid	Incompatible products. Excess heat. Combustible material. Keep away from open flames, hot surfaces and sources of ignition.
Incompatible Materials	Acids, Bases, Metals, Reducing Agent, Finely powdered metals, Strong reducing agents, Combustible material
Hazardous Decomposition Products	Carbon monoxide (CO), Carbon dioxide (CO_2), Acetone
Hazardous Polymerization	Hazardous polymerization does not occur.
Hazardous Reactions	None under normal processing.

Figure 3: DTBP SDS §10 Example B

4.2 Chemical Incompatibility Matrices

To improve the knowledge gained from the SDS, incompatibility matrices can be generated to gain insight into which chemicals may pose most challenging to the pressure-relief system and warrant further analysis.

Public domain tools are available from NOAA / US-EPA (CAMEO Software Suite) [6] and AIChE CCPS (CRW) [7] to assess chemical interactions. Where public domain software tools prove insufficient, proprietary databases are available from expert consultants. Refer to Figure 4 and Figure 5 for examples of the interaction matrices for the compounds involved in the typical production of DTBP.

	DI-TERT-BUTYL PEROXIDE		
HYDROGEN PEROXIDE, AQUEOUS SOLUTION, STABILIZED, WITH MORE THAN 60% HYDROGEN PEROXIDE	**Incompatible** ■ Flammable Generates gas Generates heat Potentially hazardous Toxic	HYDROGEN PEROXIDE, AQUEOUS SOLUTION, STABILIZED, WITH MORE THAN 60% HYDROGEN PEROXIDE	
SULFURIC ACID	**Incompatible** ■ Flammable Generates gas Generates heat Intense or explosive reaction Toxic	**Incompatible** ■ Corrosive Explosive Generates gas Generates heat Intense or explosive reaction Toxic	SULFURIC ACID
TERT-BUTYL ALCOHOL	**Caution** □ Flammable Generates heat Intense or explosive reaction	**Incompatible** ■ Corrosive Explosive Flammable Generates gas Generates heat Intense or explosive reaction Toxic	**Incompatible** ■ Explosive Flammable Generates gas Intense or explosive reaction Toxic

Figure 4: CAMEO Interaction Matrix for Compounds in Typical DTBP Production

Y : Compatible
N : Incompatible
C : Caution
SR : Self-Reactive
*: Changed by user

NFPA Health	NFPA Flammability	NFPA Instability	NFPA Special	DTBP Production Compatibility Chart	DI-TERT-BUTYL PEROXIDE	HYDROGEN PEROXIDE, AQUEOUS SOLUTION, WITH NOT LESS THAN 20% BUT	SULFURIC ACID	TERT-BUTYL ALCOHOL
3	2	4	Oxi	DI-TERT-BUTYL PEROXIDE				
3	0	1	Oxi	HYDROGEN PEROXIDE, AQUEOUS SOLUTION, WITH NOT LESS THAN	N			
3	0	2	₩	SULFURIC ACID	N	N		
2	3	0		TERT-BUTYL ALCOHOL	C	N	N	

Figure 5: CRW4 Interaction Matrix for Compounds in Typical DTBP Production

Note that the matrices clearly indicate incompatibility concerns that were not obvious from the SDS including "intense or explosive reaction."

Where inconsistencies appear to exist between the SDS's, incompatibility matrices, or scientific literature, experts may need to be consulted.

4.3 Scientific Literature

Several sources of reactivity hazards are available in scientific and academic literature including those highlighted here. Additional resources, such as the Kirk-Othmer Encyclopedia of Chemical Technology [8], Process Safety and Environmental Protection (an IChemE Journal), and the ACS Chemical Health & Safety journal provide useful insights.

4.3.1 Bretherick's Handbook of Reactive Chemical Hazards [9]

Bretherick's Handbook of Reactive Chemical Hazards (Bretherick's) Vol. 1 is organized by chemical formula. The entry for Di-tert-butyl peroxide at $C_8H_{18}O_2$ indicates additional hazards and refers to additional listings in Vol. 2 such as Dialkyl Peroxides. The DTBP entry at *3069* is:

HCS 1980, *370*

1. *Griffiths, J. F. et al., Combust. Flame, 1984,* ***56****, 135-148*
2. *Stull, 1977, 21*
3. *Sime, R. J., Chem. Eng. News, 1988,* ***66****(24), 4*
4. *Antrim, R. F. et al. Process Safety Progr., 1998,* ***17****(3), 225*

Thermal decomposition is exothermic and self-ignition may result, especially if oxygen is present [1]. Though the heat of exothermic decomposition (1.32 kJ/g) is not exceptionally high, the weakness of the peroxide link tends to ready decomposition, and the energy release, coupled with the high volume of gaseous products (7.6 mol) at the adiabatic maximum of 550°C would give a 21-fold pressure increase in a closed vessel [2]. A routine student experiment on the gas-phase decomposition of the peroxide (0.2 mL) by injection into a 250 mL flask immersed in an oil bath at 155°C. The procedure was identical to a published method except that a mercury manometer was used to measure the pressure in place of a Bourdon gauge, and the procedure had been used on 30 occasions without incident. During a run at 165°C a violent explosion occurred [3]. An explosion and fire in an earthed drum of Di-t-butyl peroxide led to an investigation which showed the material unlikely to have ignited in the drum (open cup flash point 18°C). Evidence was gained for an initial fire external to the drum leading to local overheating and thermal decomposition of the contents. Self Accelerating Decomposition Temperature is 80°C [4].

See *THERMAL EXPLOSIONS*

See *DIALKYL PEROXIDES, SELF-HEATING AND IGNITION INCIDENTS*

Figure 6: Di-tert-butyl peroxide entry in Bretherick's

4.3.2 Sax's Dangerous Properties of Industrial Materials [10]

Sax's Dangerous Properties of Industrial Materials (Sax's) lists potentially hazardous materials including drugs, ores, pesticides, lubricants extracts, and industrial intermediates and waste products from production processes. Entries contain information including physical properties, toxicity data, referenced standards and recommendations, consensus reports, and safety profiles that are summaries of the toxicity, flammability, reactivity, and incompatibility. Entries are listed in order of an alphanumeric code. The index, Volume 1, is divided into three cross-indexes organized by: DOT Guide Number, CAS Number, and Synonym. The "safety profile" for DTBP at entry BSC750 CAS: 110-05-4 says:

> ***SAFETY PROFILE:*** *Moderately toxic by intraperitoneal route. A powerful irritant by ingestion and inhalation. A mild skin and eye irritant. Questionable carcinogen with experimental tumorigenic data. Flammable liquid; see PEROXIDES, ORGANIC, for fire and explosion hazards. Warning: Water may not work to fight fire. When heated to decomposition it emits acrid smoke and fumes.*

Figure 7: Di-tert-butyl peroxide SAFETY PROFILE in Sax's

4.3.3 Process Safety Progress Archives

The AIChE publication Process Safety Progress (PSP) covers process safety for engineering professionals. Included with the online journal (https://aiche.onlinelibrary.wiley.com/journal/15475913) is an MS Excel index of all PSP articles from 1982-2019. The online article archive is searchable and provides research papers and guidance on several process safety subjects.

4.3.4 AIChE DIERS Technical Entity Knowledge Database

The AIChE Design Institute for Emergency Relief Systems (DIERS) technical entity provides an online knowledge database of presentations and papers for DIERS members: https://pse2.ioxpress.com/diers. The archive contains over 2,600 papers and presentations since the Spring 1989 meeting.

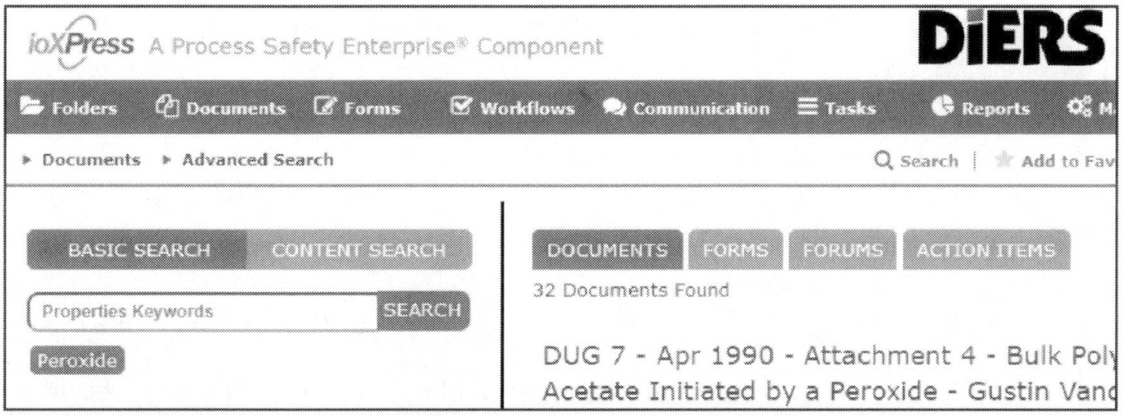

Figure 8: DIERS Searchable Online Database – "Peroxide"

4.4 Process Safety Incidents

To further investigate potential hazards, historical databases documenting incidents are available. Reactivity-related previous incidents can help identify overpressure scenarios, the main cause, and the consequences. Included here is a partial list of case history sources. The goal of reviewing process safety incidents is to be aware of and identify, which potential chemical inventory has been associated with past incidents.

4.4.1 Lees' Loss Prevention in the Process Industries

Appendix 1 of Volume 3 of Lees' Loss Prevention in the Process Industries [11] (Lees') provides case histories complied from several sources. Only significant incidents are documented in Lees'; however, the appendix contains a detailed list of incident sources with collections and databases for further research.

4.4.2 IChemE LPB

The Institution of Chemical Engineers (IChemE) publishes the Loss Prevention Bulletin (LPB) [12] bi-monthly. The LPB documents industry lessons learned and is searchable and indexed. The LPB index is available as a searchable and filterable MS Excel file.

Title	Date	Issue	Author
An approach to communicating the hazards of benzoyl peroxide	Aug-15	244	Tom Craig
Hydrogen peroxide hazards	Aug-12	226	Marcus Kneale
A chemical fire involving dicumyl peroxide	Oct-00	155	Tony Fishwick
Two explosions in the transport of organic peroxides	Jun-00	153	Ronnie Long
Hydrogen peroxide deflagration in a waste water treatment tank	Apr-94	116	Anon
Cumene hydroperoxide CHP drum explosion	Aug-93	112	Anon
Cumene hydroperoxide CHP drum explosion	Feb-92	103	Anon

Figure 9: Searchable Index Results of LPB for "Peroxide"

From the LPB search results, potential hazards and case studies provide awareness of the situations that may prove challenging to the pressure-relief system design.

4.4.3 IChemE Accident Database

In addition to the LPB, the IChemE published the Accident Database [13], which was developed in 1997, contains over 10,500 entries, with the last update made in 2000. The data is available as a free PDF download. The data is divided into five sections: Activity, Causes, Consequences, Equipment, and Substances.

4.4.4 CSB

The U.S. Chemical Safety and Hazard Investigation Board (CSB) published Incident Data [14] a reactive hazard investigation dataset of 167 serious incidents in the United States involving uncontrolled chemical reactivity from 1980 to 2001. The report is available here: https://www.csb.gov/assets/1/20/reactiveincidentdatareport.pdf.

5 Analytical Techniques

After reviewing case histories to identify which potential inventory has proven challenging to the industry, and prior to performing potentially costly calorimetry tests, analytical techniques can screen and rank reaction hazards.

5.1 CART and Minimization of Gibbs Free Energy

The Calculated Adiabatic Reaction Temperature (CART) is obtained from direct minimization of Gibbs free energy, which is a useful technique for estimating product stoichiometry for unintended chemical reactions including self-reactivity [15]. The potential reaction hazards are estimated based on the gas phase and an equilibrium final state. The gas phase is reasonable because condensed media reactions do not differ significantly from the gas phase reactions.

CART values can be calculated via minimization of Gibbs free energy with: commercial process simulators using a Gibbs Reactor unit operation; the AIChE software "SuperChems™ for DIERS; or the NASA Chemical Equilibrium and Applications software (CEA2) [16] [17]. The NASA CEA2 software was used to estimate the CART for a slate of organic peroxides. In accordance with CART estimation guidance [18] the equilibrium heat of reaction estimate is calculated at 298 K and 1 bara in the gas phase. Based on the heat of reaction estimate and heat capacity, a constant pressure adiabatic reaction temperature is calculated at 1 bara (CART).

An example is shown in Figure 10 for the decomposition of dicumyl peroxide. From the CEA2 results we can see the reaction stoichiometry based on the estimated mole fractions of the products (Equation 1).

$$DCP \rightarrow Methane + Carbon\ Dioxide + Ethyl\ Benzene$$

. 1

$$C_{18}H_{22}O_2 \rightarrow 0.273CH_4 + CO_2 + 2.09C_8H_{10}$$

Furthermore, the heat of reaction is estimated from the specific enthalpy of the products and reactants (Equation 2) and the reaction temperature rise is then estimated from the heat capacity and heat of reaction (Equation 3).

$$\left(-1{,}299.50\ \frac{kJ}{kg}\right) - \left(\frac{kgmol}{270.4\ kg} \times -79{,}100\ \frac{kJ}{kgmol}\right) = 1{,}006.97\ \frac{kJ}{kg} \quad 2$$

$$1{,}006.97\ \frac{kJ}{kg} \times \frac{kg \cdot K}{1.1583\ kJ} = 869\ K \quad 3$$

```
              THERMODYNAMIC EQUILIBRIUM PROPERTIES AT ASSIGNED

                         TEMPERATURE AND PRESSURE

 CASE = DCP

             REACTANT                    MOLES         ENERGY      TEMP
                                                      KJ/KG-MOL      K
 NAME        DCP                       1.0000000    -79100.000   298.000

 O/F=    0.00000  %FUEL=  0.000000  R,EQ.RATIO=23.500000  PHI,EQ.RATIO= 0.000000

 THERMODYNAMIC PROPERTIES

 P, BAR            1.0000
 T, K             298.00
 RHO, KG/CU M    3.2441 0
 H, KJ/KG        -1299.50
 U, KJ/KG        -1330.33
 G, KJ/KG        -2448.62
 S, KJ/(KG)(K)     3.8561

 M, (1/n)          80.379
 (dLV/dLP)t      -1.00000
 (dLV/dLT)p        1.0000
 Cp, KJ/(KG)(K)    1.1583
 GAMMAs            1.0981
 SON VEL,M/SEC      184.0

 MOLE FRACTIONS

 CH4              0.08107
 *CO2             0.29730
 C2H6             0.00001
 C8H10,ethylbenz  0.62162
```

Figure 10: CEA2 Output for Dicumyl Peroxide

A useful secondary output from the minimization of Gibbs free energies calculation is an estimate of the reaction products. The reaction product estimates provide a useful insight into the expected relief fluid and safe disposal requirements for any liquids, toxics, or flammables generated at relief conditions.

Although the examples herein are representative of self-reactive decomposition, CART and minimization of Gibbs Free Energy are applicable techniques for other reactive systems.

Table 1: CART from CEA2 Minimization of Gibbs Free Energy

Name	Unit	Hydrogen Peroxide	t-Butyl Hydroperoxide	di-t-Butyl Peroxide	Dicumyl Peroxide	tert-Butyl Peroxyacetate
Generic Acronym		H2O2	TBHP	DTBP	DCP	TBPA
Chemical Formula		H_2O_2	$C_4H_{10}O_2$	$C_8H_{18}O_2$	$C_{18}H_{22}O_2$	$C_6H_{12}O_3$
Molecular Weight	kg/kg-mol	34.015	90.12	146.23	270.4	132.2
Heat of Decomposition (ΔH_{rx})	kJ/kg	3,106	3,577	1,282	1,006	1,908
Heat of Decomposition (ΔH_{rx})	cal/g	742	855	306	240	456
Heat Capacity (C_p)	kJ/kg-K	1.419	1.470	2.677	1.158	2.377
Adiabatic Temperature Rise (ΔT_a)	K	2,188	2,433	479	868	802
CART	K	2,486	2,731	777	1,166	1,100
eCART	K	2,846	3,939	1,259	1,747	1,705

Reactions that generate more moles of products than reactants may generate additional mechanical energy. The CART is therefore scaled based on the following relationship [18]:

$$eCART = CART \times \left(\frac{N_p}{N_r}\right)^{1/3} \quad (4)$$

Where:
N_p is the reaction products number of moles (mol)
N_r is the starting number of moles in the reactant (mol)
eCART is the effective computed adiabatic temperature (K)

Based on the eCART (Equation 4), the decomposition of H2O2 and TBHP may be expected to dictate the vent area.

Common organic peroxide decomposition products are carbon dioxide, ethane, and acetone [19]. However, depending on the initial conditions and the peroxide under consideration, the minimization of Gibbs free energies provides estimates of the decomposition products [15].

From the CART conditions estimate (performed at 298 K, 1 bara), and a forced gas phase reaction in concert with estimate guidance, CEA2 calculates the decomposition products. Note that the conditions most appropriate to estimate the heat of reaction and CART may not most accurately represent the conditions representative of the expected decomposition products; the decomposition product estimates may require iteration on pressure, temperature, and compound constraints based on expert consultant selectivity. To demonstrate expected decomposition products predicted by minimization of Gibbs free energy, the standard condition decomposition products predicted by CEA2 are reported in Table 2. The reaction products can be constrained to specific compounds or opened to a broad slate of potential compounds. The examples here are based on a constrained product matrix limited to typical peroxide decomposition products.

Table 2: Predicted Gas Phase Decomposition Products from CEA2 at 298K and 1bara

Name		Hydrogen Peroxide	t-Butyl Hydroperoxide	di-t-Butyl Peroxide	Dicumyl Peroxide	tert-Butyl Peroxyacetate
Generic Acronym		H2O2	TBHP	DTBP	DCP	TBPA
Chemical Formula		H_2O_2	$C_4H_{10}O_2$	$C_8H_{18}O_2$	$C_{18}H_{22}O_2$	$C_6H_{12}O_3$
Product Mixture Molecular Weight	kg/kg-mol	22.676	30.038	34.312	80.379	35.568
Decomposition Equilibrium Products	mol-frac					
Water	H_2O	0.667				
Oxygen	O_2	0.333				
Methane	CH_4		0.485	0.180	0.081	0.169
Carbon Monoxide	CO		tr	0.003		0.003
Carbon Dioxide	CO_2		0.333	0.233	0.297	0.402
Ethylene	C_2H_4			tr		tr
Ethane	C_2H_6		0.029	0.291	tr	0.236
Propylene	C_3H_6		tr	0.293		0.189
Propane	C_3H_8		0.152			
Ethylbenzene	C_8H_{10}				0.622	
Mole noncondensable gas / Mole peroxide (N_{ncg})		0.5	1	1	1	1.5

In addition to potentially hazardous reaction products that challenge the effluent handling systems and may require special consideration, the minimization of Gibbs free energy calculations indicate how many moles of noncondensable gas are formed. The noncondensable gas generation is critical to estimating the vent area pressure index.

5.2 Melhem Index

Extending the usefulness of the CART, the Melhem Hazard Index [20] couples the equilibrium heat of reaction with the CART to develop a more intuitive screening method. Reactions with a Melhem index of 'A' indicate a highly energetic reaction that may be explosive. It may be necessary to dilute the high hazard index systems should be diluted with a high-boiling point solvent that acts as a sufficient heat sink.

Systems with an index of 'B' or 'C' are energetic, but may be protected by typical safeguards such as pressure relief systems.

Note that the eCART and heat of reaction reported in Table 1 are based on 100% peroxide concentrations and do not consider the typical solvents like water, toluene, and isopropyl alcohol. Due to the 100% peroxide concentration and number of moles in the decomposition product, the eCART calculation pushes all peroxides except DTBP into the High Hazard Index 'A' category.

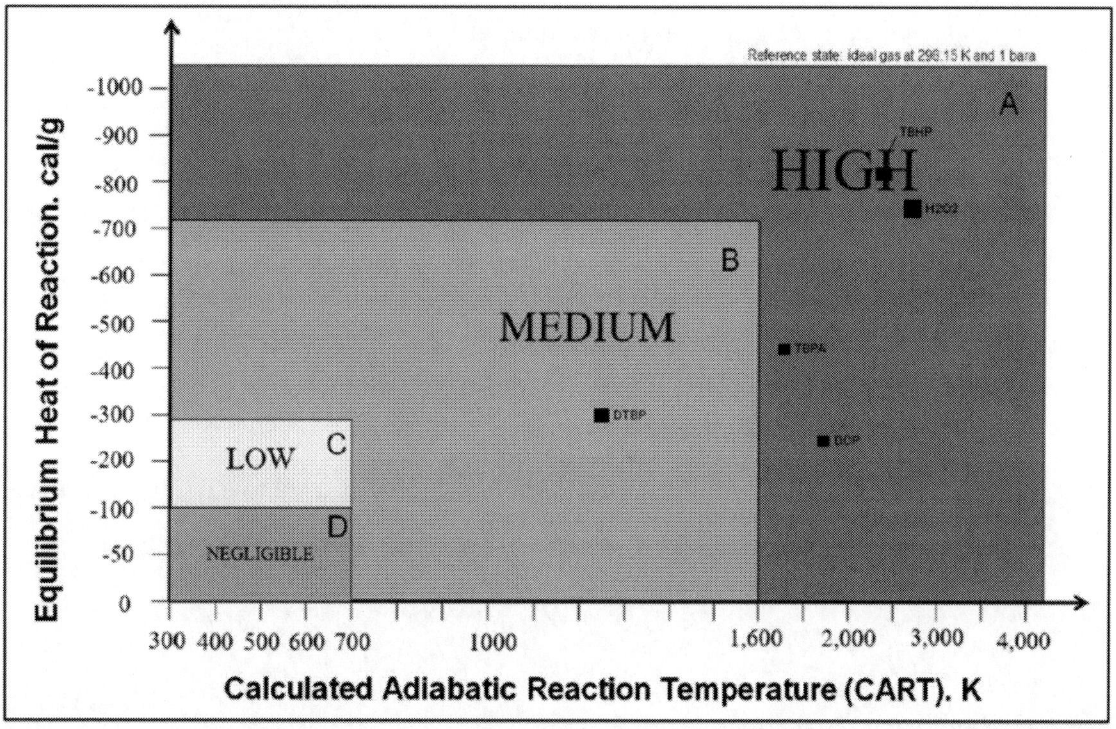

Figure 11: Melhem Chemical Reactivity Hazard Index with Select Peroxides

Coupling the CART with the equilibrium heat of reaction, the Melhem Index indicates that in addition to H2O2 and TBHP, both TBPA and DCP are high hazards.

5.3 Vent Area Pressure Index

The Melhem Hazard Index provides a simple guideline for classification. To better refine the classification and gain insight into the magnitude of the difference in required vent size, the Vent Area Pressure Index [19] (VAPI) can be calculated. Although developed specifically for organic peroxide decomposition reactions, the method can be extended to reactive systems that evolve gaseous products. VAPI is an analytical classification based on the mass specific energy and volumetric gas generation rate.

The VAPI can be calculated for all-gas vent flow or homogeneous gas-liquid vent flow depending on the degree of disengagement and expected venting regime. The all-gas VAPI is calculated using Equation 5.

$$VAPI = \frac{C_d A}{m_0} P = N_{ncg} \frac{MW_g}{MW_p} \frac{x_{mr} \, {}^{k_{mr}}\!/_{60}}{F(g,h)} \sqrt{\frac{RT_{mr}}{MW_g}} \qquad 5$$

Where:
N_{ncg} is the ratio of moles of non-condensable gas generated per mol of peroxide
MW_g is the molecular weight of the generated gas
MW_p is the molecular weight of the peroxide
x_{mr} is the concentration of the peroxide at the maximum reaction rate
k_{mr} is the rate constant at the maximum reaction rate
R is the gas constant (8.314 J/mol·K)
T_{mr} is the temperature at the maximum reaction rate

and

$$F(g,h) = \left(\frac{2g}{g-1}\left[1 - h^{(g-1)/g}\right] h^{2/g}\right)^{1/2} \qquad 6$$

Where:
g is the gas specific heat capacity ratio (C_P/C_V)
h is the ambient pressure ratio (P_{amb}/P)

For homogeneous gas-liquid mixture vent flow, Equation 7 is used to calculate the VAPI:

$$VAPI = \frac{C_d A}{m_0} P = \frac{N_{ncg}}{MW_p} \frac{x_{mr} \, {}^{k_{mr}}\!/_{60}}{P^{3/2}} RT_{mr} \sqrt{\rho_l} \qquad 7$$

Where:
ρ_l is the liquid density (kg/m^3)
P is the relief pressure (Pa)

The above relationships were used to estimate the VAPI for the six peroxides under consideration. The assumed container is a 50% full 2.6 m^3 vessel, subject to external fire heating with a design pressure >15 psig. Refer to Grolmes [19] for intermediate calculation steps and supplemental information.

Table 3: Vent Area Pressure Index for Select Peroxides

Name		Hydrogen Peroxide	t-Butyl Hydroperoxide	di-t-Butyl Peroxide	Dicumyl Peroxide	tert-Butyl Peroxyacetate
Generic Acronym		H2O2	TBHP	DTBP	DCP	TBPA
Chemical Formula		H_2O_2	$C_4H_{10}O_2$	$C_8H_{18}O_2$	$C_{18}H_{22}O_2$	$C_6H_{12}O_3$
Molecular Weight	kg/kg-mol	34.015	90.12	146.23	270.4	132.2
Heat of Decomposition (ΔH_{rx})	kJ/kg	3,106	3,577	1,282	1,006	1,908
Heat Capacity (C_p)	kJ/kg-K	1.419	1.470	2.677	1.158	2.377
Decomposition Products (MW_g)		22.676	30.038	34.312	80.379	35.568
Noncondensable Gas Parameter (N_{ncg})	mol-gas/mol-p	0.5	1	1	1	1.5
Activation Energy (E_a)	kcal/gmol	24.2	34.4	38.8	36.6	32.4
Reduced Activation Energy (B)	K	12,181	17,321	19,545	18,423	16,295
Preexponential Factor (A)	/min	7.36E+09	9.14E+13	1.48E+18	3.72E+17	8.45E+15
Onset Temperature (T_{onset})	K	413	449	415	402	384
Peroxide density (rho)	kg/m^3	1443	886	790	1029	949
External Heating Rate ($(dT/dt)_{ext}$)	°C/min	4	6.5	7.5	5.5	6
Maximum Reaction Temp (T_{max})	K	2622	2900	907	1284	1199
Maximum Reaction Rate Temp (T_{mr})	K	2218	2530	868	1205	1122
Concentration at Max Rate (x_{mr})	wt frac	0.184	0.152	0.080	0.090	0.096
Rate Constant at Max Rate (k_{mr})		3.04E+07	9.72E+10	2.48E+08	8.51E+10	4.15E+09
All-Gas Vent Area Pressure Index (C_d A P / m_0)		1.29E+06	3.34E+09	1.79E+06	6.91E+08	6.84E+07
Homogeneous two-phase Vent Index (C_d A P / m_0)		1.55E+01	5.52E+04	2.10E+01	5.53E+03	8.64E+02

Based on the VAPI the decomposition of TBHP and DCP appear to be the worst-case inventory for both all-gas and homogeneous two-phase relief.

Note that the interactions of the solvents used to dilute the inventory may participate in the reaction and alter the stoichiometry, reaction rate, reaction products, and temperature rise. For example, TBHP is often in solution with DTBP and H2O, which together generate different decomposition products at a lower decomposition heat of reaction and adiabatic temperature rise. To ensure safe relief system design, it is critical to include the solvent in the reactivity evaluation.

6 Calorimetric Testing

After screening for reactive chemistry hazards and identifying the worst-case suspects, appropriate calorimetric experiments can be designed and performed to ensure safe pressure-relief system design. Depending on the types of reactivity hazards, a selection of tests and apparatuses are available.

6.1 Screening

Tests to screen for thermal stability include blasting cap tests, thermogravimetric analysis (TGA), differential scanning calorimetry (DSC), and the (advanced) reactive systems screening tool ((A)RSST). Screening tests allow comparison between reactive inventories.

Blasting cap tests provide means for evaluating self-reactivity hazards and sensitivity to relatively intense impact and are typically used to evaluate new compositions compared to known materials. Blasting cap tests also provide semi-quantitative, rudimentary insight into propagation. Select blasting cap test results from Melhem [21] are shown in Figure 12.

Blasting Cap Test Results, December 1997

Run #	Composition	D. max. cm	delta D	Volume ml	delta V ml	D g/cc	Condition of Witness tube	delta H(r) kJ/g Gas phase	CART K reactants
20	Cumene hydroperoxide 80%	4.52	0.29	209	4	1.02	Ditto	-3.13	956
21	Cumene hydroperoxide 80%	4.57	0.34	209	4	1.02	Ditto	-3.13	956
22	Di t-butyl peroxide	4.52	0.29	210	5	0.79	Ditto	-2.72	847
23	Di t-butyl peroxide	4.50	0.27	208	3	0.79	Ditto	-2.72	847
24	Benzoyl peroxide	4.30	0.07	206	1	0.71	Ditto	-3.47	1016
25	Benzoyl peroxide	4.50	0.27	206	1	0.64	Ditto	-3.47	1016
26	H2O2 40%+EtOH, balanced	4.55	0.32	209	4	1.12	Ditto	-3.39	1874
27	H202 40%+EtOH, balanced	4.49	0.26	210	5	1.12	Ditto	-3.39	1874

Figure 12: Blasting Cap Tests for Select Peroxides

The thermogravimetric analysis (TGA) technique measures the mass of a substance as a function of temperature, which quantifies several characteristics including loss of water, loss of solvent, oxidation, and decomposition. Refer to Table 4 and Figure 13 for TGA results indicating onset temperature of DCP decomposition. [22]

Table 4: TG T_o Results for Varying DCP Concentrations and Heat Rates [22]

Heating Rate (K/min)	Mass %	T_o (°C)
10	30	150
	50	167
	70	165
	100	162
5	30	139
	50	152
	70	159
	100	152
3	30	140
	50	144
	70	139
	100	147

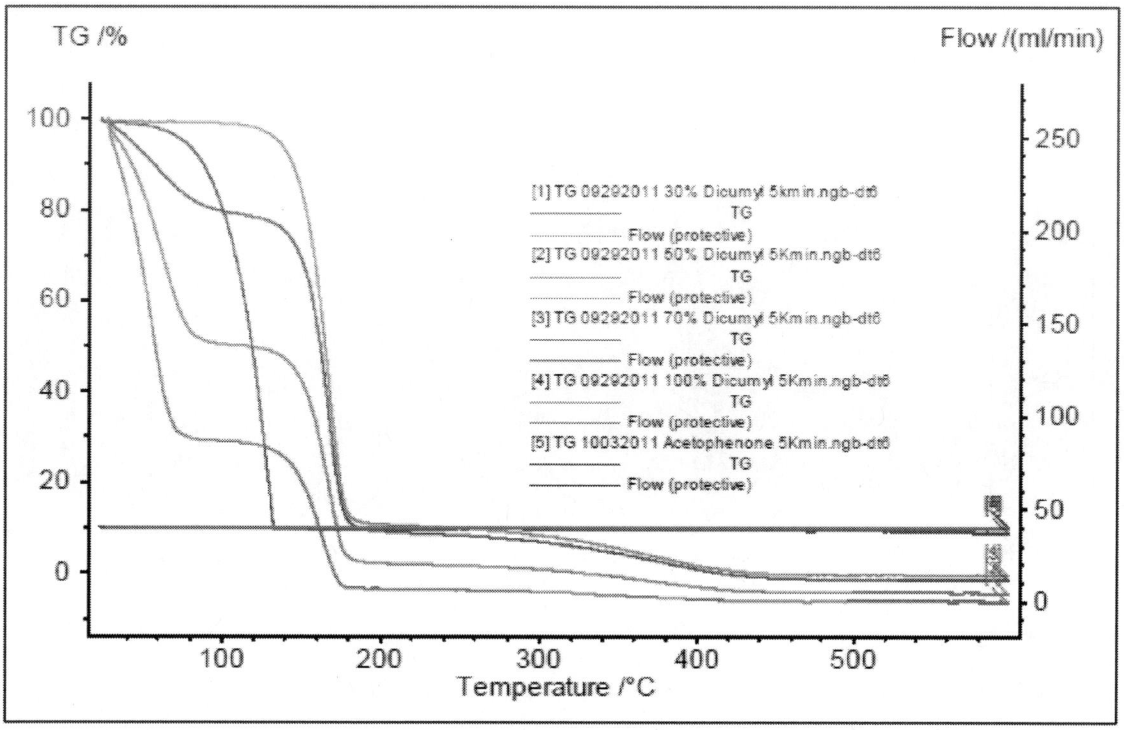

Figure 13: TGA Values for Varying DCP Concentrations at 5K/min [22]

The differential scanning calorimeter (DSC) measures the difference in heat input between the sample and a reference. Results from the DSC include heat capacity (C_p) measurements used to calculate adiabatic temperature rise and thermal inertia for use in further adiabatic reactive calorimeter (ARC) tests. Refer to Figure 14 for a typical DSC thermogram [22] and to Table 5 for tabulated decomposition characteristics tested by DSC. [23]

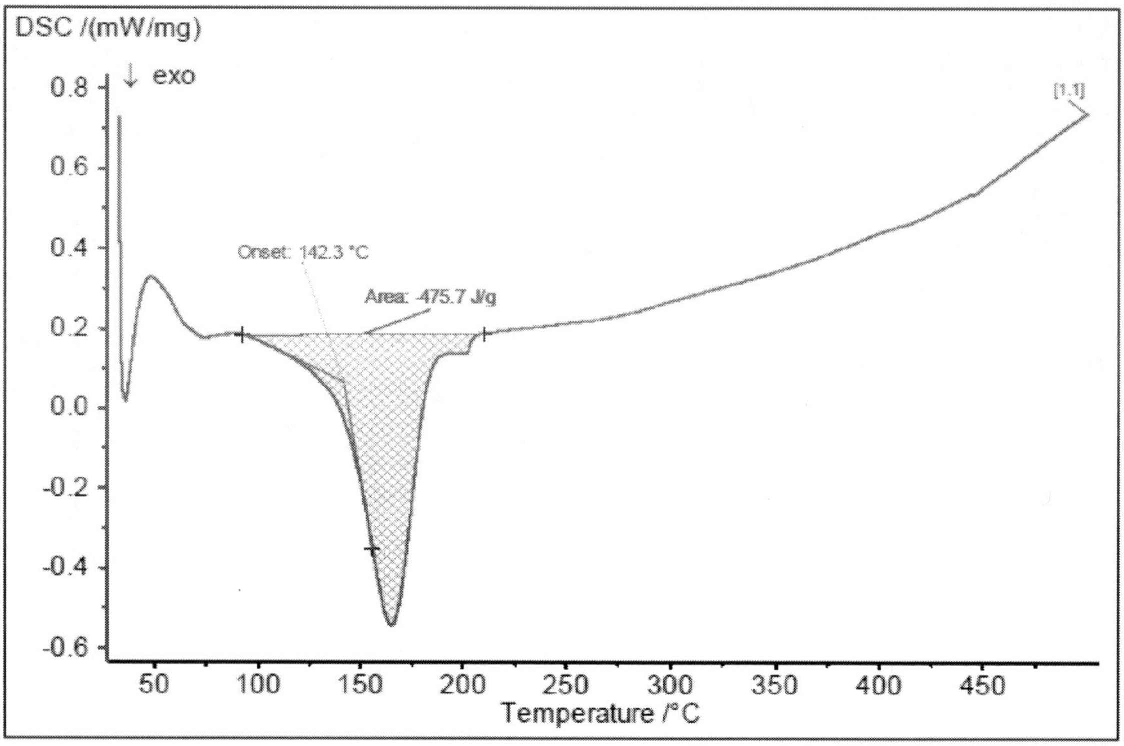

Figure 14: DSC Scan 70% dicumyl peroxide at 3K/min [22]

Table 5: Self-reactive decomposition of organic peroxides by DSC [23]

Organic Peroxide	Weight (mg)	Scanning Rate (°C/min)	Onset Temp (T_{onset}, °C)	Heat of Decomposition (J/g, 100 wt % Organic Peroxide)
CHP 12%	4.62	10	130	1,527
CHP 16%	5.55	10	130	1,692
CHP 20%	4.15	10	130	1,571
CHP 30%	4.40	10	110	1,743
CHP 35%	4.50	10	100	1,790
CHP 80%	4.90	2	100	1,825
DTBP	2.70	10	125	1,197
DTBP	1.33	4	115	1,202
DTBP	2.5	2	110	1,312
DTBP	3.9	1	105	1,288
MEKPO	3.08	4	35+155	724 + 740
MEKPO	3.74	4	40+160	705 + 772
MEKPO	8.55	1	35+145	652 + 831
TBHP	1.83	4	65+225	1,185 + 707
TBHP	1.71	4	65+225	1,408 + 892
BPO	3.87	4	105	1,034
LPO	3.82	4	70	809
LPO	3.00	4	70	787
LPO	4.67	1	70	766
TBPBZ	9.0	10	105	1,435
TBPBZ	1.43	4	100	1,446
TBPBZ	4.52	4	95	1,477
TBPBZ	2.70	2	95	1,522
TBPBZ	7.50	1	90	1,515
DCPO	2.1	4	110	993
DCPO	2.82	4	110	942

The ARSST can be used to directly simulate overpressure scenarios and to estimate parameters used in vent sizing including temperature and pressure rise rates, adiabatic temperature rise, and heat of reaction. The ARSST can also be used for sizing required vents by scale-up.

6.2 Kinetic Evaluation

Useful apparatuses to evaluate reaction rates and kinetics for relief system design include the Accelerating Rate Calorimeter (ARC), the Vent Sizing Package (VSP2), and the Automatic Pressure Tracking Adiabatic Calorimeter (APTAC).

The ARC is particularly effective due its sensitivity and high adiabaticity. The ARC measures both temperature and pressure during exotherms. ARC datasets include pressure and temperature as a function of time, pressure versus temperature, and the derivatives of self-heat rate (dT/dt), and pressure rise rate (dP/dt) as functions of temperature. From these datasets, kinetic parameters can be derived, and the models adjusted to match experimental results. A typical pressure-rise rate plot from ARC testing is shown in Figure 15.

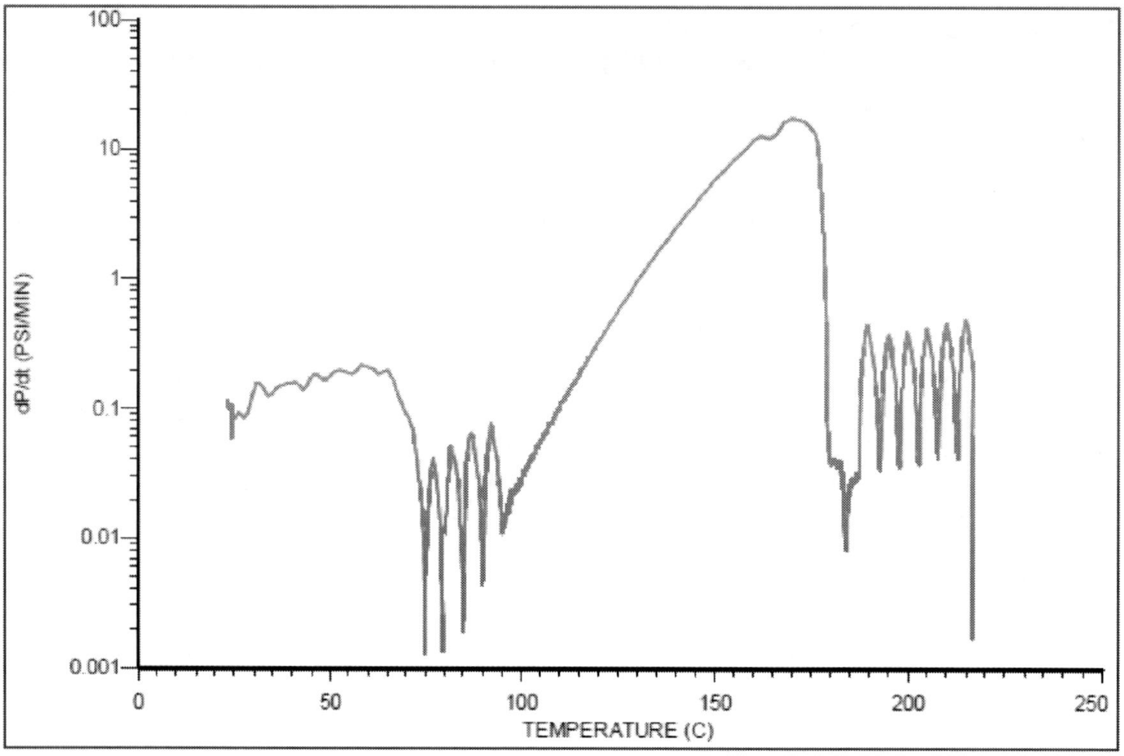

Figure 15: ARC data plot of pressure rise rate vs. temperature for 50% dicumyl peroxide in toluene [22]

The ARSST can also be used to develop reaction kinetic parameters. From the temperature rise rate vs. temperature data, the Arrhenius expression can be fit to the data. Note that for constant heat input experiments, rather than heat-wait-search, the data must be corrected to include only the self-heat temperature rise rate. Consider Figure 16, which shows ARSST data used to determine reaction rate parameters for neat dicumyl peroxide. [24]

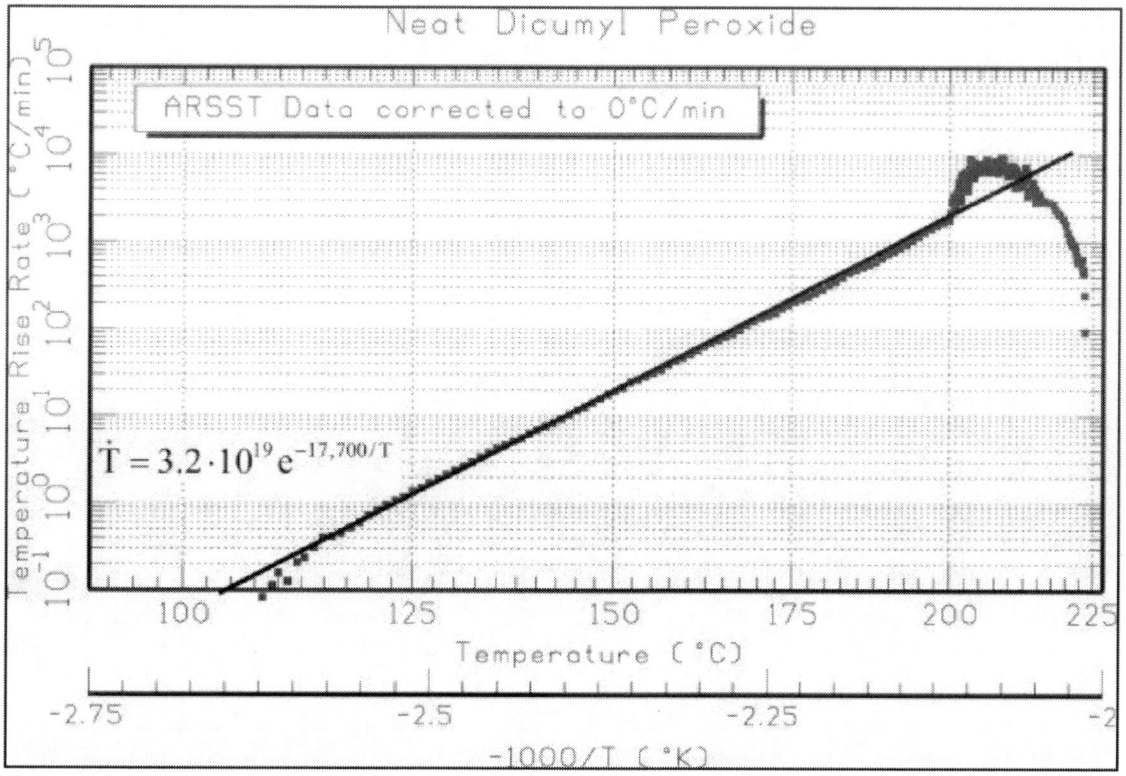

Figure 16: Temperature rise rate curve with kinetic data

7 Size Required Vents

With the worst-case inventories identified through analytical screening and appropriate calorimetry performed, the required vent size can be calculated.

When sufficient knowledge is gained about the chemistry and reactivity through appropriate testing and analyses, detailed kinetic models and stoichiometry may be used to estimate vent sizes with dynamic modeling. The AIChE DIERS Software SuperChems™ for DIERS and commercial process simulators include dynamic sizing options.

When such software is unavailable, sizing can be performed with DIERS technology using scale-up relationships directly correlated with the calorimetric testing. Equations are available for tempered, gassy, and hybrid systems [25] that can be used in conjunction with DIERS estimates for vapor/liquid disengagement.

Based on the Vent Area Pressure Index (Table 3), TBHP and DCP appear to be the most challenging to a vent design. A prudent design would consider at least these two systems for additional calorimetry and vent sizing. For the DCP inventory, the gassy system vent sizing equation (Equation 8 [25]) is appropriate.

$$\frac{A}{V} = \frac{1}{0.61\, C_D} \frac{\rho \nu \dot{P}}{m_t P} \left(\frac{M_{w,g}}{RT}\right)^{1/2} \qquad 8$$

Where:
A is the vent area (m^2)
V is the volume of reactants (m^3)
C_D is the vent system discharge coefficient
ρ is the loading density (kg/m^3)
ν is the test freeboard volume (m^3)
$\dot{P}$ is the maximum pressure rise rate (Pa/s)
m_t is the test sample mass (kg)
P is the relief pressure (Pa)
$M_{w,g}$ is the gas molecular weight
T is the relief temperature (K)
R is the gas constsant (8,314 Pa-m^3 / K-kgmol)

RSST data for neat dicumyl peroxide [26] indicates a peak pressure rise rate of ~4,000 psi/min (Figure 17). Applying this to Equation 8 results in an unreasonably large vent area to volume ratio (10^3 m^{-1}) demonstrating that despite initial two-phase relief, sizing for all-gas venting is typically most conservative.

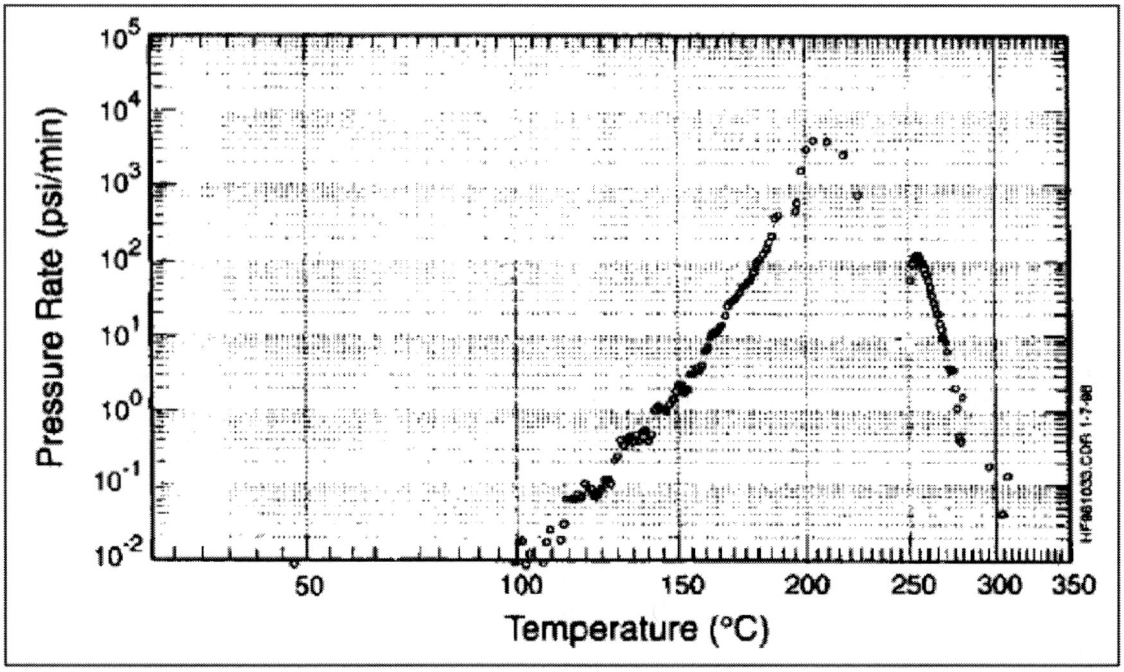

Figure 17: RSST test with 100 wt% dicumyl peroxide with 1°C/min heating [26]

Experimental test results support the suspected behavior that initial two-phase venting leads to a smaller vent size compared to all gas venting, even for gassy systems. This effect can be applied to the Fauske generalized sizing equation for gassy systems resulting in Equation 9 [27]:

$$\frac{A}{V} = \frac{3.5 \times 10^{-3}}{C_D P \left[1 + \frac{1.98 \times 10^3}{P^{1.75}}\right]^{0.286}} \dot{P}(1 - \bar{\alpha}) \qquad 9$$

Where:
A is the vent area (m^2)
V is the vessel volume (m^3)
$\dot{P}$ is the maximum pressure rise rate in RSST with 10g sample (psi/min)
C_D is the vent system discharge coefficient
P is the venting pressure (psig)
$\bar{\alpha}$ is the average void fraction (initial fill fraction otherwise)

Consider a 15 psig vertically oriented 6 ft tall, 4 ft diameter vessel with 2:1 ellipsoidal heads (690 gal, 2.6 m^3) with typical rupture disks ($C_D = 0.62$) and applying Equation 9 to the DCP inventory:

$$\frac{A}{V} = \frac{3.5 \times 10^{-3}}{0.62 \cdot 18 \left[1 + \frac{1.98 \times 10^3}{18^{1.75}}\right]^{0.286}} 4{,}000\ (1 - 0.5)$$

$A/V = 0.297\ m^{-1}$

$A = 0.78\ m^2$

Which is consistent with full-scale experiments. Variants of Equations 8 and 9 have also been developed for vapor (tempered) and hybrid systems [27].

With the vent size established, consideration can be given to other layers of protection to avoid the runaway or reduce the required vent size reduction measures including storing with sufficient high-boiling solvent. Note that some runaway reactions may be so energetic that relief design is impractical and safe handling must be ensured through proper risk reduction measures.

Regardless of the vent size, consideration must be given to the safe discharge of the effluent. In the case of two-phase venting, the effluent will be a mixture of the initial inventory and reaction products. Adequate effluent handling equipment including knockout drums, quench vessels, and ultimate disposal must be designed.

8 Conclusion

A practicable approach to pressure-relief system design for reactive hazards is a multistep procedure.

After obtaining the PSI, the potential reactive chemistry is screened starting with a review of the safety data sheets (SDS) and reference literature including Sax's Dangerous Properties of Industrial Materials, Bretherick's Handbook of Reactive Chemical Hazards, the "*Process Safety Progress*" archives, and the AIChE Design Institute for Emergency Relief Systems (DIERS) online knowledgebase. Chemical incompatibility databases are referenced, and reactivity matrices are constructed from readily available resources (U.S. EPA, NOAA, and Coast Guard, for example). With the potential reactions identified, a review of relevant process safety incidents specific to the chemical inventories is reviewed for potential upset scenarios and consequences.

The potential reactivity is then classified according to characteristics such as the calculated adiabatic reaction temperature (CART), the Melhem Index, and the Vent Area Pressure Index. In addition to the severity of the reaction, equilibrium estimates are performed to identify potentially hazardous reaction products, further potential side reactions, and toxic or flammable mixtures that require special effluent handling. Based on the identified worst-case reactive inventories, a minimum set of calorimetry tests can be performed to assess pressure-relief system vent size. DCP and TBHP would be primary candidates for calorimetry testing in this example slate of peroxides whereas DTBP appears less hazardous.

9 References

1. Seggerman B. Ensuring Process Safety in Batch Tolling. *Chemical Engineering Progress*. December 2017:34-40.

2. Hirsh C. Lamart Corporation. *What is Toll Manufacturing? Everything You Should Consider* [Blog]. February 7, 2022. Available at: https://lamartcorp.com/blog/what-is-toll-manufacturing/. Accessed January 23, 2024.

3. Melhem GA. Relief System's Last Line of Defense, Only Line of Defense? *Process Safety Progress*. December 2006;25(4):290-297.

4. NIST. ThermoData Engine User's Guide. *NIST-Modified UNIFAC Prediction Method*. Available at: https://trc.nist.gov/TDE/Help/TDE103b/NIST-Modified-UNIFAC-AC-Model.htm. Accessed February 2, 2024.

5. Joback K. Molecular Knowledge Systems. *Physical Property Estimation and Chemical Product Design*. Available at: https://www.molecularknowledge.com/index.html. Accessed February 2, 2024.

6. NOAA. CAMEO Chemicals. *Database of Hazardous Materials*. Available at: https://cameochemicals.noaa.gov/. Accessed February 1, 2024.

7. CCPS. Chemical Reactivity Worksheet. *Download/Install | AIChE*. Available at: https://www.aiche.org/ccps/resources/downloadinstall. Accessed February 1, 2024.

8. Cummings J. *Kirk-Othmer Encyclopedia of Chemical Technology*. Hoboken, NJ: J. Wiley; 2007.

9. Bretherick L. *Bretherick's Handbook of Reactive Chemical Hazards*. Vol 1&2. 7th ed. Burlington, MA: Academic Press; 2007.

10. Lewis RJ. *Sax's Dangerous Properties of Industrial Materials*. Vol II. 12th ed. Hoboken, NJ: John Wiley & Sons, Inc.; 2012.

11. Lees FP. *Lees' Loss Prevention in the Process Industries*. Vol 3. 3rd ed. Burlington, MA: Elsevier Butterworth-Heinemann; 2005.

12. Institute of Chemical Engineers. Loss Prevention Bulletin - Archive - IChemE. *IChemE LPB Archive*. 1974-2024. Available at: https://www.icheme.org/knowledge-networks/loss-prevention-bulletin/lpb-archive/index/. Accessed February 1, 2024.

13. Institute of Chemical Engineers. IChemE Accident Data. *ICehmE Safety Centre - Accident Database - IChemE* [Database]. 2000. Available at: https://www.icheme.org/knowledge-networks/safety-centre/resources/accident-data/. Accessed February 02, 2024.

14. U.S. Chemical Safety and Hazard Investigation Board. *Incident Data - Reactive Hazard Investigation*: CSB; 2003. 2003-15-D.

15. Melhem GA. An Advanced Method for the Estimation of Reaction Stoichiometry and Rates from ARC Data. Paper presented at: International Symposium on Runaway Reactions and Pressure Relief Design, 1995; Boston, MA.

16. Gordon S, McBride BJ. *Computer Program for Calculation of Complex Chemical Equilibrium Compositions and Applications - I. Analysis*: National Aeronautics and Space Administration; 1994. 1311.

17. McBride BJ, Gordon S. *Computer Program for Calculation of Complex Chemical Equilibrium Compositions and Applications - II. Users Manual and Program Description*: National Aeronautics and Space Administration; 1996. 1311.

18. Melhem GA. *Managing Chemical Reactivity Hazards in Multipurpose Equipment*: ioMosaic Corporation; 2023. Process Safety and Risk Management Practices.

19. Grolmes MA. Pressure Relief Requirements for Organic Peroxides and Other Related Compounds. Paper presented at: International Symposium on Runaway Reactions, Pressure Relief Design, and Effluent Handling, 1998; New Orleans, LA.

20. Melhem GA, Shanley ES. On the Estimation of Hazard Potential for Chemical Substances. *Process Safety Progress*. Fall 1996;15(3):168-172.

21. Melhem GA. An Experimental Study of Organic Peroxide Hazards. Paper presented at: 22nd DIERS Users Group Meeting, 1998; Charleston, SC.

22. Murphy MR. New Calorimetry Data for Dicumyl Peroxide (30-100%). Paper presented at: DIERS 48, 2011; Chicago, IL.

23. Duh YS, Wu XH, Kao CS. Hazard Ratings for Organic Peroxides. *Process Safety Progress*. June 2008;27(2):89-99.

24. Theis AE, Burlebach JP, Askonas CF. Safely Scale-up Processes and Accommodate Recipe Changes. *Process Safety Progress*. June 2009;28(2):135-140.

25. Fauske HK. Properly Size Vents for Nonreactive and Reactive Chemicals. *Chemical Engineering Progress*. February 2000:17-29.

26. Fauske HK. The Reactive System Screening Tool (RSST): An Easy, Inexpensive Approach to the DIERS Procedure. *Process Safety Progress*. 1998;17(3):190-195.

27. Fauske HK. Revisiting DIERS' Two-Phase Methodology for Reactive Systems Twenty Years later. *Process Safety Progress*. September 2006;25(3):180-188.

Considerations for the Safe Handling and Processing of Unstable Materials

Brian Ott
Exponent, Inc.
5401 McConnell Ave, Los Angeles, California 90066
bott@exponent.com

Nick Welchert, Laurent Delafontaine, Matevz Frajnkovic, Ali Reza
Exponent, Inc.

Keywords: Decomposition, polymerization, self-accelerating, unstable chemicals, SADT, SAPT

Abstract

We evaluate the instability hazards of initiators, monomers, and other unstable materials during storage and handling in individual containers, pallets of containers, and storage and process vessels. We propose a simplistic model to analyze the temperature-time profile of an unstable chemical, given heat transfer from/to the environment, heat generation due to decomposition and other reaction kinetics, and heat generated from auxiliary equipment (e.g., agitators/mixers, pumps, vessel jackets). Lumped capacitance-based heat transfer and decomposition kinetics are integrated to predict self-accelerating decomposition temperature (SADT), self-accelerating polymerization temperature (SAPT), and the temperature profile given exposure conditions (e.g., ambient temperature and additional sources of heat). Case studies related to the storage and processing of unstable liquid monomers (i.e., methyl methacrylate, MMA), and organic peroxides, (i.e., tert-butyl peroxybenzoate, TBPB), are presented. We also provide recommendations on determining safe storage and processing temperatures and steps to prevent emergency situations and include an overview of thermal analysis and data treatment procedures which might affect chemical process safety recommendations.

1 Introduction

The chemical industry handles and processes an increasing amount and variety of unstable chemicals. For example, certain chemicals used to manufacture plastics, including peroxide-based initiators and olefinic monomers, are known to be unstable. The quantity of plastics produced worldwide has increased 12% over the 5 years from 2017 to 2021 and is expected to grow at a rate of at least 5% per year through 2030.[1,2] Technological advancements are also expected to yield new products, including sustainable polymers, performance fibers used in textiles, biological-based medical-grade polymers, and polymers developed for use in the next generation of automobiles and aircraft. Plastics manufacture will also see an increase in the variety and quantity of unstable chemicals due to limited and tailored formulations demanded by end customers. Therefore, production and use of the unstable initiators and monomers is also expected to increase, and manufacturers will need to develop a methodology to assess and mitigate increased risks from these chemicals.

Unstable chemicals typically generate heat at an accelerating rate as the storage and processing temperature increases. They can also release flammable and toxic gases that can create greater risk of fires, explosions, and/or poisoning as they react and/or decompose. Therefore, chemical manufacturers and plant operators add chemical inhibitors and/or implement temperature control measures (i.e., refrigeration or vessel jacket quenching) to prevent or mitigate unwanted heating and runaway exothermic reactions. The screening method presented in this paper will allow better assessment of the risks associated with producing, handling, storing, and transporting unstable materials and identifying processes that require additional analysis.

2 Determining Whether a Material is Unstable

The temperature profile of unstable monomers and organic peroxides (which are the primary focus of this study) has been evaluated by several researchers. Kinetic models are often used to estimate the thermal stability of the material to calculate self-accelerating decomposition temperature (SADT) or self-accelerating polymerization temperature (SAPT), the critical temperature (T_{CR}), time to maximum rate (TMR), temperature at which time to reach maximum rate is 24 hours (T_{D24}), temperature of no return (TNR), and other relevant kinetic parameters.[3]

The United Nations "Recommendations on the Transport of Dangerous Goods, Manual of Tests and Criteria" (UNRTDG) includes four tests to determine SADT (test series H), which are used to ascertain whether the substance falls within the hazard class of self-reactive substances as delimited by the globally harmonized system (GHS):[4]

- H1: The United States SADT test (US SADT Test)
- H2: Adiabatic storage test (AST)
- H3: Isothermal storage test (IST)
- H4: Heat accumulation storage test (Dewar test)

The US SADT Test (H1) determines SADT, which is defined as "the minimum constant air temperature air environment at which auto-accelerative decomposition occurs for a substance in a specific package" (characterized by the middle of the packaging reaching a temperature 6 °C higher than 2 °C below the air temperature in 7 days or less).[4] The critical temperature, T_{CR}, is the highest temperature at which the substance as packaged does not show self-accelerating decomposition.[4] Test series H is summarized in Table 1.

While the H1 test unequivocally determines the SADT, it is constrained by the cost and volume requirements of ~ 220 L of the material. The H2-H4 tests do not require large amounts of reactive product(s) and therefore are often used for lab-scale experiments. For example, the H2 and H3 tests define SADT as "the critical ambient temperature (°C) rounded to the next higher multiple of 5 °C" determined graphically from a plot of heat generation versus heat loss.[4] The H2 test utilizes adiabatic calorimetry (accelerating rate calorimetry (ARC)) while the H3 test utilizes isothermal calorimetry at different temperatures to determine the temperature dependency of the reaction rate.[3] The Dewar test (H4) uses the same definition of SADT as the US SADT Test (H1), but is based on isothermal storage of a smaller quantity of the reactive product. However, the accuracy of the H4 test depends on whether the Dewar vessel has similar heat loss per unit mass characteristics to the transport packaging.

Test series H for determining SADT allows classification of a substance as *self-reactive* if its SADT is less than 75 °C (167 °F) and the heat of decomposition is greater than 300 J/g.[5] Self-reactive substances are defined as "thermally unstable liquid or solid substances or mixtures liable to undergo a strongly exothermic decomposition even without participation of oxygen (air)". Temperature control is stipulated for certain organic peroxides with a SADT ≤ 45 °C (113 °F) and all self-reactive substances with a SADT ≤ 55 °C (131 °F). SAPT is defined as the lowest temperature at which self-accelerating polymerization may occur and is also determined by test series H.[6] The two values are used interchangeably to determine control and emergency temperatures for self-reactive substances that require temperature control.[6]

Table 1. Summary of test series H

Test	Sample Size and Container Type	Test Conditions	Determining SADT/SAPT
H.1 – U.S. SADT Test	Representative of substance and packaging being shipped.	Test chamber kept at constant temperature for minimum of 10 days. Conducted at 5 °C intervals.	Sample temperature exceeds test chamber temperature by 6 °C or more.
H.2 – Adiabatic storage test (AST)	1.0 L or 1.5 L Dewar vessel.	Heat vessel at 5 °C intervals until self-heating is detected. Monitor heat generation rate as a function of temperature.	Draw line tangential to heat generation curve with the slope of the heat loss of the packaging. Round temperature where tangential line intersects x-axis to next higher multiple of 5 °C.
H.3 – Isothermal storage test (IST)	70 cm^3 sample holder.	Determine the maximum heat generation rate of seven tests conducted at 5 °C intervals.	Draw line tangential to heat generation curve with the slope of the heat loss of the packaging. Round temperature where tangential line intersects x-axis to next higher multiple of 5 °C.
H.4 – Heat accumulation storage test	Dewar vessel with heat loss per unit mass similar to package offered for transport.	Test chamber kept at constant temperature for up to 7 days. Conducted at 5 °C intervals.	Sample temperature exceeds test chamber temperature by 6 °C or more.

Under the globally harmonized system (GHS), separate classifications are assigned to explosives (Chapter 2.1), oxidizing liquids/solids (Chapters 2.13 and 2.14), and organic peroxides/organic peroxide mixtures (Type A through Type G) containing less than 5% combustible organic substances (Chapter 2.15).[5] These classifications require additional tests such as UN Test Series Type 1(a-c) for explosive substances, and UN Test O.1/O.2/O.3 (test for oxidizing solids/liquids).[6] Requirements for materials that meet the GHS criteria for explosive substances but are excluded from the GHS criteria for self-reactive substances are subject to additional hazards, and the heat transfer and chemical kinetics models developed in this paper should not be used for them.

The additional explosive hazards of self-reactive substances can include mass explosion (affecting the entire quantity of material virtually instantaneously) for division 1.1 explosive substances,[5] shock sensitivity, friction sensitivity, and temperature sensitivity. For explosive materials, United Nations test series E should be performed first to assess the influence of heating under confinement. Test series A and test series C may also be required to determine whether they propagate a detonation or deflagration.[6] Test series E prescribes three methods to assess sensitiveness to intense heat under defined confinement: E.1: Koenen test, E.2: Dutch pressure vessel test, and E.3: USA pressure vessel test.[6] These tests are different from test series H because they do not incorporate decomposition kinetics and heat loss models to account for different packaging amounts or configurations. A summary of the unstable materials reviewed in this paper and their associated GHS hazard classes and test series from the UN Manual of Tests and Criteria is provided in Table 2.[6]

Table 2. Test series for analyzing thermal stability of chemical/substance.

Chemical	Chemical Type	GHS Hazard Class	Applicable Classification Test Series	Maximum SADT/SAPT Requiring Temperature Control per Hazard Class
Tert-butyl peroxybenzoate (TBPB)	Organic Peroxide	Organic Peroxide Type C	A, C, E, H	50 °C (122 °F)
Methyl methacrylate (MMA)	Monomer	Self-reactive substance	A, C, E, H	Portable Tank: 45 °C (113 °F) IBC: 50 °C (122 °F)

2.1 Characterization Techniques

Thermal analysis (TA) techniques used to determine decomposition kinetics of unstable substances include thermogravimetric analysis (TGA), differential scanning calorimetry (DSC), and accelerated rate calorimetry (ARC). The self-heating rate is calculated according to recommendations by the Kinetics Committee of the International Confederation for Thermal Analysis and Calorimetry (ICTAC).[7] The methods to determine thermal stability, in particular, the self-heating rate of solids (e.g., sodium dichloroisocyanurate), and liquids (e.g., polymer initiators such as tert-butyl 2-ethylperoxyhexanoate) which decompose via different mechanisms, are also considered.

Xu et al. utilized differential scanning calorimetry (DSC), and accelerating rate calorimeter (ARC) in conjunction with chemical kinetic models to analyze the thermal stability and decomposition of solid materials, e.g. sodium dichloroisocyanurate (NaDCC).[8] These methods have also been used to study the thermal stability and decomposition of liquid organic peroxides (OPs), including tert-butyl peroxy-2-ethyl hexanoate (TBPO),[9,10] which is widely used in the chemical industry (under the brand name Trigonox 21S) as an initiator for (co)polymerization of ethylene, styrene, acrylonitrile, (meth)acrylates, and for acrylic resin production.[11]

Analysis of ARC data to determine SADT values depends on the kinetic model used. Mores et al provide general and simplified models for a solid slab with a steady non-uniform temperature distribution from self-heating.[12] The general model (Thomas model) accounts for heat transfer resistance within the substance from thermal conductivity and heat transfer at the boundary layer.[16] The simplified model (Semenov model) assumes the effects of thermal conductivity are negligible and the reacting substance remains isothermal.[13] This is a reasonable assumption for a well-stirred low-viscosity liquid material. A separate simplification, known as the Frank-Kamenetskii model, assumes no resistance to heat transfer at the surface boundary while a temperature gradient is established within the reacting substance.[16] Frequently, the Semenov model is used for well-stirred liquids which can be approximated as a lumped system while the Frank-Kamenetskii model is more appropriate for solids (especially those in bulk packaging) which develop significant temperature gradients.[3,14]

3 Case Studies

Preventing thermal runaway, which represents a major hazard in the chemical processing industry, requires a comprehensive understanding of the chemicals involved in the process and implementing a systematic approach to their use and storage. Barton and Nolan reviewed over 180 industrial incidents in the UK between 1962 and 1987 that were caused by thermal-runaway chemical reactions. The incidents occurred due to several factors:[15]

1) inadequate understanding of reaction chemistry
2) inadequate material quality control
3) process temperature control
4) agitation
5) mischarging the reactants
6) equipment maintenance
7) human error

Interestingly, only 6% of the incidents were attributed to human error. The leading causes were mischarging of the reactants (21%), inadequate understanding of reaction chemistry (20%), and process temperature control issues (19%).

A more recent study by Zhang et al. evaluated thermal runaway incidents in China between 1984 and 2019.[16] They identified over 270 incidents out of which over 180 caused at least one fatality. The annual number of incidents and fatalities has trended upward since 2000. Over half of the incidents (57.2%) occurred in oil, gas, and coal industries (i.e., organics), followed by production of inorganic chemicals (10.7%). The remaining industries with incidents include pharmaceuticals (7%), agricultural chemicals (6.6%), dyes (3%), hazardous materials transport (2.6%), metallurgy (1.8%), plastics, rubbers, and resins (1.5%), and other (9.6%). Most incidents occurred during production (175 incidents with 530 fatalities). Although chemical storage had a relatively low number of incidents (31), these accounted for 375 (34.3%) of total fatalities. Overall, the possible causes were attributed to:

1) physical and technical issues
2) human and organizational causes
3) natural causes
4) other causes

In contrast to the study by Barton and Nolan, Zhang et al. reported that operator error was the most common cause of thermal-runaway incidents (35.4%).

Some representative incidents that demonstrate the risk of runaway chemical decomposition are described below:

3.1 1976 Seveso, Italy – Chemical release involving dioxin.

A chemical release occurred during the production of 2,4,5-trichlorophenol from 1,2,4,5-tetrachlorobenzene reaction with sodium hydroxide at elevated temperatures.[17,18] The batch process utilized a 10 m^3 stainless steel reactor equipped with an agitator that was normally charged with 7,000 kg of chemicals. The reaction mixture temperature was increased to the point where it caused a rapid thermal runaway reaction.[19] The event released 2,900 kg organic matter, including 600 kg of sodium trichlorophenate and an unknown quantity of a dioxin chemical, 2,3,7,8-tetrachlorodibenzo-p-dioxin (TCDD). While there was no immediate loss of life, the release caused widespread, long-term, adverse health consequences.[18]

3.2 2001 Toulouse, France – Explosion of ammonium nitrate and chlorinated compounds

A warehouse holding about 300 metric tons of ammonium nitrate exploded in the AZF/Grande Paroisse chemical plant, causing 31 deaths.[20] Although the cause was not determined, it is suspected that chemical contamination from sodium dichloroisocyanurate (SDIC), trichloroisocyanurate acid (ATCC), or other incompatible chemicals contributed to the initiation of the event.[21,22] When SDIC is exposed to moisture at ambient temperatures, a rapid runaway reaction can be initiated and produce unstable trichloramine (NCl_3), which may have contributed to the explosion.[22]

3.3 2002 Pascagoula, MS – MNT explosion caused by contact with steam

An explosion and subsequent fire involving ~1,200 gallons of mononitrotoluene (MNT) occurred inside a distillation column at First Chemical Corporation plant in Pascagoula, Mississippi. Steam entered the column through leaking reboiler steam supply valves and heated the MNT, which decomposed over several days and eventually experienced a thermal runaway and explosion. Several post-explosion fires were ignited, including a fire in a MNT storage vessel.[23]

3.4 2003 Gnadenhutten, OH – BPO explosion during drying

An explosion and fire occurred within the Catalyst Systems, Inc. production facility while drying benzoyl peroxide (BPO). A double cone vacuum dryer was being used to convert 75% BPO to 98% BPO over multiple drying sessions that typically lasted 2 to 2.5 days. The dryer had a jacket filled with heated water at 82°C, which would heat the BPO to 42°C before automatically shutting off. A 200 lb. batch of BPO, which completed two days of drying sessions, was allowed to remain in the dryer over an extended period (~6 days). A final drying session was started within the batch just before the explosion. It is possible that the BPO was heated above its SADT because the dryer's high temperature limit was set based on the SADT for a much smaller quantity (68°C for a 1 lb. package) than what was present in the dryer.[24]

3.5 2013 West, TX – Fire and explosion involving FGAN

A fire and explosion of fertilizer grade ammonium nitrate (FGAN) occurred at the West Fertilizer Company in West, Texas. Under normal conditions, FGAN is a stable substance. However, if exposed to temperatures greater than its melting point, it can rapidly decompose, forming toxic gasses and nitrous oxide (N_2O). High pressure, rapid heating, and heat retention decreases the temperature at which FGAN can explode.[25] The density, packing, particle size, moisture content, presence of organic contaminants, and low pH environments are also relevant, and decomposing or contaminated oxidizer can also detonate when exposed to a strong impulse force.[25,26]

The facility stored between 40 to 60 tons of FGAN inside the building and 100 tons were also stored in adjacent railcars. Approximately 30 tons detonated during the incident. The primary factors contributing to the fire development and detonation were determined to be (i) contamination of the material with potential fuels and (ii) heat buildup within the material.[23] In response to the incident, the U.S. Environmental Protection Agency (EPA) recommended that ammonium nitrate should not be stored in confined spaces, contamination with combustible, organic, and/or inorganic materials should be minimized, and localized heating should be avoided.[27]

3.6 2017 Crosby, TX – Explosion of organic peroxides

The Arkema plant in Crosby manufactured organic peroxides, which were packaged in one-gallon/five-gallon plastic containers, shrink-wrapped together, stacked onto pallets, and stored in refrigerated containers (-20 to 0 °F (-29 to -18 °C)).[28,29] Due to flooding from Hurricane Harvey, failure of multiple refrigeration systems caused the ambient temperature to increase above the SADT of the organic peroxides.[29] The resulting runaway reactions caused an explosion that exposed 21 people to toxic fumes, requiring them to seek medical attention.[28]

3.7 2020 Westlake, LA – Exothermic decomposition and fire involving TCCA

TCCA is a chemical commonly used as a sanitizer to prevent algae and bacteria growth when added to swimming pools and hot tubs. At dilute concentrations, it dissolves slowly and releases chlorine into the water. However, if TCCA is wetted by a small volume of water, it can decompose exothermically, generating toxic chlorine gas and explosive nitrogen trichloride.

An exothermic decomposition of trichloroisocyanuric acid (TCCA) occurred at the Bio-Lab, Inc. Lake Charles facility in Westlake, Louisiana. The facility stored granulated TCCA in 2,750 lb. bulk bags. During Hurricane Laura, part of the roof of the facility was damaged and rainwater wetted the stored TCCA, initiating a decomposition reaction. The impacted part of the Bio-Lab plant housed between 70,000 and 100,000 lb. of TCCA while an adjacent warehouse held over 1,000,000 lb. The heat released during the decomposition ignited a fire which quickly destroyed the building while also releasing a large volume of toxic gases into the environment.

Approximately 9 hours later, the warehouse, which was also damaged by the hurricane, started to emit white smoke. However, the stored TCCA did not ignite.[30]

3.8 2020 Belle, WV – Explosion involving NaDCC dihydrate

An explosion occurred within a 165 ft^3 Hastelloy® C-276 rotary double cone dryer at Optima Belle LLC while drying (dehydrating) 8,820 lbs. of sodium dichloroisocyanurate dihydrate (NaDCC dihydrate). The cause of the over-pressurization and subsequent explosion was self-accelerated decomposition of NaDCC dihydrate. Analysis of process data revealed an exponential temperature increase within the dryer, consistent with self-accelerating exothermic activity. The decomposition reaction started to accelerate above 83°C, which is well below the decomposition temperature of 240-250°C listed in the manufacturer's safety data sheet (SDS). However, published SADT values for NaDCC dihydrate range from 45-65 °C, and post-incident testing determined thermal runaway could occur at 81°C.

The U.S. Chemical Safety Board (CSB) concluded that the incident occurred because Optima Belle had insufficient knowledge regarding the thermal characteristics of NaDCC dihydrate and did not perform an effective thermal hazard assessment. They recommended revisions to safety information for NaDCC dihydrate within the database hosted by the National Center for Biotechnology Information (NCBI), which Optima Belle had relied upon for their process hazard analysis. Specifically, they asked for the SADT information to be updated, and DSC and ARC results from their testing be included. The manufacturer was also advised to update their SDS to provide clarity on thermal requirements, and Optima Belle was urged to establish a written thermal and reactive hazards evaluation and management program, including an analysis of the thermal characteristics of the processed material.[31]

In the rest of this paper, we introduce a heat transfer and kinetics model that can be used in tandem to assess the performance of unstable chemicals and discuss how to integrate potential hazards into a risk assessment methodology to prevent thermal runaway.

4 Heat Loss Rate and UA of Packaging

The heat transfer rate is driven by temperature difference and is proportional to the heat transfer coefficient. Although the temperature difference can change drastically over the course of a day (e.g., daytime, and nighttime temperatures), the overall heat transfer coefficient, U, remains relatively constant per unit area, A, over which the heat transfer occurs. The heat transfer rate per unit area can be expressed as:

$$\dot{Q} = U \cdot A_S \cdot \Delta T \tag{1}$$

Here, $\dot{Q}$ is the heat transfer rate (in W), U is the overall heat transfer coefficient (in W/m^2K), As is the wetted surface area (in m^2), and ΔT is the temperature difference (in K). The overall heat transfer coefficient, U, can be viewed as a reliable heat transfer rate metric and it can be calculated for a wide range of geometries using the thermal circuit analogy:[32]

$$U = \frac{1}{\sum R_i \cdot A_i} \tag{2}$$

Here, R_i is the thermal resistance of an individual region of heat transfer (i.e., convection inside the container and convection outside the container) or layer of the container (i.e., conduction through container wall and conduction through insulation layer of the container) and A is the area.

The product of the overall heat transfer coefficient, U, and the surface area, A, over which the heat transfer occurs can be determined experimentally by measuring the temperature change during cooling of the vessel.[33] To do so, it is recommended to use the so-called "half-time of cooling" method by fitting the cooling data with a line according to:[34]

$$\ln(\theta) = \ln(\theta_i) - \frac{U \cdot A_S}{m \cdot c_p} t \quad (3)$$

Where $\theta = T - T_\infty$ is the temperature difference between the material and the environment at time t, $\theta_i = T_i - T_\infty$ is the initial temperature difference between the material and the environment (i.e., at t = 0), m is the mass of the material, and c_p is the specific heat of the material.

The Nusselt number is a dimensionless group that compares the heat transfer by conduction to that of convection and is defined as:

$$Nu = \frac{h \cdot L}{k} \quad (4)$$

Where h is the convective heat transfer coefficient, L is the characteristic length dimension of the object, and k is the thermal conductivity of the fluid. Based on Eqn. 4, at an arbitrary fixed Nusselt number, a more thermally insulative fluid (i.e., lower thermal conductivity) will exhibit lower convective heat transfer. Furthermore, the thermal conductivity of a gas is typically at least an order of magnitude lower than the thermal conductivity of its liquid counterpart.[32] Therefore, when storing liquids in containers, it is reasonable to assume that the heat transfer occurs through the wetted container surface only, and the heat transfer in the ullage of the container can be neglected.

4.1 1-Gallon drum

Sheng et al.[33] measured the temperature change of a 1-gallon drum of water as it cooled. A 17 cm diameter and 22 cm high drum was filled to about 90%, or 3.49 L, with water. The drum was heated in the oven at 76 °C and was left to cool at ambient conditions. The cooling curve for water digitized from data published by Sheng et al. is shown in Figure 1.[33]

Based on the linear fit of the data (Figure 1), the UA value for the 1-gallon container filled to 90% of its volume with water (ρ = 1000 kg/m^3, c_p = 4.181 J/gK) is equal to 1.182 W/K. When accounting only for the internal area wetted by water (A_S = 0.105 m^2), the U value is equal to 11.2 W/m^{2}K, which is less than the value reported by Sheng et al., but consistent with previously published values (11.3 W/m^{2}K).[33]

Figure 2 shows the cooling curve for MMA plotted from the data published by Sheng et al.[33] Based on a linear fit, the UA value for the one-gallon container filled to 90% of its volume with MMA is 0.911 W/K. The overall heat transfer coefficient, U, computed from the linear fit of the data is 8.68 W/m^{2}K. Although the overall heat transfer coefficient U is lower than that reported by Sheng et al. (9.0 W/m^{2}K), the value of 8.68 W/m^{2}K is within the 10% experimental uncertainty reported in the publication.[33]

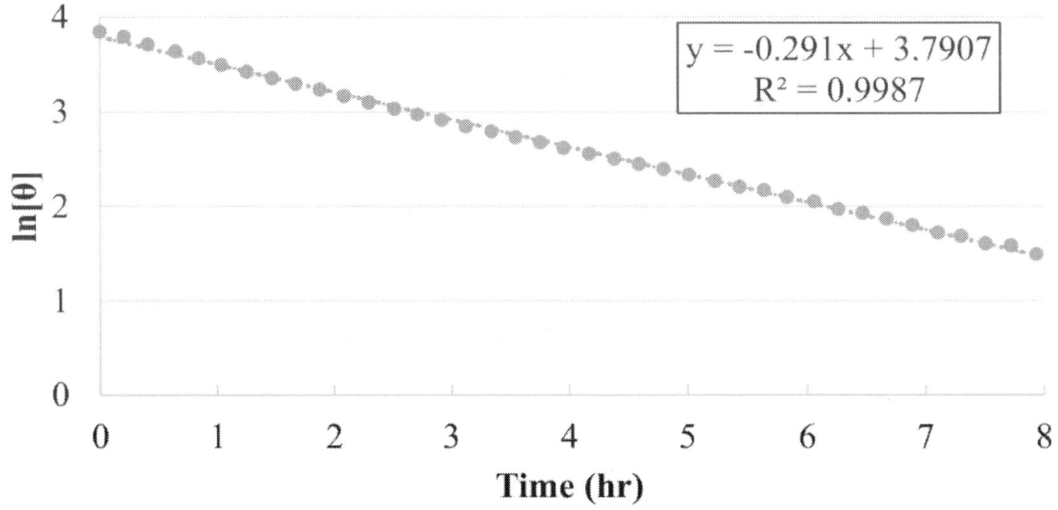

Figure 1: Cooling curve of a 1-gallon container filled with water to 90% of its volume (data digitized from Sheng et al., Figure 2a[33]).

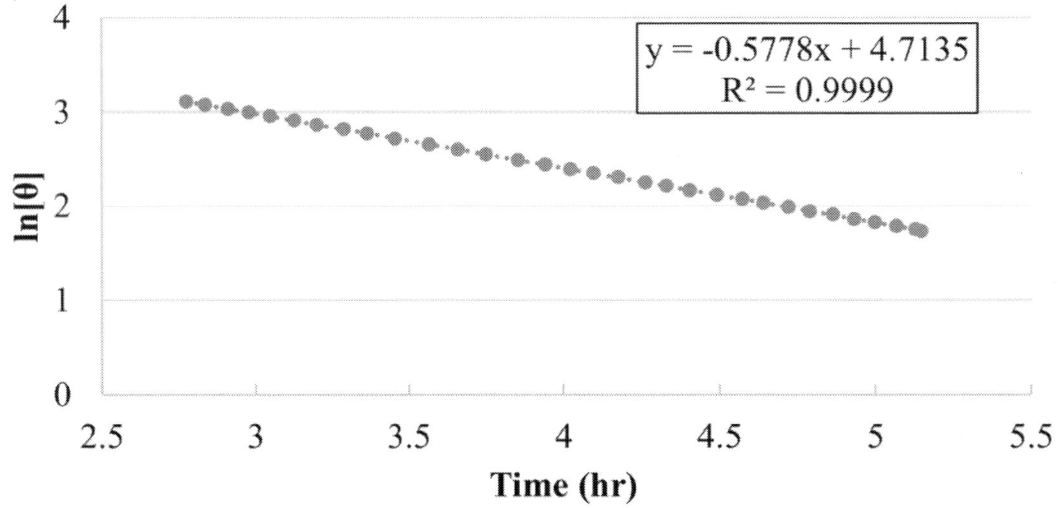

Figure 2: Cooling curve of a 1-gallon container filled with MMA to 90% of its volume (data digitized from Sheng et al., Figure 3[33]).

4.2 55-gallon insulated drum

If experimental heat transfer data is not readily available (i.e., in case of large or non-standardized containers), the product of the overall heat transfer coefficient and the heat transfer area, UA, can be estimated from geometry and typical convective heat transfer coefficient values.

For a 55-gallon insulated drum, the UA can be estimated according to Equation (5), where R_{Cyl} is the resistance through the wetted shell of the drum and R_{Bottom} is the resistance through the wetted bottom of the drum.

$$UA_S = \frac{1}{R_{Cyl}} + \frac{1}{R_{Bottom}} \tag{5}$$

The overall thermal resistance through the shell of the drum, R_{Cyl}, consists of four thermal resistances representing (i) the external convection, $R_{e,c}$; (ii) conduction through the insulation, R_{ins}; (iii) conduction through the container wall, R_{cond}; and (iv) internal convection, $R_{i,c}$. These can be written as:[32]

$$R_{cyl} = R_{e,c} + R_{insul} + R_{cond} + R_{i,c} \tag{6}$$

$$R_{e,c} = \frac{1}{2 \cdot \pi \cdot (r_i + t_{shell} + t_{ins}) \cdot z \cdot h_e} \tag{7}$$

$$R_{ins} = \frac{\ln((r_i + t_{shell} + t_{ins})/(r_i + t_{shell}))}{2 \cdot \pi \cdot z \cdot k_{ins}} \tag{8}$$

$$R_{cond} = \frac{\ln((r_i + t_{shell})/r_i)}{2 \cdot \pi \cdot z \cdot k_{shell}} \tag{9}$$

$$R_{i,c} = \frac{1}{2 \cdot \pi \cdot r_i \cdot z \cdot h_i} \tag{10}$$

Where, r_i is the internal drum diameter (11.25 in), t_{shell} and t_{ins} are the thickness of the container (0.25 in) and insulation (2 in), respectively, z is the wetted drum height (28.758 in at 90% full), h_e (10 W/m^{2}K) and h_i (100 W/m^{2}K) are representative values of the external and internal convective heat transfer coefficients, and k_{shell} (60.5 W/mK) and k_{ins} (0.042 W/mK) are the thermal conductivity of the drum material and insulation, respectively.[32]

The resistance through the bottom of the drum, R_{bottom}, is calculated as follows:

$$R_{bottom} = \frac{\frac{1}{h_e} + \frac{t_{ins}}{k_{ins}} + \frac{t_{shell}}{k_{shell}} + \frac{1}{h_i}}{\pi (r_i + t_{shell} + t_{ins})^2} \tag{11}$$

The UA calculated for a 55-gallon insulated drum filled to 90% capacity is 1.387 W/K. Given the total wetted surface area is 1.943 m^2, the overall heat transfer coefficient is 0.714 W/m^{2}K.

Based on data availability, a more accurate estimation of convective heat transfer coefficients can also be obtained by employing the appropriate Nusselt number correlations on a case-by-case basis.

5 Kinetic Properties of Unstable Chemicals

As discussed above, determining the heat transfer properties of packaging used to transport unstable materials, is necessary to properly classify them for shipment. For example, the UNRTDG considers self-heating solids and monomers to be class 4, division 4.1 substances if their SADT or SAPT is 75 °C (167 °F) or less within the packaging used, the heat of reaction is greater than 300 J/g, and they do not meet other criteria in classes 1-8. Further, the substances must be temperature controlled if the SADT or SAPT is less than 50°C (113 °F) within IBC or packaging, and less than 45 °C (131 °F) within a portable tank. These criteria can also be incorporated into a kinetic model to assess risks associated with the manufacture, storage, and handling of unstable materials as discussed below.

5.1 Calculation of dT/dt from ARC data

Accelerating rate calorimetry (ARC) data was extracted for liquid TBPB and liquid MMA with 25 ppm MEHQ inhibitor to determine the relationship between temperature rise (dT/dt) and temperature and assess chemical instability. A zeroth order Semenov model, which is applicable to low-viscosity self-reactive monomer or organic peroxide systems, was assumed to apply.[3,9,33] For reaction kinetics to be considered zeroth order, the natural log of the change in temperature with respect to time (dT/dt) versus inverse temperature should be linear. Although MMA will transition to a high viscosity system as it self-polymerizes and this can create a large temperature gradient within the liquid, the Semenov model is applicable to the initial rate of reaction (i.e., at low conversion).[33]

Although zeroth-order reaction kinetics are a conservative approximation for reactions that follow n^{th}-order Arrhenius kinetics, many systems follow more complex reaction mechanisms, including autocatalytic and multi-stage reaction systems. Therefore, the method presented herein should be considered as a screening tool, and the application of more complex reaction mechanisms should be applied to systems known to significantly deviate from n^{th}-order Arrhenius-type reaction rates.

The rate change in concentration (dC/dt) is shown in Eqn. 12, where C is concentration, t is time, k(T) is the rate constant as a function of temperature, and n is the order of reaction.[35] The rate constant, k(T), is expressed by the Arrhenius equation, Eqn. 13, where A is the pre-exponential factor, E_A is the activation energy, R is the ideal gas constant, and T is temperature. The rate change in temperature (dT/dt) is equated to the rate change in concentration in Eqn. 14, where m is mass of the sample, C_p is specific heat of the sample, V is the sample volume, and ΔH_R is the heat of reaction, where exothermic reactions (i.e., decomposition and polymerization reactions) have a negative heat of reaction. Substituting Eqn. 12 and Eqn. 13 into Eqn. 14 and solving for dT/dt results in Eqn. 15, where ρ is density. Taking the natural log of both sides of Eqn. 15 results in Eqn. 16. Given that the ARC data was collected while the reactants were being consumed (i.e., when the concentration was decreasing), Eqn. 16 will yield a linear relationship only when the reaction rate is zeroth order (n=0), as shown in Figure 3.

$$\frac{dC}{dt} = -k(T) \cdot C^n \qquad (12)$$

$$k(T) = A \exp\left(\frac{-E_A}{RT}\right) \qquad (13)$$

$$mC_p \frac{dT}{dt} = V \frac{dC}{dt} \Delta h_R \qquad (14)$$

$$\frac{dT}{dt} = -\frac{\Delta h_R}{\rho C_p} \cdot A \exp\left(\frac{-E_A}{RT}\right) \cdot C^n \qquad (15)$$

$$Ln\left(\frac{dT}{dt}\right) = Ln\left(A\frac{\Delta h_R}{\rho C_p} \cdot C^n\right) + \left(\frac{-E_A}{R}\right)\frac{1}{T} \tag{16}$$

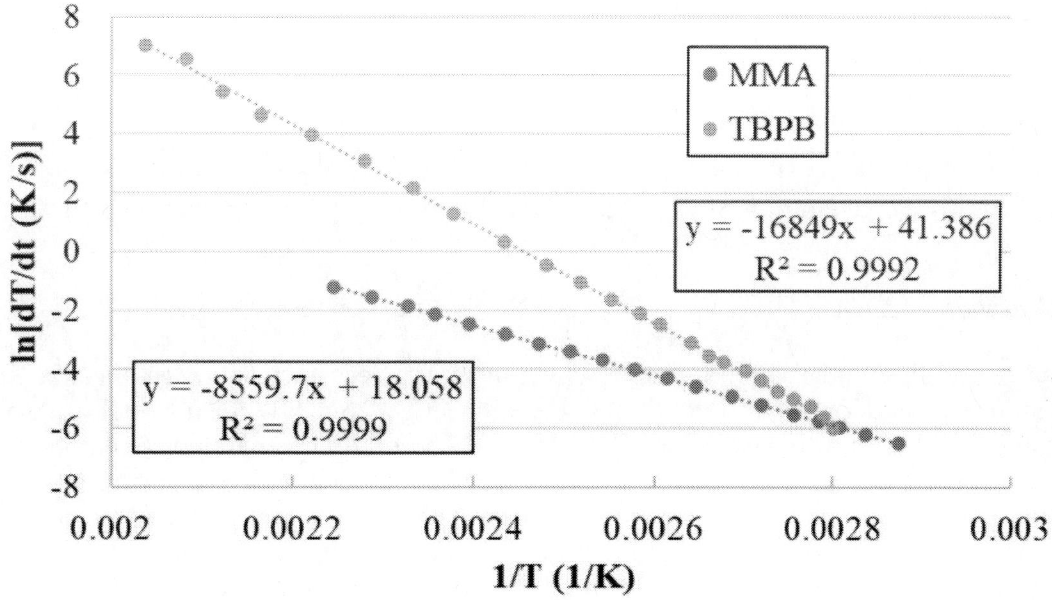

Figure 3. Linear regression of ln (dT/dt) versus inverse temperature for MMA with 25 ppm MEHQ inhibitor (blue) and TBPB (orange).

The natural log of dT/dt versus inverse temperature for MMA and TBPB is plotted in Figure 3. The values of dT/dt have been corrected to account for the heat absorbed by the ARC vessel using the Fisher method presented by Kossoy et al.[3] Adiabatic values used in the kinetic evaluation below were extracted from ARC data for MMA published by Sheng et al. and for TBPB by Gong et al.[36,37]

First, the adiabatic onset temperature ($T_{on,ad}$) is calculated from the onset temperature in the ARC test (T_{on}) using Eqn. 17:

$$T_{on,ad} = \left[\frac{1}{T_{on}} + \frac{R\ ln\phi}{E_a}\right]^{-1} \tag{17}$$

Where ϕ is the thermal inertia of the ARC test (provided in Table 3). Then, the adiabatic temperature ($T_{ad}(t)$) is calculated from the temperature versus time ARC data (T(t)) using Eqn. 18:

$$T_{ad}(t) = T_{on,ad} + \phi[T(t) - T_{on}] \tag{18}$$

Finally, the adiabatic change in temperature with respect to time (dT/dt) as a function of adiabatic temperature (T_{ad}) used in Eqn. 12-15 (and elsewhere in this paper) is calculated from the change in temperature with respect to time values of the ARC data ($(dT/dt)_{ARC}$), using Eqn. 19.

$$\frac{dT}{dt} = \phi \left(\frac{dT}{dt}\right)_{ARC} exp\left[\frac{E_A}{R}\left(\frac{1}{T} - \frac{1}{T_{ad}}\right)\right] \tag{19}$$

The activation energy is iteratively determined such that the E_A in Eqns. 17-19 is equal to the E_A derived from the linearization from Figure 3 and Eqn. 16.

The R^2 values of the linear regression in Figure 3 are essentially uniform and therefore, reaction kinetics for MMA and TBPB can be approximated by a zeroth order reaction rate during the initial stages of the reaction. Per Eqn. 16, the slope of the regression (m_G) is given by $-E_A/R$ with units of K and the intercept (b_G) is given by ln $(A \cdot \Delta h_R/\rho C_p)$ with units of ln(K/s) since C^n=1 for zeroth order reactions. The activation energy, E_A, calculated from the slope in Figure 3 is 71,165 J/mol for MMA and 139,509 J/mol for TBPB.

5.2 Calculation of TMR

TMR represents the amount of time at a set temperature necessary to reach the maximum decomposition rate under adiabatic conditions (i.e., all heat generated by the material is absorbed by the material and no heat is lost to the environment). Therefore, this parameter can be used to evaluate the minimum time necessary for exothermic decomposition given an initial starting temperature.[38] Townsend and Tou[39] developed equations for self-heating rate, TMR, and TNR based on ARC data. Wilberforce[40] and Whitmore[41] expanded on this methodology and proposed a TMR curve developed with ARC data to calculate SADT. Their method of determining TNR and SADT has been widely used and accepted.[10,42] For reactions with high activation energy (> 83,680 J/mol),[39] TMR can be expressed as:

$$TMR = \frac{RT^2}{E_A\frac{dT}{dt}} = \frac{RT^2}{E_A * \exp\left(b_G + m_G\frac{1}{T}\right)} \tag{20}$$

Where R is the universal gas constant (8.314 J/mol·K), T is the temperature of the material, and b_G is the intercept, m_G is the slope, and E_A is the activation energy calculated in Figure 3 using Eqn. 16. The form of Eqn. 20 is also acceptable for reactions with an activation energy slightly below 83,680 J/mol (e.g., MMA) for temperatures where TMR > 1 hr, since the neglected term is on the order of a few minutes.

Parameters used to calculate TMR for MMA and TBPB extracted from previously published work and publicly available thermodynamic data are summarized in Table 3.[9,33,43] Zeroth order reaction kinetics (i.e., n=0) were assumed.[44] The 24-hr TMR (TMR_{24}), which is used as a reference value for comparison of data on thermal stability, is extrapolated from the TMR versus temperature plot shown in Figure 4.[45] MMA and TBPB have TMR_{24} values of 44.4 °C and 57.7 °C, respectively. Figure 4 demonstrates that neither chemical will readily exhibit runaway exothermic conditions at typical ambient temperatures, i.e., 15 °C to 25 °C (59 °F to 77 °F). However, additional precautions are warranted in hot weather, such as tropical climates and/or regions with high solar insolation. For example, the Methacrylate Producers Association recommends storing MMA below 40 °C (104 °F) and acknowledges that temperatures exceeding 45 °C (113 °F) are indicative of an ongoing polymerization reaction, which is consistent with the TMR_{24} value calculated by the method presented in this paper.[46]

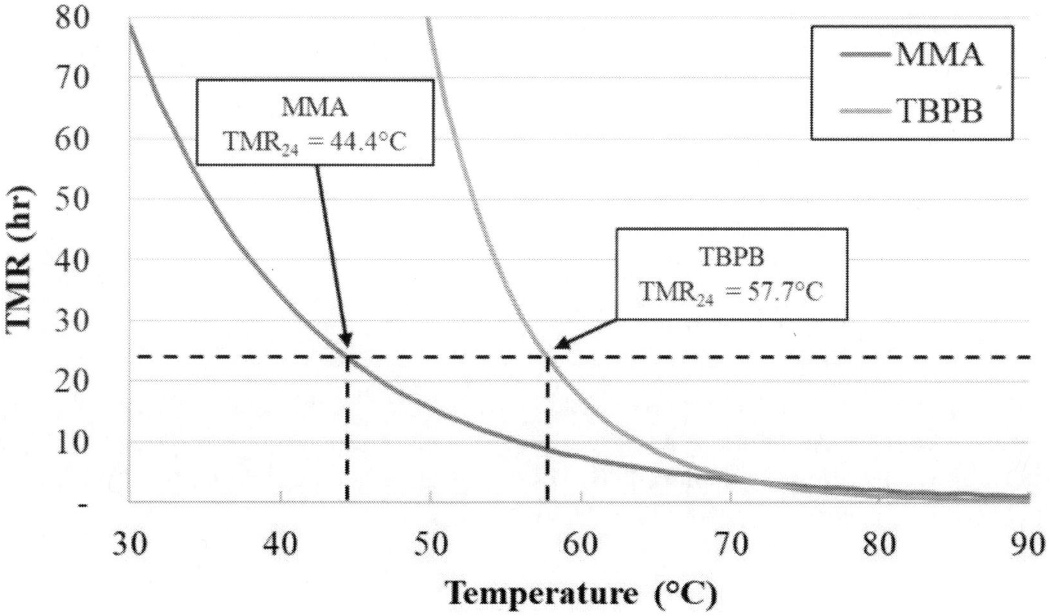

Figure 4. TMR graph for MMA (blue line) and TBPB (orange line) indicating TMR_{24} for both materials (intersection with black dashed line).

Table 3. MMA and TBPB parameters and TMR_{24}.

Chemical	Φ	T_{on} (°C)	C_p (J/g°C)	ρ (kg/m³)	E_A (J/mol)	TMR_{24} (°C)
MMA	2.04	113	1.791	908.59	71,165	44.4
TBPB	2.82	89.44	1.290	1040	139,509	57.7

5.3 Calculation of SADT or SAPT

To calculate the SADT or SAPT using the Semenov model, the heat loss curve (Q_L), which is a linear line with the slope of the heat loss rate, is drawn tangential to the heat generation curve (Q_G) developed from the ARC data (Figure 5). The heat loss curve is dependent on the sample mass (ρV) and the overall heat transfer coefficient (U) and wetted surface area (A) of the packaging, given by Eqn. 21.[6]

$$Q_L = \frac{UA_s}{\rho \cdot V} T + b_L \tag{21}$$

The heat generation curve, Q_G, is calculated by multiplying the temperature rate of change (Eqn. 15) with the specific heat, as given by Eqn. 22. Substituting the slope and intercept of the linear regression plotted in Figure 1 into Eqn. 22 yields Eqn. 23.

$$Q_G = C_p \frac{dT}{dt} = C_p \tag{22}$$

$$Q_G = C_p \exp\left(b_G + m_G \frac{1}{T}\right) \qquad (23)$$

The temperature of no return (TNR) is the temperature at which the heat generation curve has the same slope as the heat loss curve. The slope of the heat generation curve is determined by taking the derivative of Eqn. 23 with respect to temperature, as expressed in Eqn. 24. The TNR is the temperature at which the slope of Eqn. 21 is equal to Eqn. 24, as shown in Eqn. 25.[3]

$$\frac{dQ_G}{dT} = \frac{-m_G C_P}{T^2} \exp\left(b_G + m_G \frac{1}{T}\right) \qquad (24)$$

$$\frac{-m_G C_P}{(TNR)^2} \exp\left(b_G + m_G \frac{1}{TNR}\right) = \frac{UA}{\rho \cdot V} \qquad (25)$$

The critical temperature (T_{CR}) is determined from where the heat loss curve is zero (i.e., intercepts the x-axis). The intercept of the heat loss curve (b_L) is determined by setting Eqn. 21 equal to Eqn. 23 when the temperature is equal to TNR (Eqn. 26). T_{CR} is then determined by setting Eqn. 21 equal to 0 and solving for temperature, given by Eqn. 27.

$$b_L = C_p \exp\left(b_G + m_G \frac{1}{TNR}\right) - \frac{UA_S}{\rho V} TNR \qquad (26)$$

$$T_{Cr} = TNR - \frac{\rho V C_P}{UA_S} \exp\left(b_G + m_G \frac{1}{TNR}\right) \qquad (27)$$

The SADT or SAPT is determined by rounding T_{CR} up to the next higher multiple of 5 °C.[6] Figure 5 demonstrates the process of calculating TNR and T_{CR} from the heat generation curve for MMA and the heat loss curves for 1-gallon and 55-gallon packages. SAPT for MMA in a 1-gallon drum reported here is consistent with values reported by Sheng, et al.[33] The process for calculating these parameters for TBPB is shown in Figure 6.

Since the heat loss rate is proportional to U and the ratio of the wetted surface area of the package to the volume of the material (A/V), decreasing U (e.g., by adding insulation) and/or decreasing A/V (e.g., by increasing sample size) decreases the SADT. For example, SADT/SAPT for MMA and TBPB in a 1-gallon drum is 70 °C and 75 °C, respectively (Table 4), which is above the temperature control threshold.[6] However, the SADT/SAPT for MMA and TBPB in a 55-gallon insulated drum is 30 °C and 50 °C, and temperature control becomes necessary.

Table 4. TNR, T_{CR}, and SADT/SAPT calculated for MMA and TBPB from heat loss curve parameters for 1-gallon and 55-gallon packages.

Package	A_S (m^2)	V (m^3)	A_S/V (m^{-1})	U W/m^2K	Chemical	TNR (°C)	T_{CR} (°C)	SADT/ SAPT (°C)
1-gallon drum	0.105	0.00349	30	8.68	MMA	82.0	67.2	70
					TBPB	78.3	71.0	75
55-gallon insulated drum	1.943	0.187	10.4	0.714	MMA	33.0	22.0	25
					TBPB	53.1	46.8	50

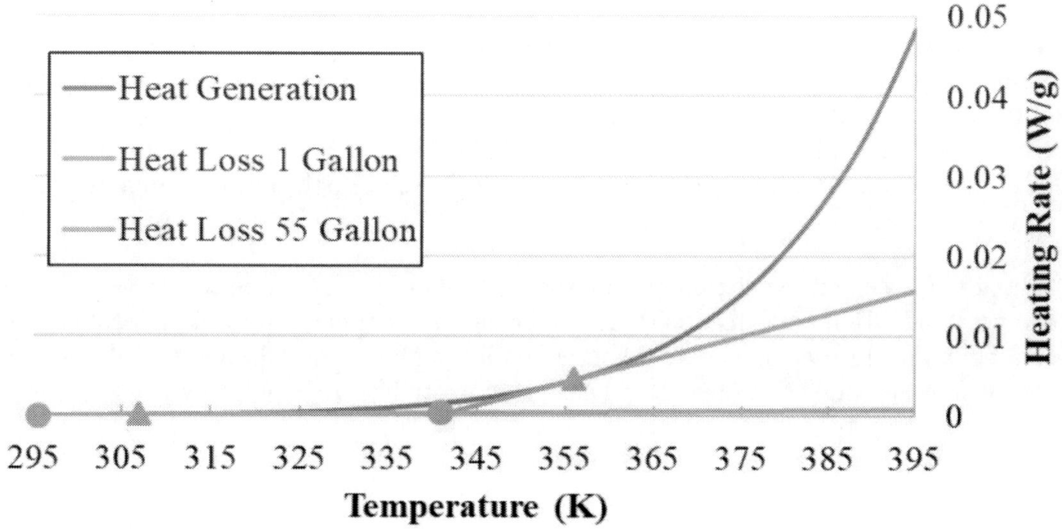

Figure 5. Determining TNR (triangles) and T_{CR} (circles) of MMA packaged in UN 1A1 1-gallon drum and UN 1A2 55-gallon drum.

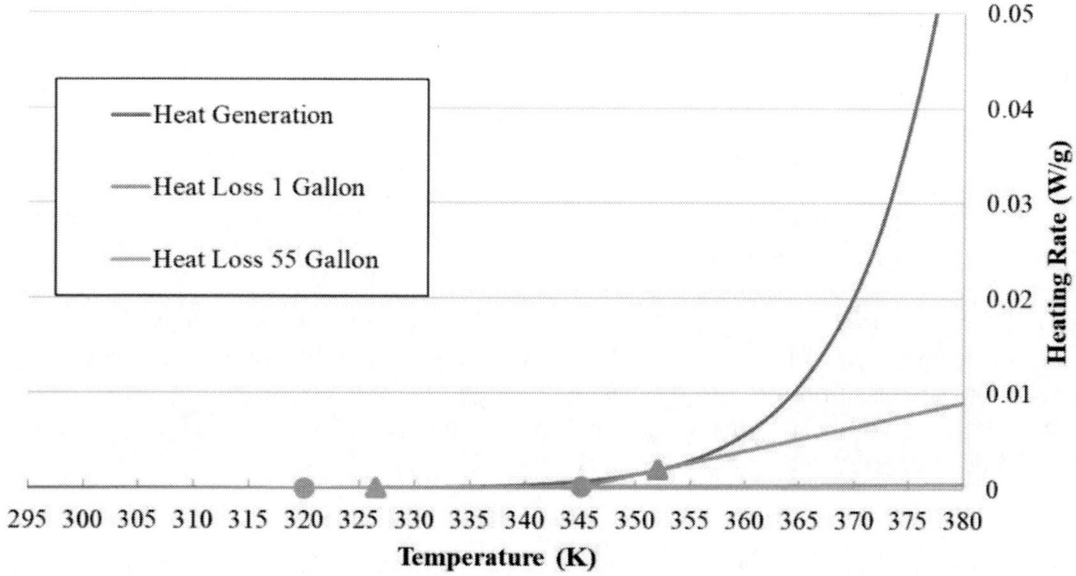

Figure 6. Determining TNR (triangles) and T_{CR} (circles) of TBPB packaged in UN 1A1 1-gallon drum and UN 1A2 55-gallon drum.

6 Heat Transfer Model

Once the adiabatic behavior (e.g., change rate of temperature with respect to time and TMR_{24}) of an unstable material is determined, a heat transfer model can be applied to understand how it will respond to heat gained or lost to the environment. Lower heat transfer rates (e.g., within insulated containers) will cause the material to slowly heat up when exposed to warmer ambient temperatures, which might allow longer storage times. However, if the material also generates significant heat from an exothermic reaction, heat loss to the environment is also slower. This implies that once the material self-heats above ambient conditions it will more readily achieve run-away conditions. For high heat transfer rates (e.g., uninsulated containers), the opposite is true. Chemical handlers must understand these tradeoffs to select appropriate container and storage conditions.

Heat transfer between stored chemicals and the environment occurs continuously and is based on their temperature difference. Heat conduction through the container as well as heat convection at the container walls must be considered. Factors including heat generation within the container due to decomposition or side reactions as well as agitation are also relevant.

The heat transfer model can be separated into three main contributions:

1. Heat transfer with the environment,
2. Heat generation due to equipment, and
3. Heat generation due to decomposition and/or side reactions.

The total temperature change of the container can be expressed as the sum of the individual contributions according to:

$$\Delta T = \Delta T_{HT} + \Delta T_{AG} + \Delta T_{CR} + \Delta T_{other} \tag{28}$$

Where, ΔT_{HT} is the temperature change due to the heat transfer with the environment, ΔT_{AG} is the temperature change due to heat generated by equipment (e.g., agitation), ΔT_{CR} is the temperature change due to any decomposition reactions, and ΔT_{other} is the temperature change due to parasitic/side reactions.

6.1 Heat transfer with the ambient (UA)

Temporal temperature change can often be modeled via a lumped capacitance model,[32] which assumes uniform temperature within an object. Assuming negligible temperature gradients within is an appropriate assumption for objects with high thermal conductivity, small mass, or in containers with agitation. In general, the temperature change, ΔT_{HT}, using the lumped capacitance model can be expressed as:[32]

$$\Delta T_{HT} = (T_i - T_\infty) \cdot \left[e^{-\left(\frac{h \cdot A_S}{m \cdot c_p} \cdot \Delta t\right)} - 1 \right] \tag{29}$$

Where, T_i is the initial temperature of the object, T_∞ is the ambient temperature, and Δt is the time change. Moreover, h is the convective heat transfer coefficient at the surface.

In the case of liquid in a container, the lumped capacitance model can be used for relatively small masses and for agitated liquids where the temperature is relatively uniform throughout. However, the convective heat transfer inside as well as outside of the container and heat conduction through container walls must be considered. Therefore, the lumped capacitance model from Eqn. 29 can be written as:

$$\Delta T_{HT} = (T_i - T_\infty) \cdot \left[e^{-\left(\frac{UA_S}{m \cdot c_p} \cdot \Delta t\right)} - 1 \right] \qquad (30)$$

The convective heat transfer coefficient, h, is replaced by the overall heat transfer coefficient, U, which is discussed above (Eqn. 3 and Eqns. 5-11):[32]

6.2 Heat generated from equipment

While equipment such as agitators and circulators are often used in chemical processes, they can introduce unintended heat to the process. These sources are often neglected during process design and may cause thermal-runaway incidents.[15] The temperature change due to heat dissipation from equipment can be expressed as:

$$\Delta T_{AG} = \frac{\dot{W}}{m \cdot c_p} \Delta t \qquad (31)$$

Where, $\dot{W}$ is the power of equipment dissipated as heat within the material. During startup, heat dissipation from the equipment might be reduced because part of the energy is used to reach steady-state operating conditions.

6.3 Heat generated from decomposition reaction

For a decomposition or polymerization reaction that can be approximated as a zeroth order reaction, the Arrhenius-type linearization expressed in Eqn. 16 can be used to determine heat generation and the associated temperature change according to:

$$\Delta T_{CR} = \exp\left(b_G + m_G \frac{1}{T}\right) \Delta t \qquad (32)$$

6.4 Heat generated from unintended side reactions (contamination)

In some instances, the presence of contaminants can also generate heat if they cause parasitic and/or side exothermic reactions. A simplified case of heating due to a side reaction with a constant heat generation rate, $\dot{r}$, can be expressed as:

$$\Delta T_{other} = \frac{\dot{r}}{m \cdot c_p} \Delta t \qquad (33)$$

6.4 Risk assessment application

Consider a condition where MMA is stored in a 55-gallon insulated tank. The MMA was removed from refrigeration at 10 °C and exposed to an ambient temperature of 20 °C, which is below the

SADT. Considering the effects of heat transfer (ΔT_{HT}) and heat generated from the self-polymerization reaction (ΔT_{Cr}), the temperature of the MMA will increase until the heat generated from the polymerization reaction equals heat loss to the environment. The temperature approaches an asymptote of ~25.6 °C, which is between the T_{CR} (22.0 °C) and TNR (33.0 °C), and therefore is not expected to run away (Figure 7, Unagitated MMA).

Now consider the same condition, but the MMA is also exposed to mechanical agitation (ΔT_{AG}) from a stirrer that imparts 20 W of frictional heat energy (Figure 7, Agitated MMA). The additional heat causes the MMA to undergo a self-accelerating polymerization reaction. However, if the stirrer is programmed to turn off after the temperature of the MMA reaches the TNR (Figure 7, Agitated MMA with Shutoff), the MMA begins to cool to the same temperature as the unagitated MMA.

As part of the risk assessment, it may be determined that the agitator should be turned off when the MMA reaches a temperature that is 5 °C below the TNR to incorporate a margin of safety. Other process-specific scenarios should also be considered during a risk assessment. For example, evaluating whether refrigeration would be effective at preventing a runaway reaction and determining effective temperatures of the heat transfer medium within the storage tank jacket. Acceptable storage times might also be evaluated based on the analysis presented to ensure stored chemicals do not undergo unintended runaway reactions.

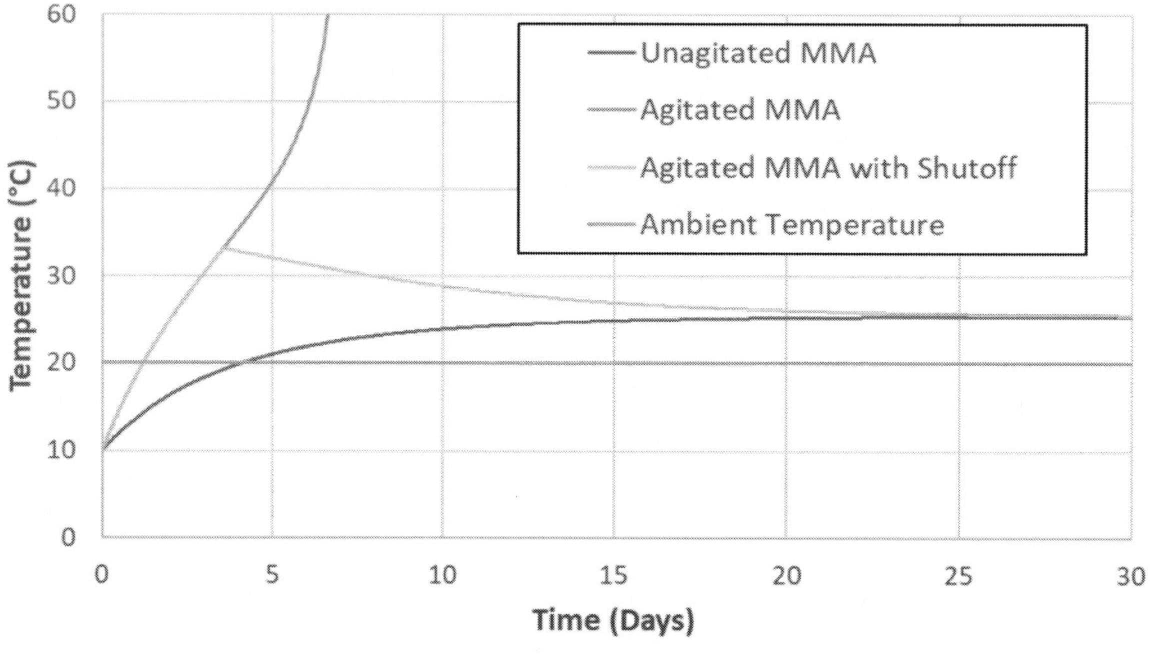

Figure 7. Temperature profile of MMA in 55-gallon insulated drum exposed to an ambient temperature of 20 °C (grey line) without agitation (blue line), with agitation (orange line), and with agitation that is turned off when the temperature of the MMA reaches the TNR (yellow line).

7 Conclusion

In this paper, we have reviewed historical case studies where unstable materials underwent a self-accelerating decomposition or polymerization reaction. Additional study of unstable materials is necessary because the number of incidents has continued to increase since 2000.

We developed a methodology to evaluate chemical stability of unstable materials by calculating heat transfer coefficients via a lumped capacitance model based on experimental data (i.e., cooling curves and other temperature vs. time data) and the physical properties of the container (i.e., conductive and convective heat transfer coefficients, shell and insulation thicknesses, and wetted surface area). We determined heat generation rates by deriving adiabatic self-heating rates from ARC data using the Fisher method and using zeroth-order reaction kinetics to approximate the self-heating rate during the initial stages of a decomposition or polymerization reaction. We described how to calculate the temperature that causes a runaway reaction under adiabatic conditions (TMR_{24}). We also developed a method to combine the heat transfer and heat generation models to assess the temperature at which the heat generation rate is greater than the heat loss to the environment (TNR and SADT). Finally, we combined the heat transfer and chemical kinetics models to analyze various scenarios (including exposure to ambient temperatures, heat generation e.g., by agitation and other mechanical processes, and unintended exothermic reactions) that can be used as a screening tool during the risk assessment.

8 References

[1] Annual production of plastics worldwide from 1950 to 2022. https://www.statista.com/statistics/282732/global-production-of-plastics-since-1950. Accessed February 1, 2024.
[2] Global Polymer Market Size, Share, Growth Analysis … Industry Forecast 2023-2030. https://www.skyquestt.com/report/polymer-market. Accessed February 1, 2024.
[3] Kossoy AA, Sheinman IY. Comparative analysis of the methods for SADT determination. *J. Hazard Mater* 2007; **142(3)**: 626-638, DOI: 10.1016/j.jhazmat.2006.06.068.
[4] UN Recommendations on the Transport of Dangerous Goods, Manual of Tests and Criteria, 6th revised edition 2015, United Nations.
[5] UN Globally Harmonized System of Classification and Labelling of Chemicals, 9th revised edition 2021, United Nations.
[6] UN Recommendations on the Transport of Dangerous Goods, Manual of Tests and Criteria, 7th revised edition 2019, United Nations.
[7] Koga N, Vyazovkin S, Burnham A, Favergeon L, Muravyev N, Pérez-Maqueda L, Saggese C, Sánchez-Jiménez P. ICTAC kinetics committee recommendations for analysis of thermal decomposition kinetics. *Thermochim Acta* 2023; **719**: 179384, DOI: 10.1016/j.tca.2022.179384.
[8] Xu H, Zhang H, Qi M, Bai M, Xu Y, Li L, Liu Y. Thermal hazard evaluation of sodium dichloroisocyanurate via TG-MS, DSC, and ARC. *Process Saf Environ Prot* 2022; **166**: 68-77, DOI: 10.1016/j.psep.2022.08.010.
[9] Gong LQ, Li L, Pan Y, Jiang JJ, Jiang JC. Thermal hazards and initial decomposition mechanisms study of four tert-butyl organic peroxides combining experiments with density functional theory method. *Thermochim Acta* 2022; **708**: 179142, DOI: 10.1016/j.tca.2021.179142.

[10] Yao H, Jiang J, Li B, Ni L, Ni Y, Yao X. Investigation of pyrolysis kinetics, mechanism and thermal stability of tert-butyl peroxy-2-ethyl hexanoate. *Process Saf Environ Prot* 2022; **160**: 734-748, DOI: 10.1016/j.psep.2022.02.059.
[11] Trigonox 21S tert-Butyl peroxy-2-ethylhexanoate (nouryon.com). Accessed Feb 1, 2024.
[12] Mores S, Nolan FP, O'Brien G. Determination of the self-accelerating decomposition temperature (SADT) from thermal stability data generated using accelerating rate calorimetry. *IChemE Symp Series No. 134* 1994.
[13] Semenov NN. *Some Problems of Chemical Kinetics and Reactivity, Vol 2* (Translated : J.E.S. Bradley). Pergamon Press; 1959.
[14] Fierz H. Influence of heat transport mechanisms on transport classification by SADT-measurement as measured by the Dewar-method. *J. Hazard Mater* 2003; **96**; 121-126), DOI: 10.1016/S0304-3894(02)00202-9.
[15] Barton JA, Nolan PF. Incidents in the chemical industry due to thermal-runaway chemical reactions. *IChemE Symp Series No. 115* 1989.
[16] Zhang H, Bai M, Wang X, Gai J, Shu CM, Roy N, Liu Y. Thermal Runaway Incidents - A Serious Cause of Concern: An Analysis of Runaway Incidents in China, *Process Saf Environ Prot* 2021; **155**: 277-286. DOI: 10.1016/j.psep.2021.09.027.
[17] Health and Safety Executive (HSE). Icmesa chemical company, Seveso, Italy. 10th July 1976. Control of Major Accident Hazards (COMAH). 2015.
[18] Bertazzi PA, Di Domenico, A. Health consequences of the Seveso, Italy, accident. *Dioxins and Health* 2nd Edition, Schecter A, Gasiewicz TA (Editors); Hoboken, NJ John Wiley & Sons 2003; 827-853.
[19] Kletz T. *What Went Wrong? Case Histories of Process Plant Disasters*; 4th Edition. Gulf Publishing Company 1999; 376-378.
[20] Guiochon, G. On the catastrophic explosion of the AZF plant in Toulouse. *Process Saf Prog* 2020; **39(4):** e12197, DOI: 10.1002/prs.12197.
[21] Toulouse 2001- Incident Summary. *Process Safety Integrity* 2024. https://processsafetyintegrity.com/incidents/summary/. Accessed February 1, 2024.
[22] Dechy N, Gyenes Z, Merad M. A review on Toulouse accident trials: some lessons to learn despite uncertainty on direct causes. IChemE *Loss Prevention Bulletin 269* 2019; 25-28.
[23] U.S. Chemical Safety and Hazard Investigation Board, *Explosion and Fire, First Chemical Corporation*; October 2003. 2003-01-I-MS.
[24] U.S. Chemical Safety and Hazard Investigation Board. *Fire and Explosions: Hazards of Benzoyl Peroxide;* October 2003. 2003-03-C-OH.
[25] Davis, ROE. *Explosibility and Fire Hazard of Ammonium Nitrate Fertilizer*. U.S. Department of Agriculture; March 1945. Circular No. 719.
[26] Van Dolah R, Mason C, Perzak F, Hay J, Forshey D. *Explosion Hazards of Ammonium Nitrate Under Fire Exposure.* U. S. Department of the Interior, Bureau of Mines; 1966.
[27] U.S. Environmental Protection Agency. *Explosion Hazard from Ammonium Nitrate.* Office of Solid Waste and Emergency Response; December 1997. EPA 550-F-97-002d.
[28] CSB Releases Arkema Final Report. News release. U.S. Chemical Safety and Hazard Investigation Board; May 24, 2018. https://www.csb.gov/csb-releases-arkema-final-report/. Accessed February 1, 2024.
[29] U.S. Chemical Safety and Hazard Investigation Board. *Organic Peroxide Decomposition, Release, and Fire at Arkema Crosby Following Hurricane Harvey Flooding;* May 2018. 2017-08-I-TX.

[30] U.S. Chemical Safety and Hazard Investigation Board. *Trichloroisocyanuric Acid Reaction, Decomposition, and Toxic Gas Release at Bio-Lab, Inc;* April 2023. 2020-05-I-LA.
[31] U.S. Chemical Safety and Hazard Investigation Board. *Fatal Chemical Decomposition Reaction and Explosion at Optima Belle LLC*; July 2023. 2021-02-I-WV.
[32] Bergman TL, Lavine AS. *Fundamentals of Heat and Mass Transfer*; 8th Edition. Hoboken, NJ John Wiley & Sons; 2017.
[33] Sheng M, Dan F, Horsch S, Bellair R, Holsinger M, Scholtz T, Weinberg S, Sopchik A. Calorimetric method to determine self-accelerating polymerization temperature (SAPT) for monomer transportation regulation: a heat balance approach. *Org Process Res Dev* 2019; **23(5)**: 750-761. DOI: 10.1021/acs.oprd.9b00012.
[34] UN Recommendations on the Transport of Dangerous Goods, Manual of Tests and Criteria, 5th revised edition 2009, United Nations.
[35] Vyazovkin S. *Isoconversional Kinetics of Thermally Stimulated Processes*. Heidelberg Springer; 2015. DOI: 10.1007/978-3-319-14175-6.
[36] Sheng M, Dan F, Horsch S, Bellair R, Holsinger M, Scholtz T, Weinberg S, Sopchik A. Calorimetric method to determine self-accelerating polymerization temperature (SAPT) for monomer transportation regulation: Ki9netics and Screening Criteria. *Org Process Res Dev* 2019; **23(5)**: 737-749. DOI: 10.1021/acs.oprd.9b00011.
[37] Gong L, Li L, Yue P, Jiang J, Jiang J. Thermal hazards and initial decomposition mechanisms study of four tert-butyl organic peroxides combining experiments with density functional theory method. Thermochim. Acta 708 (2022). DOI: 10.1016/j.tca.2021.179142.
[38] Keller A, Stark D, Fierz H, Heinzle E, Hungerbühler K. Estimation of the time to maximum rate using dynamic DSC experiments. *J. Loss Prev Process Ind* 1997; **10(1)**: 31-41.
[39] Townsend DI, Tou JC. Thermal hazard evaluation by an accelerating rate calorimeter. *Thermochim Acta* 1980; **37(1)**: 1-30. DOI: 10.1016/0040-6031(80)85001-5.
[40] Wilberforce JK. The use of the accelerating rate calorimeter to determine the SADT of organic peroxides. Columbia Scientific Industries Corp July 1981; Internal Report.
[41] Wilberforce JK, Whitmore MW. Use of the accelerating rate calorimeter and the thermal activity monitor to estimate stability temperature. *J. Loss Prev Process Ind* 1993; **6(2)**: 95-101. DOI: 10.1016/0950-4230(93)90006-J.
[42] Fisher HG, Goetz DD. Determination of self-accelerating decomposition temperature using the accelerating rate calorimeter. *J. Loss Prev Process Ind* 1991; **4(5)**: 305-316. DOI: 10.1016/0950-4230(91)80044-U.
[43] Green W, West R. RMG - Reaction Mechanisms Generator. Massachusetts Institute of Technology 2024. https://rmg.mit.edu/. Accessed February 1, 2024.
[44] Sun J, Li Y, Hasegawa K. A study of self-accelerating decomposition temperature (SADT) using reaction calorimetry. *J. Loss Prev Process Ind* 2001; **14(5)**: 331-336. DOI: 10.1016/S0950-4230(01)00024-9.
[45] UN Sub-Committee of Experts on the Transport of Dangerous Goods. *Proposal to amend the proper shipping name of Crotonaldehyde (UN 1143)*. Transmitted by the International Council of Chemical Associations (ICCA; April 2004. ST/SG/AC.10/C.3/2004/33, United Nations. https://digitallibrary.un.org/record/521371?ln=en. Accessed February 1, 2024.
[46] MPA. *Methacrylate Esters Safe Handling Manual*. Methacrylate Producers Association, Inc. and Methacrylates Sector Group of the European Chemical Industry Council, 2019.

I've identified overpressure caused by tube failure and subsequent chemical reaction as a credible scenario during my PHA, now what?

Elizabeth Meegan, Alec Slazas, Sara Townsend, Ben Doup, Ken Kurko, Billie Spreadbury, Emily Ziemba, Alai Musleh

Keywords: tube rupture, heat exchanger, reactive system vent sizing, API 521

Abstract

Identifying and then mitigating potential sources of overpressure is a crucial aspect of a robust safety program. API STD 521, "Guide for Pressure-Relieving and Depressuring Systems," provides a convenient checklist defining common causes of overpressure for consideration. Potential scenarios include closed outlets, overfilling, exterior fire, chemical reaction, and heat transfer equipment failure. When focusing on the common scenario of heat transfer equipment failure, API STD 521 and other standards provide some guidance as to things to consider when assessing the consequences and mitigation strategies such as type and extent of internal failure, selection of a relieving device that reacts fast enough to prevent overpressurization, and selection of the optimal relief device location. When defining the credible scenario, API STD 521 recognizes that the extent of internal failure can vary from a pinhole leak to a complete tube rupture, and therefore in practice, a sharp break (or double-ended guillotine tube rupture) is often a conservative first point of consideration. Assessing the required relieving rate to protect the vessel from this scenario is further complicated when the inadvertent mixing of the two fluids results in a chemical reaction. The relief device on the low-pressure side will now need to relieve the overpressure from the mixing of the high-pressure fluid in addition to vapors and/or gases generated from the reaction. This paper provides an example of a double-ended guillotine tube rupture of pressure reactor's internal cooling coils where cooling water is inadvertently mixed with a water-reactive chlorine containing material. Guidance is provided on calculating the equivalent addition rate and resulting mixture composition following this scenario, as well as how to simulate the scenario using adiabatic calorimetry. The presented experimental temperature and pressure generation rates are then used to design an emergency relief system.

1 Introduction

A common hazard identified when developing a safe process is the failure of heat transfer equipment. There is some uncertainty in this hazard's failure mechanism, and the expected transient phenomena can be quite complex. In the case where a shell and tube heat exchanger is passing a high pressure gas on the tube side with a liquid filled shell side, the sudden failure of a tube will rapidly (on the scale of milliseconds) impact the liquid resulting in a sudden pressure rise creating a shock wave. As the wave travels through the shell and tube heat exchanger, reflection of the wave can further increase the pressure rise by additional reflection of the wave as it propagates along the heat exchanger. A relief device such as a rupture disk or fast acting relief valve mounted on the low-pressure shell side then must react quickly and have enough capacity to vent the two-phase fluid. This is further complicated when the failure leads to mixing of chemically incompatible materials resulting in pressure generation from the reaction in addition to mixing. This paper will provide some background information on API STD 521's guidelines for addressing heat transfer equipment failure and chemical reactions, as well as provide a case study documenting a simplified approach to addressing this hazard.

1.1 Process Hazard Analyses

A process hazard analysis (PHA) is a rigorous and systematic approach to identifying, evaluating, and controlling potential hazards associated with processes and operations with the goal being to implement controls or actions necessary to reduce the risk(s) arising from such hazards which might have severe, extensive, or possibly catastrophic consequences. Typically, such events include but are not limited to major uncontrolled emissions, fires, or explosions that present serious danger to employees in the workplace, cause serious injury, cause a major disruption to business, and/or leave the boundaries of the property.

Qualitative or semi-quantitative assessment methodologies are typically utilized when conducting PHAs with risk quantification being presented in the form of a risk matrix. A risk matrix categorizes probability of likelihood against a category of consequence severity to rank risks and assist management with decision-making. In essence, hazard assessments utilize guidewords or questions to identify credible deviations from the process. Causes and consequences for credible deviations are identified and evaluated by the hazard review team. The combination of the worst-case credible consequence and the likelihood of the scenario are quantified by the hazard review team utilizing qualitative categories defined by a risk matrix. Occasionally, quantitative methods are also used to help prioritize risk reduction.

Loss-of-containment is a common hazard identified and addressed during a PHA. These situations can result in delays in production, toxic releases resulting in injury, environmental damage, fires and/or explosions. LOC can be caused directly by equipment failure or indirectly by a process deviation such as an unexpected buildup of pressure. Equipment and piping failures are commonly identified in hazard assessments, including tube failure in heat exchangers. The severity of the failure can vary depending on what is determined to be the cause of the failure. Many common initiating events for equipment and piping failures, such as pump failure, above ground piping full-breach failure, above ground piping leak, pressure regulator failure, and vessel failure, have been provided typical generic initiating event frequencies (IEFs) with associated criteria for each value readily available in literature such as "*Guidelines for Initiating Events and Independent Protection layers in Layer of Protection Analysis"* or "*Guidelines for Quantitative Risk Assessment*" [1, 2]. However, the values are conservative values under conditions specified, and literature cannot capture all possible initiating event scenarios with associated frequencies. In some instances, such as those for tube failures in heat exchangers, the need for integrating the input of multiple functional areas of expertise to define credible scenarios may be required. This integrated team may determine varying likelihood of a tube leak vs. tube rupture as well as varying consequences. Further, the more likely tube leak scenario and the less likely tube rupture scenarios can have different adequate layers of protection, requiring both scenarios to be evaluated independently.

1.2 API STD 521 Guidance to Addressing Overpressure Scenarios

1.2.1 Heat Transfer Equipment Failure

API STD 521 [3] provides the primary causes of overpressure for pressure vessels, heat exchangers, operating equipment, and piping as well as guidance to minimize the effects of overpressure, and recommendations for estimating the relief requirements for mitigating the pressurization. Conditions resulting in overpressure specified in API STD 521 include scenarios such as closed outlets, accumulation of noncondensables, overfilling, exterior fire, heat transfer equipment failure, and chemical reaction. Guidance for addressing heat transfer equipment failure is outlined in API STD 521 Section 4.4.14 and includes details for heat exchanger tube rupture, double pipe, and plate and frame. API STD 521 states that "heat exchangers and similar vessels may require protection with a relieving device of sufficient capacity to avoid overpressure in case of an internal failure," noting the following key points of consideration:

- the required relieving rate on the low-pressure side of the failure
- the extent (i.e. size) of the failure
- the selection of a fast-acting relief device
- the selection of the location of the relief device

When defining the credible scenario, API STD 521 recognizes that the extent of internal failure can vary from a pinhole leak to a complete tube rupture, and therefore in practice, a sharp break (or double-ended guillotine tube rupture) is often a conservative first point of consideration. Multiple tube failures, or complete failure of the entire tube inventory is cited as a possibility in some guidance documents, but it not explicitly mentioned in API STD 521 which is the focus of this discussion. Here, a complete tube rupture scenario involves a sharp break in one tube, where the failure is assumed to occur at the back side of the tubesheet. Here, the high-pressure fluid is assumed to flow both through the tube stub remaining in the tubesheet and through the longer section of tube.

It is noted that the estimated frequency of equipment tube failure is approximately 0.0009 per unit per year (~1 per 1,100 years) [4], and there is some data available on the probability of leak size indicating smaller leaks are more common (see Figure 1 for general process equipment leak size and Table 1 for LOC frequencies specific to heat exchangers handling hydrocarbons), however there is not clear guidance or criteria for considering lesser failures. Because of this, it is recommended to first try to design the emergency relief system (ERS) to accommodate a complete tube rupture. In the event that this is not a practical design, a quantitative risk assessment can help to support the consideration of a lesser failure.

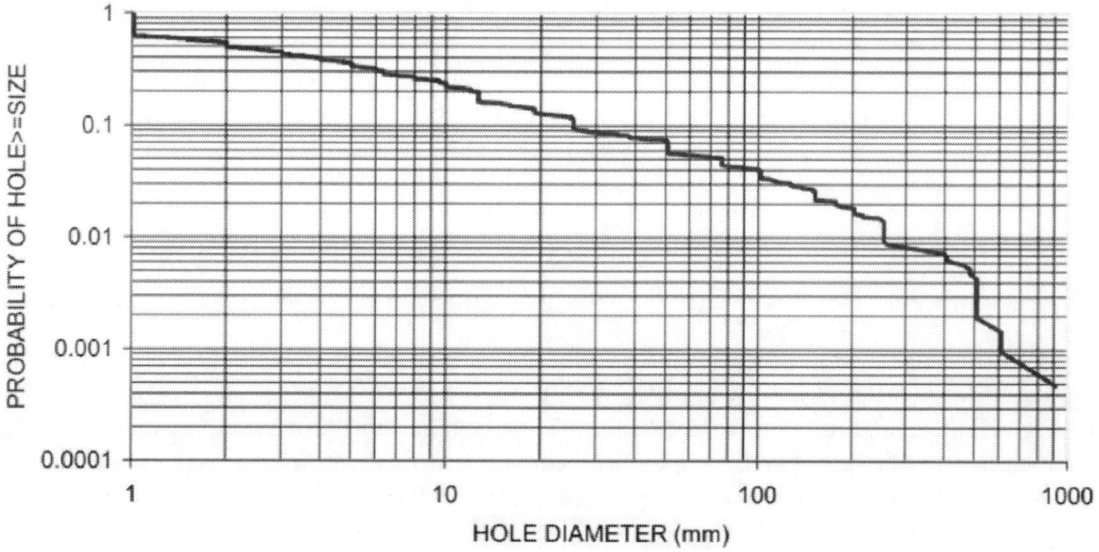

Figure 1 Leak Size Distribution for HSE Offshore Data [5, 6]

Table 1: LOC Failure Frequencies for Heat Exchangers Handling Hydrocarbons [6]

Equipment Type	Frequency of ≥ 1 mm Diameter[1]	Frequency of ≥ 50 mm Diameter[1]
Hydrocarbon in Shell	1.4E-03	1.3E-04
Hydrocarbon in Tube	1.0E-03	4.9E-05

[1] Per Equipment Item Per Year

API STD 521 recognizes that the approach for sizing should vary depending on the phase of the leaking fluid (e.g. non-flashing liquids, flashing liquids, gas/vapor, two phase), and provides two methods for determining the required size of the relief device: steady state, and dynamic analysis. Performing a dynamic analysis can provide valuable insight into the transient phenomenon occurring on the low-pressure side of the rupture such as the time at pressure, allowing for engineering judgment on if that level of pressure can lead to physical deformation or failure, and should be considered the optimal approach [7, 8]. API 521 specifically recommends this approach when there is a large pressure differential across the tube failure (e.g. ~1,000 psi or more), especially if the high-pressure side contains a gas or flashing liquid. Other sources further recommend a dynamic approach for systems with a long shell with baffles on a tight pitch or for existing units where the pressure relief devices are mounted on the utility piping. However, a simple and first-pass steady-state evaluation is economical and can provide an estimate of the requirement before investing in a more costly approach and can be used to verify or benchmark the transient modeling approach.

1.2.2 Chemical Reaction

Estimating the relief requirement following a heat exchanger failure is further complicated when the mixing of the two fluids leads to a chemical reaction generating vapor and/or gas. The API STD 521 guidance for addressing chemical reactions is outlined in Section 4.4.11 [3]. In this section, the recommended methodology is in accordance with the approach established by the Design Institute for Emergency Relief Systems (DIERS) which is as follows:

- define the upset scenario leading to the reaction
- characterize the reactive system through bench-scale tests simulating the upset scenario
- use vent-sizing equations that account for two-phase flow.

API STD 521 further recommends considering "the potential for chemical reaction in conjunction with the other overpressure scenarios in 4.4" when appropriate (e.g. fire exposure or heat exchanger failure leading to chemical reaction). This paper will focus on the overlapping situation of heat exchanger failure resulting in an unexpected chemical reaction. It will provide an example of a simplified steady-state approach to evaluating the relief requirement on the low-pressure side of a heat exchanger failure and document examples of how the extent of the failure can impact the required relieving rate.

1.3 Relief System Sizing for Runaway Chemical Reactions

1.3.1 Fundamentals

During a reactive upset scenario, the source of pressurization is often modeled using three general classifications: a tempered (or vapor system), a non-tempered (or gassy system), and a gas generating tempered system (hybrid system) [9].

Vapor systems describe both reactive and nonreactive situations where pressure generation is due to an increase in the vapor pressure of a liquid. Vapor systems have the advantage that the latent heat from boiling prevents the temperature from increasing (or tempers the system). As a result, the temperature rise due to the runaway reaction can be controlled by venting. In these cases, the relief system must be designed to accommodate the vapor generated during venting. As such, the relief requirement is based on the temperature rise rate, and the required vapor generation ($\dot{Q}_v$) rate is described in Equation 1.

$$\dot{Q}_v = \frac{V \rho c \dot{T}}{\lambda \rho_v} \tag{1}$$

where
$\dot{Q}_v$ = rate of vapor generation, m^3/s
V = reactant volume, m^3
ρ = density, kg/m^3
c = heat capacity, kJ/K/kg
λ = latent heat of vaporization, kJ/kg
ρ_v = vapor density, kg/m^3
$\dot{T}$ = temperature rise rate, K/s

Gassy systems describe a situation where noncondensable gas generation ($\dot{Q}_g$), for example oxygen or carbon dioxide reaction products, is the cause of the pressure buildup. In this system, the latent heat of vaporization is not available, and the reaction temperature cannot be controlled by venting. Therefore, the source term is a function of pressure rise rate alone. The gas generation term is described in Equation 2.

$$\dot{Q}_g = \frac{V \rho v}{m_t} \frac{\dot{P}}{P} \tag{2}$$

where
$\dot{Q}_g$ = rate of gas generation, m^3/s
v = experimental void volume, m^3
P = venting pressure, Pa
m_t = experimental mass, kg
$\dot{P}$ = pressure rise rate, Pa/s

Finally, hybrid systems describe situations where both vapor and gas generation coincide in the venting region ($\dot{Q}_v + \dot{Q}_g$). The latent heat of cooling is available at the relief pressure, and the reaction temperature can be controlled by venting, but noncondensable gas generation is also occurring simultaneously. The hybrid (summation of vapor and gas) generation term is described in Equation 3.

$$\dot{Q}_{v,g} = \frac{V \rho c \dot{T}}{\lambda \rho_v} + \frac{V \rho v}{m_t} \frac{\dot{P}}{P} \tag{3}$$

It is typically difficult to predict the outcome of process deviations leading to unexpected chemical reactions, and even more difficult to quantify the rate of vapor and/or gas generation, and oftentimes the most economical approach for assessing the hazard is by experimentally simulating the process deviation.

1.3.2 Experimental Techniques

Adiabatic calorimetry is the preferred method for simulating reactive hazards when the goal is designing an emergency relief system (ERS). Calorimeters are designed to measure the amount of heat associated with chemical or physical processes, and an adiabatic calorimeter is designed to measure temperature and pressure rise rates during a reaction and ensures that there is minimal heat transferred between the system and its surroundings. This technique allows all of the exothermic heat produced by a reaction to increase the temperature of the system. As the temperature increases, the rate of reactions often exponentially increases. This appropriately simulates what can occur at the plant scale where the amount of available cooling is fixed. When the reaction generates enough heat to surpass the available cooling, the batch temperature will start to increase further, increasing the rates of reaction.

A key parameter used to describe adiabatic calorimetry studies is the phi-factor (φ). The phi-factor is the ratio of the thermal inertia of the sample to the thermal inertia of the sample holder (i.e. how much heat is absorbed in heating the test cell as opposed to heating the sample). Phi-factor is calculated according to Equation 4.

$$\varphi = \frac{C_{p,s} m_s + C_{p,b} m_b}{C_{p,s} m_s} \tag{4}$$

where

	$C_{p,s}$	=	heat capacity of sample (J/g/K)
	m_s	=	sample mass (g)
	$C_{p,b}$	=	heat capacity of test cell (J/g/K)
	m_b	=	test cell mass (g)

Low phi-factor experiments are ideal for direct scale up of the data as a correction is not needed (or the extrapolation of the data is small) to account for the heat loss to the test cell. This maximizes the quality of the data collected over the temperature rise by more closely simulating the phi-factor observed in a large-scale process vessels where the phi-factor approaches 1. Collecting adiabatic data close to this value minimizes or eliminates the need for data extrapolation. This is best illustrated in Figure which provides simulations of temperature rise rate as a function of temperature for the thermally initiated polymerization of styrene with various phi-factors considered.

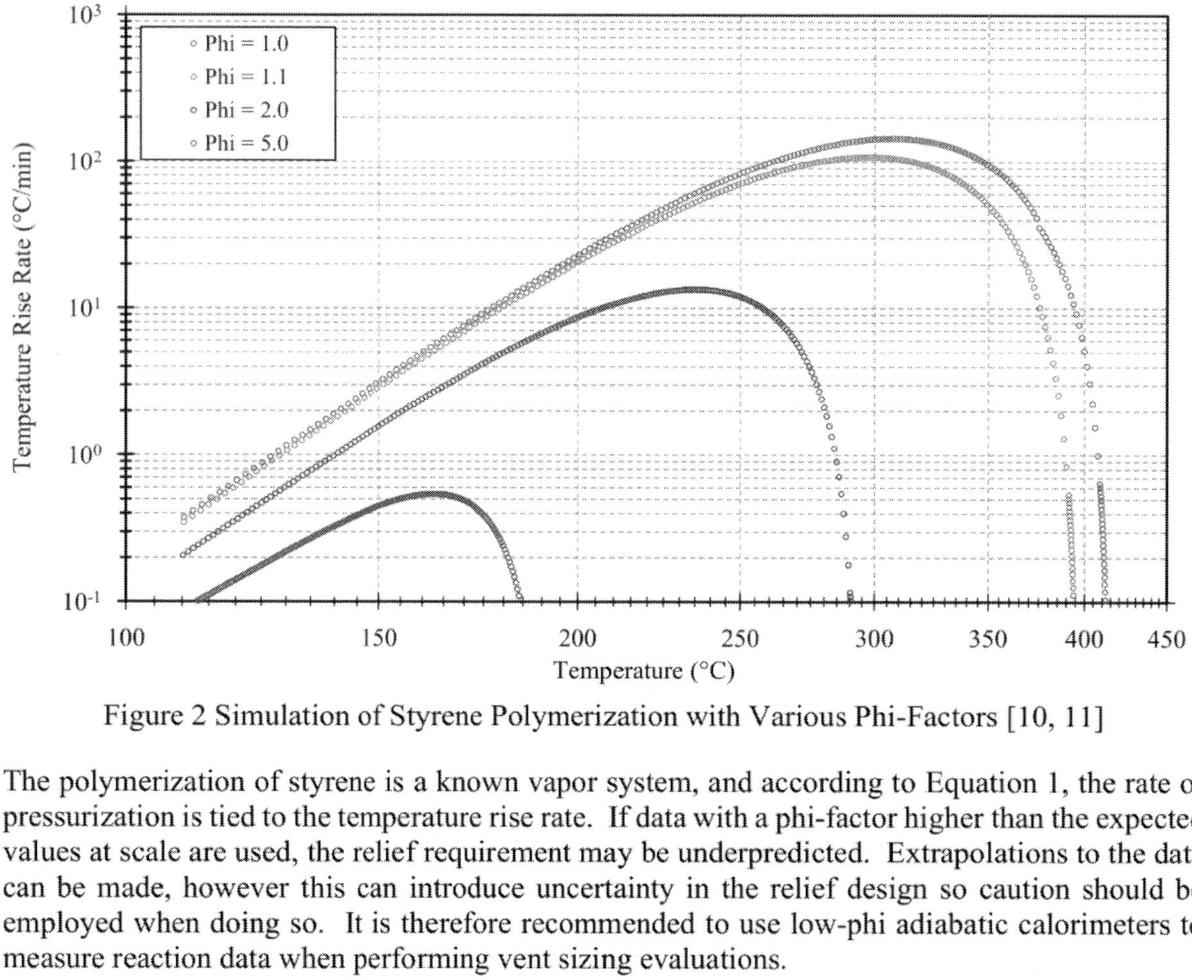

Figure 2 Simulation of Styrene Polymerization with Various Phi-Factors [10, 11]

The polymerization of styrene is a known vapor system, and according to Equation 1, the rate of pressurization is tied to the temperature rise rate. If data with a phi-factor higher than the expected values at scale are used, the relief requirement may be underpredicted. Extrapolations to the data can be made, however this can introduce uncertainty in the relief design so caution should be employed when doing so. It is therefore recommended to use low-phi adiabatic calorimeters to measure reaction data when performing vent sizing evaluations.

1.3.3 VSP2 Adiabatic Calorimeter

There are a few commercially available adiabatic calorimeters, and the present work will utilize the Vent Sizing Package 2 (VSP2) [12]. The VSP2 apparatus is a low-phi adiabatic calorimeter designed to acquire thermal and vent sizing data and is based on DIERS research results. This apparatus and typical results have been discussed in relevant publications [13-16]. Key features of the apparatus are summarized in Figure 3.

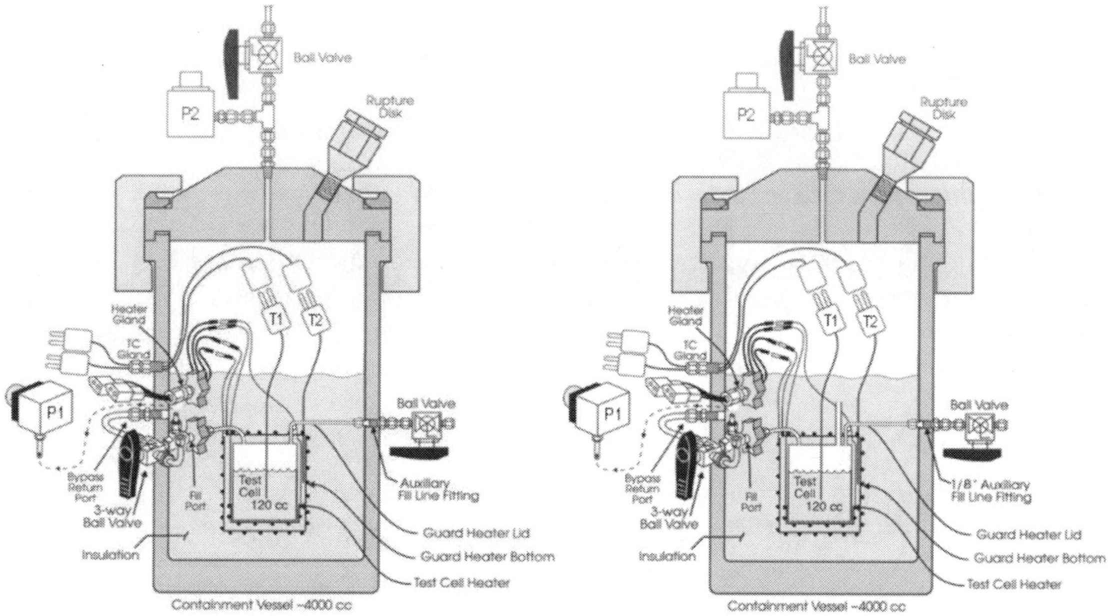

Figure 3 VSP2 Apparatus in Closed (Left) and Open (Right) Cell Configurations

The VSP2 commonly uses cylindrical test cells that are thin walled and thus very lightweight, allowing the test to have a low phi-factor. Typical values of phi for VSP2 experiments are around 1.1, which is common of many full-scale applications, validated by the comparison of VSP2 data against large scale experiments. There is a wide variety of materials of construction and test cell configurations (e.g. baffles for immiscible solutions, dip tube for subsurface additions, a pluggable vent line for adding solids prior to performing a closed cell test, etc.) allowing for experimental customization. The approximately 110-120 ml test cell is fitted with a guard heater (for maintaining adiabatic conditions) and auxiliary heater (for imposing a background heating rate or heating to a process temperature), wrapped in thermal insulation, and is placed in a 3.8 liter high pressure containment vessel.

The apparatus measures sample temperature (T1) and pressure (P1), and external (guard) temperature (T2) and containment vessel pressure (P2), with additional temperature and pressure measurement ports available depending on the experimental need. For a normal test setup the reactions can be tracked to above 400°C and up to approximately 1,500 psi, with alternative options available for high pressure reactions.

For closed (non-vented) test cells, the pressure balancing system maintains equilibrium between the pressure in the containment vessel and the pressure in the test cell by either adding or removing nitrogen. This pressure control feature makes the utilization of the thin wall (low ϕ-factor) test cell design possible. For open or vented test cells, no pressure regulation is necessary since the vent provides adequate pressure equalization. Pressure control allows for tests to be carried out under constant volume (containment volume), constant pressure, or depressurization operation modes. In most cases an initial back pressure usually determined by the process vessel specifications is imposed on the system. The heaters then function in a manner similar to the closed non-vented tests.

Stirring is typically performed using a Super Magnetic Stirrer (capable of stirring up to 10,000 cps) but can also be performed using a mechanical overhead agitator. The VSP2 can be combined with auxiliary equipment such as an additional containment vessel, cooling coil, syringe pump, or piston to perform a wide variety of tests.

2 Case Study: Tube Rupture Leading to Unexpected Chemical Reaction

2.1 Description of Upset Scenario

A PHA identifies a potential upset scenario involving a cooling coil rupture inside a pressure reactor at a pilot plant. The cooling coil utilizes city water, and the vessel is used for processes involving titanium tetrachloride, a known water reactive material [17]. It is identified that the worst-case scenario would be tube rupture while titanium tetrachloride is inside the vessel at the start of processing. The details of the internal coils and pressure vessel are outlined in Table 2. Both fluids are considered to be at approximately the same temperature at the start of the upset scenario.

Table 2: Vessel and Cooling Coil Dimensions

Parameter	Value
Vessel Diameter, m	0.91
Vessel Length, m	1.27
Vessel Head Type, -	2:1 Ellipsoidal
Vessel Maximum Allowable Working Pressure, Pa	583,958
Relief Device Set Pressure, Pa	583,958
Reactor Operating Pressure, Pa	104,772
Cooling Coil Inner Diameter, m	0.032
Cooling Coil Length, m	55.8
Initial Titanium Tetrachloride Charge, kg	249.6
Maximum Quantity of Water in Cooling Coils, kg	44.2

Three potential scenarios are considered:

1. double-ended guillotine tube rupture with high pressure differential
2. double-ended guillotine tube rupture with low pressure differential
3. pinhole leak with high pressure differential

A summary of the parameters considered for each scenario is provided in Table 3.

Table 3: Summary of Upset Scenarios

Parameter	Scenario 1	Scenario 2	Scenario 3
Water Pressure, Pa	273,700	104,797	273,700
Leak Area, m^2	7.92E-04	7.92E-04	2.66E-06

2.2 *Experimental Test Design*

When simulating a tube rupture scenario resulting in a chemical reaction, there are two crucial parameters to assess so that the experiment can be designed to best simulate the scenario: 1. The addition rate (and direction) of fluid flow, and 2. The expected mass fraction of the two components.

2.2.1 Gas/Vapor Flow Rates

To estimate the leak flow rate of a gas or vapor, the first thing that should be identified is the flow type, i.e. if the flow is critical (choked) or subcritical (unchoked). To do so, the critical pressure ratio should be determined using Equation 5.

$$\frac{P_c}{P_o} = \left(\frac{2}{k+1}\right)^{\left(\frac{k}{k-1}\right)} \tag{5}$$

where P_c = absolute critical pressure, Pa
P_o = absolute cooling water pressure, Pa
k = isentropic coefficient, -

Determination of the presence of subcritical flow is done using the inequality,

$$\frac{P_b}{P_o} > \frac{P_c}{P_o} \tag{6}$$

where P_c = absolute backpressure, Pa

Here, the venting pressure is taken to be the pressure at the higher-pressure side, and the backpressure is taken to be the static initial reactor pressure. If this inequality holds true, then the flow is subcritical. If not, critical flow must be considered. The generalized equation for critical gas/vapor flow is written as:

$$u = \left(\frac{P_o}{\rho}\right)^{0.5} \left[k\left(\frac{2}{k+1}\right)^{\left(\frac{k+1}{k-1}\right)}\right]^{0.5} \tag{7}$$

where	ρ	=	density, kg/m^3
	u	=	vapor velocity, m/s

In cases of ideal subcritical flow, Equation 8, can be used.

$$u = \left(\frac{P}{\rho}\right)^{0.5} \left\{\left(\frac{2k}{k-1}\right)\left[\left(\frac{P_b}{P_o}\right)^{\frac{2}{k}} - \left(\frac{P_b}{P_o}\right)^{\frac{k+1}{k}}\right]\right\}^{0.5} \tag{8}$$

where	ΔP	=	pressure differential between tube break opening, Pa

The mass flow is then estimated using Equation 9.

$$w = A * u * \rho \tag{9}$$

where	A	=	effective leak area, m^2
	w	=	leak flow rate, kg/s

In the case of a double-ended guillotine tube rupture, the effective leak area can be described by the following equation.

$$A = (1 + C_D)A_t \tag{10}$$

where	A_t	=	tube leak area, m^2
	C_D	=	discharge coefficient of flow through tube, -

The flow area is derated by the discharge coefficient, which is obtained assuming a Fanning friction factor, f, of 0.005 and Equation 11 for subcritical flow and Equation 12 for critical flow.

$$C_D = \frac{1}{\sqrt{1 + 4f\frac{l}{d}}} \tag{11}$$

$$C_D = \frac{1}{\left(1 + 4f\frac{l}{d}\right)^{0.39}} \tag{12}$$

where	d	=	inner tube diameter, m
	f	=	Fanning friction factor, 0.005
	l	=	tube length, m
	C_D	=	discharge coefficient of flow through tube, -

In other leak scenarios, the flow area and discharge coefficient may vary. For the present work, the pinhole leak was assumed to have a diameter of 0.07" and a discharge coefficient of 0.6.

2.2.2 Liquid Flow Rates

In the event of a single-phase incompressible flow like the present case, the flow rate can be estimated using a simplified version of Equation 8 provided in Equation 13 with the same approach for estimating the mass flow rate according to equations 9-11.

$$u = \left(2\frac{\Delta P}{\rho}\right)^{0.5} \tag{13}$$

If flashing flows are expected, an alternative methodology is recommended [18].

2.2.3 Composition

It is assumed for this case study that there are safeguards installed so that in the event of a leak the supply is isolated from the vessel, and there are systems in place to prevent backflow into the vessel. Therefore, the final mass fraction of water introduced can be determined by the following:

$$x_{water} = \frac{m_{water}}{m_{TiCl4} + m_{water}} \tag{14}$$

where

m_{TiCl_4}	=	mass of batch loaded $TiCl_4$, kg
m_{TiCl_4}	=	mass of water contained in cooling coils, kg
x_{water}	=	maximum mass fraction of water, -

Considering the starting quantity of titanium tetrachloride, and the water amount available in the cooling coils, the resulting mass fraction of water is calculated to be 0.15. In instances where the source of water cannot be isolated, the additional quantity of water pumped into the system could also be simulated. When there is uncertainty in the possible composition, testing near the stochiometric composition is recommended to generate the highest expected reaction rates. Considering this, the approximate VSP2 recipe studied in the experiments is outlined in Table 4.

Table 4: VSP2 Sample Charge

Chemical Name	Mass (g)	Moles (mol)	Density (g/mL)	Volume (ml)	Composition (wt. %)
DI Water	15	0.83	1.0	15	15.00
Titanium Tetrachloride	85	0.45	1.7	50	85.00
Total	**100**	-	**1.031**	**65**	**100.00**

The resulting leak rates can be scaled to the VSP2 by identifying the total leak time at the reactor scale and then taking the VSP2 quantity of water and dividing by the leak time. The estimates for leak rate for the presented upset scenarios are available in Table 5.

Table 5: Summary of Upset Scenarios

Parameter	Scenario 1	Scenario 2	Scenario 3
Water Density, kg/m^3	1,000	1,000	1,000
Discharge Coefficient, -	0.17	0.17	0.60
Water Velocity, m/s	18.4	0.2	18.4
Water Leak Rate, kg/s	16.97	0.20	0.03
Leak Time, s	2.6	217.9	1,504.8
VSP2 Addition Rate, ml/min	345.6	4.1	0.6

It is noted that there are some practical limitations at the laboratory scale. For example, the starting volume inside the test cell must be large enough to ensure proper contact with the internal thermocouple. If the upset scenario indicates a starting composition below the thermocouple line, early reaction rate information can be missed. Possible remedies include considering a reverse addition, or performing multiple experiments to extract a kinetic model so that the reaction rates of the upset scenario can be simulated. Additionally, there are limitations with how fast and how slow quantities can be added at a controlled rate at the laboratory scale. In this instance, the fastest that material can be introduced is using a piston injection system. Care can be taken to time the rate of the addition, however there is uncertainty in the true addition rate into the test cell. For rapid additions, alternative approaches can also be employed such as scaling the observed reaction rates to the expected leak rates, or developing a kinetic model to better predict the behavior when the expected conditions cannot be simulated experimentally.

2.3 Experimental Procedure

Each scenario was tested inside the VSP2. Significant quantities of HCl noncondensable gas were expected, and therefore the experiments were conducted as open system experiments. A description of the test cell used is available in Table 6. The material of construction was selected to minimize corrosion due to the expected HCl generation.

Table 6: VSP2 Test Cell Configuration

Parameter	Value
Material of Construction, -	All-Welded Hastelloy C
Fill Line Diameter, in	1/16"
Pressure Line Diameter, in	1/8"
Vent Line Diameter, in	2 x 1/4"
Thermowell Location, -	1/2" from Bottom of Test Cell
Stir Bar Type, -	1" Teflon-Coated
Test Cell Volume, ml	112
Approximate Test Cell Mass, g	40

The general test procedure involved inserting the test cell into the standard heater and insulation assembly, and then the entire assembly was installed inside the VSP2 containment vessel. The containment vessel and test cell were pressurized with 50 psig of nitrogen five times to purge the system of air and moisture leaving the containment vessel with approximately 1-2 psig of nitrogen. The titanium tetrachloride was added to the test cell through the auxiliary line. The containment vessel and test cell were then pressurized with 80 psig of nitrogen. The Super Stirrer was turned on and set to 350 rpm. For scenario 1, deionized water was loaded into a stainless-steel piston and connected to the auxiliary fill line. The addition was completed within a two seconds indicating an addition rate on the order of 450 ml/min or greater. An improvement to the experimental procedure would be to time the addition to better quantify the expected addition rate. Adjustments to the backpressure on the piston or flow area could be adjusted to somewhat vary the fast addition rates. For scenarios 2 and 3, deionized water was loaded into a stainless-steel syringe and installed in a high-pressure syringe pump connected to the auxiliary fill line. The pump was used to control the addition rate at the specified values. After data acquisition was initiated and the guard heater was enabled, the specified quantity of water was injected or metered into the test cell at the rates specified in Table 5. The reactions were tracked adiabatically, and once the temperature and pressure signals stabilized, the heaters were disabled, and cooldown data were collected.

2.4 Experimental Results

A summary of the test results is available in Table 7, with comparison plots of temperature rise rate, pressure rise rate, and pressure as a function of temperature data from the experiments available in Figures 4-6. In all cases, exothermic activity was observed immediately with noncondensable gas generation. The presented rates are based on an eleven-point regression of the collected data. During scenario 1, choked flow was observed indicating insufficient venting capacity, and during scenario 3, some brief plugging in the vent lines around 65°C was observed. Therefore, in all instances the presented pressure rise rates are based on a volume weighted average of the test cell pressure and containment vessel pressure.

Table 7: Summary of Test Results

Parameter	Scenario 1	Scenario 2	Scenario 3
Mass of Titanium Tetrachloride	85.75 g	81.87 g	87.44 g
Moles of Titanium Tetrachloride	0.45 mol	0.43 mol	0.46 mol
Mass of Water	14.21 g	14.01 g	14.98 g
Moles of Water	0.79 mol	0.78 mol	0.83 mol
Tested Addition Rate	>450 ml/min	4.1 ml/min	0.6 ml/min
Peak Measured Temperature	105.1°C	90.7°C	103.4°C
Peak Measured Temperature Rise Rate	1,869°C/min at 48.0°C	110.2°C/min at 79.2°C	5.1°C/min at 73.1°C
Peak Measured Pressure	122 psig	115 psig	125 psig
Peak Measured Pressure Rise Rate	719.6 psi/min at 29.7°C	41.5 psi/min at 75.1°C	3.5 psi/min at 63.8°C
Quantity of Noncondensable Gas Generated/Mole of Titanium Tetrachloride	0.7 mol/mol	0.5 mol/mol	0.6 mol/mol

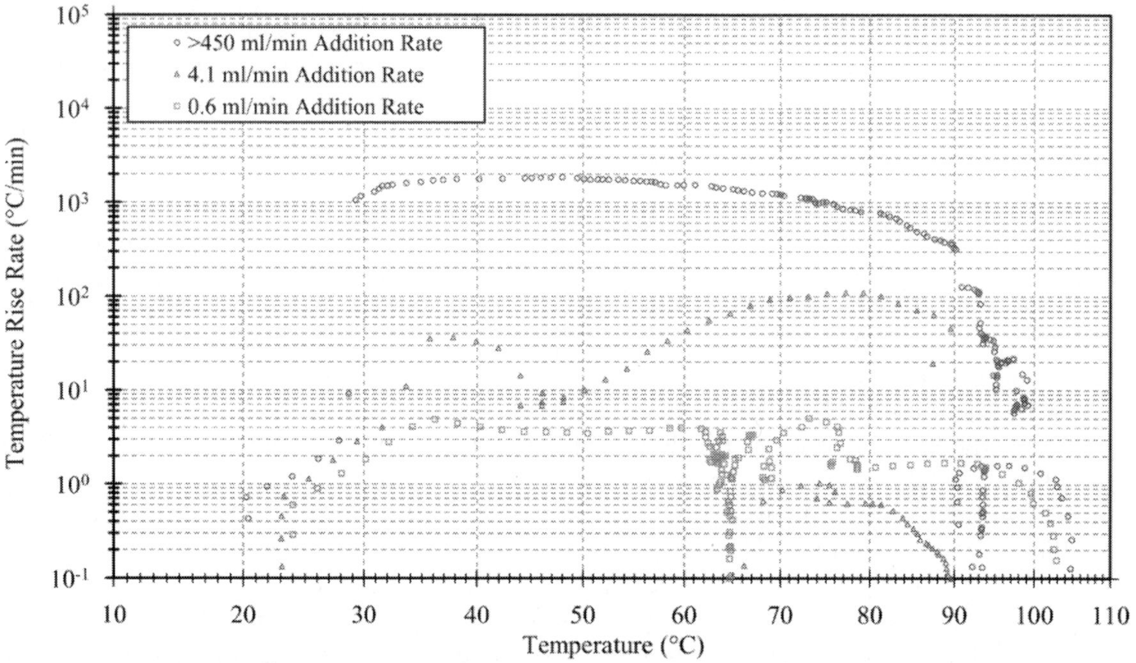

Figure 4 Comparison of Temperature Rise Rate vs. Temperature Data

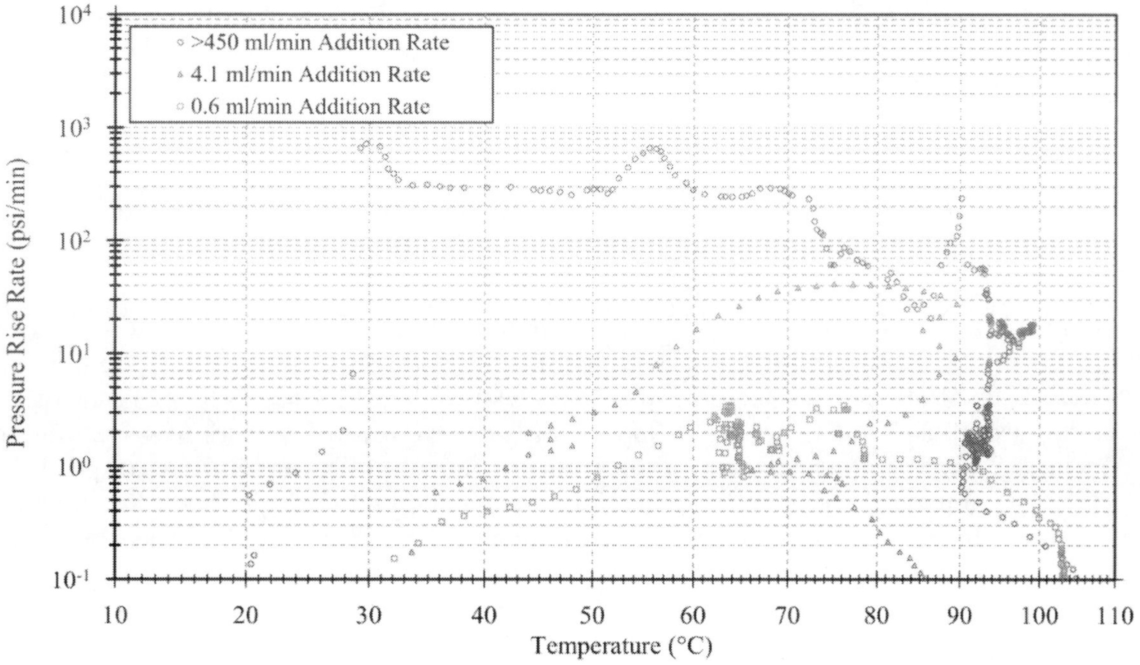

Figure 5 Comparison of Pressure Rise Rate vs. Temperature Data

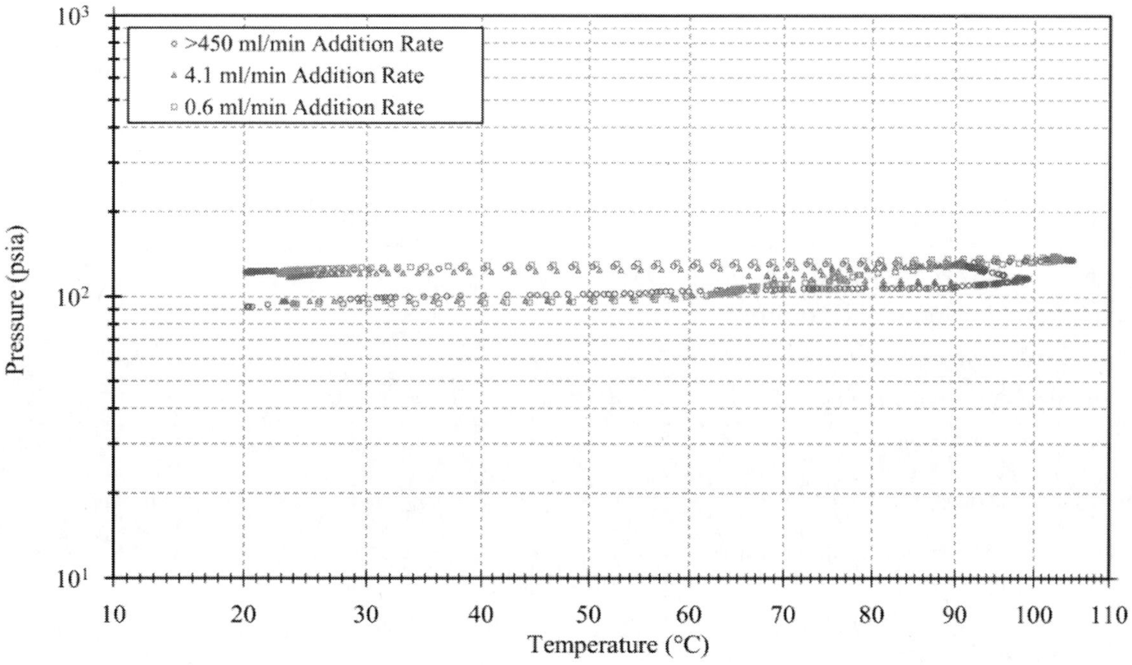

Figure 6 Comparison of Pressure vs. Temperature Data

2.5 Vent Sizing Results

These upset scenarios result in titanium tetrachloride reacting with water, producing hydrochloric acid that could cause overpressure. The venting temperatures and pressure are expected to be below the vapor pressure of water indicating that the system can be classified as a reactive non-tempered system (or gassy system). The ideal vent size evaluations were conducted using the Leung Omega methodology [19-21] within FERST powered by CHEMCAD [22] assuming a homogeneous flow regime. The batch liquid was treated as a mixture of titanium tetrachloride and water. Early mass loss was not considered. The resulting FERST inputs and outputs are available in Tables 9-11 with a summary provided in Table 8.

The results indicate that the venting requirement is directly related to the addition rate. In the event of a complete tube rupture resulting in an addition rate of approximately >450 ml/min, it is recommended to install at least a vent(s) with an equivalent area of 483 cm^2. This can be achieved through the use of a 14" diameter relief line. As the addition rate decreases to 4.1 ml/min or 0.6 ml/min, the corresponding relief requirements decrease to 29.1 cm^2 and 3.3 cm^2, respectively. This shows the importance of identifying not only a credible upset scenario, but also a probable upset scenario. In the absence of historical data or other archives eliminating the possibility of a complete tube rupture, it is clear why this is the recommended first step as it represents the worst credible internal failure with the largest relief requirement. In the event that a double-ended guillotine tube rupture resulting in a chemical reaction is unventable, next steps may be such as changing the cooling medium or effectively changing the leak rate by altering the pressure differential.

Table 8: Summary of Vent Sizing Results

Parameter	Scenario 1	Scenario 2	Scenario 3
Tested Addition Rate	>450 ml/min	4.1 ml/min	0.6 ml/min
Ideal Vent Area[1], cm^2	483	29.1	2.3
Ideal Vent Diameter, m	0.25	0.06	0.02
Ideal Vent Diameter, in	9.8	2.4	0.7
Recommended Nominal Rupture Disk Diameter[2], in	14	3	1
Allowable Frictional Losses[3], K, -	4.5	3.3	10.8

[1] Based on Schedule 40 piping
[2] Based on Schedule 40 piping
[3] Provided K value does not account for entrance loss (typically 0.5) or pressure loss coefficient for the rupture disk

Table 9: FERST Vent Sizing Results – >450 ml/min

```
Vessel
Volume: 1.023278 m3
Max Allowable Working Pressure: 583958.000000 Pa
Diameter: 0.910001 m
Length: 1.270000 m
```

Orientation: Vertical
Head Type: Ellipsoidal

Vent Path
Path Type: Rupture Disk
Relief Set Pressure: 583958.000000 Pa
Back / Atm. Pressure: 101325.000000 Pa

Inlet Pipe
Fanning friction Factor: 0.005000
Elevation Change from Vessel to Highest Point: 0.000000 m
Pipe Schedule: Schedule 40

System Type: Reactive-Non-Tempered (Gassy)

Summary of Input to Leung Omega Method

Mode: Design
Vessel Flow Regime: Homogeneous

Relief Set Pressure: 583958.000000 Pa
Back Pressure: 101325.000000 Pa
Maximum Venting Pressure: 632222.000000 Pa
Vessel Volume: 1.023278 m3
Vessel X-Sectional Area: 0.513625 m2
Total Mass in Vessel: 293.800018 kg
Liquid Density: 1494.030029 kg/m3
Liquid Constant Press Specific Heat: 0.842512 kJ/kg-K
Liquid Constant Volume Specific Heat: 0.842512 kJ/kg-K
Gas Molecular Weight: 36.459999 kg/kmol
Gas Compressibility Factor: 1.000000
Gas Constant Press Specific Heat: 1.000000 kJ/kg-K
Gas Constant Volume Specific Heat: 0.990099 kJ/kg-K
T @ dP/dt, max: 29.700003 C
dP/dt, max: 82691.155556 Pa/sec
Vapor/Gas Space Temperature: 29.700003 C

Test Freeboard Volume: 0.003800 m3
Test Sample Mass: 0.099960 kg

Summary of Results from Leung Omega Method

Overpressure (gauge): 10.000146%
Turnaround Time: 0.331571 sec
Peak Venting Pressure: 632222.000000 Pa
Temperature Correction: 1.000000
Gas Generation Rate: 13.372822 kg/sec
Freeboard Volume Fraction: 0.807824
alpha0/k2p: 0.807587
Exit Quality: 0.025110
Density of Discharge Flow: 287.116486 kg/m3
Discharge Mass Flow Rate: 419.427277 kg/sec
Discharge Mass Flux: 8675.824106 kg/sec-m2
Ideal Vent Area: 0.048344 m2
Ideal Vent Diameter: 0.248101 m

Rupture Disk/Tailpipe Sizing Program

Note: Assumes turbulent (non-viscous) flow regime.
Calculates max allowable pressure loss coefficient for RD tailpipe.
Evaluation based upon Results from Leung Omega Method.

Mass Flow Rate: 411.344604 kg/sec
Reference Pressure: 608090.000000 Pa
Reference Density: 287.116486 kg/m3
Back Pressure: 101325.000000 Pa
Elevation Above Reactor: 0.000000 m
Ideal Vent Area Required: 0.048344 m2
Pipe Schedule: Schedule 40
Fanning friction Factor: 0.005000

Pipe Dia (m)	Flow Area (m2)	4fL/D* (allowable)
0.254508	0.050874	0.138756
0.303225	0.072214	2.260987
0.333350	0.087275	4.527369

* Allowable total piping resistance factor

Note: Relief set pressure > 50% of allowable pressure limit *** Transient mass loss effect has NOT been credited ***

Table 10: FERST Vent Sizing Results – 4.1 ml/min

Vessel
Volume: 1.023277 m3
Max Allowable Working Pressure: 583958.000000 Pa
Diameter: 0.910000 m
Length: 1.270000 m
Orientation: Vertical
Head Type: Ellipsoidal

Vent Path
Path Type: Rupture Disk
Relief Set Pressure: 583958.000000 Pa
Back / Atm. Pressure: 101325.000000 Pa

Inlet Pipe
Fanning friction Factor: 0.005000
Elevation Change from Vessel to Highest Point: 0.000000 m
Pipe Schedule: Schedule 40

System Type: Reactive-Non-Tempered (Gassy)

Summary of Input to Leung Omega Method

Mode: Design
Vessel Flow Regime: Homogeneous

Relief Set Pressure: 583958.000000 Pa

Back Pressure: 101325.000000 Pa
Maximum Venting Pressure: 632221.000000 Pa
Vessel Volume: 1.023277 m3
Vessel X-Sectional Area: 0.513624 m2
Total Mass in Vessel: 293.800018 kg
Liquid Density: 1494.030029 kg/m3
Liquid Constant Press Specific Heat: 0.842512 kJ/kg-K
Liquid Constant Volume Specific Heat: 0.842512 kJ/kg-K
Gas Molecular Weight: 36.459999 kg/kmol
Gas Compressibility Factor: 1.000000
Gas Constant Press Specific Heat: 1.000000 kJ/kg-K
Gas Constant Volume Specific Heat: 0.990099 kJ/kg-K
T @ dP/dt, max: 79.199989 C
dP/dt, max: 4768.866667 Pa/sec
Vapor/Gas Space Temperature: 79.199989 C

Test Freeboard Volume: 0.003800 m3
Test Sample Mass: 0.095880 kg

Summary of Results from Leung Omega Method

Overpressure (gauge): 9.999939%
Turnaround Time: 5.514686 sec
Peak Venting Pressure: 632221.000000 Pa
Temperature Correction: 1.000000
Gas Generation Rate: 0.691084 kg/sec
Freeboard Volume Fraction: 0.807824
alpha0/k2p: 0.807619
Exit Quality: 0.021658
Density of Discharge Flow: 287.116913 kg/m3
Discharge Mass Flow Rate: 25.218102 kg/sec
Discharge Mass Flux: 8675.729502 kg/sec-m2
Ideal Vent Area: 0.002907 m2
Ideal Vent Diameter: 0.060836 m

Rupture Disk/Tailpipe Sizing Program

Note: Assumes turbulent (non-viscous) flow regime.
Calculates max allowable pressure loss coefficient for RD tailpipe.
Evaluation based upon Results from Leung Omega Method.

Mass Flow Rate: 24.732140 kg/sec
Reference Pressure: 608090.000000 Pa
Reference Density: 287.116913 kg/m3
Back Pressure: 101325.000000 Pa
Elevation Above Reactor: 0.000000 m
Ideal Vent Area Required: 0.002907 m2
Pipe Schedule: Schedule 40
Fanning friction Factor: 0.005000

Pipe Dia (m)	Flow Area (m2)	4fL/D* (allowable)
0.062713	0.003089	0.170580
0.077927	0.004769	3.258032
0.090119	0.006379	8.155478

* Allowable total piping resistance factor

Note: Relief set pressure > 50% of allowable pressure limit *** Transient mass loss effect has NOT been credited ***

Table 11: FERST Vent Sizing Results – 0.6 ml/min

```
Vessel
Volume: 1.023278 m3
Max Allowable Working Pressure: 583958.000000 Pa
Diameter: 0.910001 m
Length: 1.270000 m
Orientation: Vertical
Head Type: Ellipsoidal

Vent Path
Path Type: Rupture Disk
Relief Set Pressure: 583958.000000 Pa
Back / Atm. Pressure: 101325.000000 Pa

Inlet Pipe
Fanning friction Factor: 0.005000
Elevation Change from Vessel to Highest Point: 0.000000 m
Pipe Schedule: Schedule 40

System Type: Reactive-Non-Tempered (Gassy)
------------------------------------------------------------------
Summary of Input to Leung Omega Method

Mode: Design
Vessel Flow Regime: Homogeneous

Relief Set Pressure: 583958.000000 Pa
Back Pressure: 101325.000000 Pa
Maximum Venting Pressure: 632222.000000 Pa
Vessel Volume: 1.023278 m3
Vessel X-Sectional Area: 5136.250977 cm2
Total Mass in Vessel: 293.800018 kg
Liquid Density: 1494.030029 kg/m3
Liquid Constant Press Specific Heat: 0.842512 kJ/kg-K
Liquid Constant Volume Specific Heat: 0.842512 kJ/kg-K
Gas Molecular Weight: 36.459999 kg/kmol
Gas Compressibility Factor: 1.000000
Gas Constant Press Specific Heat: 1.000000 kJ/kg-K
Gas Constant Volume Specific Heat: 0.990099 kJ/kg-K
T @ dP/dt, max: 73.099998 C
dP/dt, max: 402.195000 Pa/sec
Vapor/Gas Space Temperature: 73.099998 C

Test Freeboard Volume: 0.003800 m3
Test Sample Mass: 0.102420 kg
------------------------------------------------------------------
Summary of Results from Leung Omega Method

Overpressure (gauge): 10.000146%
```

```
Turnaround Time: 69.848564 sec
Peak Venting Pressure: 632222.000000 Pa
Temperature Correction: 1.000000
Gas Generation Rate: 0.055524 kg/sec
Freeboard Volume Fraction: 0.807824
alpha0/k2p: 0.807616
Exit Quality: 0.022032
Density of Discharge Flow: 287.116425 kg/m3
Discharge Mass Flow Rate: 1.991020 kg/sec
Discharge Mass Flux: 0.867574 kg/sec-cm2
Ideal Vent Area: 2.294929 cm2
Ideal Vent Diameter: 0.017094 m
------------------------------------------------------------------
Rupture Disk/Tailpipe Sizing Program

Note: Assumes turbulent (non-viscous) flow regime.
      Calculates max allowable pressure loss coefficient for RD tailpipe.
      Evaluation based upon Results from Leung Omega Method.

Mass Flow Rate: 1.952652 kg/sec
Reference Pressure: 608090.000000 Pa
Reference Density: 287.116425 kg/m3
Back Pressure: 101325.000000 Pa
Elevation Above Reactor: 0.000000 m
Ideal Vent Area Required: 2.294929 cm2
Pipe Schedule: Schedule 40
Fanning friction Factor: 0.005000

       Pipe Dia       Flow Area          4fL/D*
            (m)           (cm2)     (allowable)
       0.020930        3.440422        2.295255
       0.026645        5.575814       10.762009
       0.035052        9.649737       40.018164

*  Allowable total piping resistance factor

Note: Relief set pressure > 50% of allowable pressure limit *** Transient
mass loss effect has NOT been credited ***
```

3 Other Considerations

Other considerations when mitigating a heat exchanger failure scenario include relief device type and location. API 521 STD [3] and Institute of Petroleum "Guidelines for the design and safe operation of shell & tube heat exchangers to withstand the impact of tube failure" [4] recommends installing a fast acting relief devices like a spring loaded pressure relief valve, angle pin valve, or most favorably, a rupture disk mounted directly on the shell (low-pressure side) of the heat exchanger. It is recommended to use two disks when the unit is very long or there is a very high differential operating pressures. Other considerations when addressing a heat exchanger failure include assessing the potential impact of the rupture on downstream or upstream piping and/or equipment, rapid phase changes following the failure, rapid temperature changes above or below the design temperature of the system, and momentary transient forces such as shock or surge pressures and reaction forces or structural loads during venting. It is also recommended to consider multi-phase flow even in the absence of chemical reactions. If all gas/all vapor flow is considered, the relief device(s) may be undersized. Many of these factors can be addressed during a dynamic evaluation, and otherwise are not discussed in the present work.

4 Conclusions

A common scenario identified when evaluating a process for potential hazards is failure of heat transfer equipment. When defining the credible scenario, API STD 521 recognizes that the extent of internal failure can vary from a pinhole leak to a complete tube rupture, and therefore in practice, a sharp break (or double-ended guillotine tube rupture) is often a conservative first point of consideration, although it is noted that this may not be very probable. Assessing the required relieving rate to protect the vessel from this scenario is further complicated when the inadvertent mixing of the two fluids results in a chemical reaction. This paper provides a simplified approach for assessing the relief requirement following a cooling coil failure resulting in an unintended chemical reaction. This paper provides guidance for estimating the equivalent addition rate and mixture composition following a double-ended guillotine tube rupture with two high and low side pressure differentials as well as a pinhole leak. Direct simulation of the scenario using low phi-factor adiabatic calorimetry then allows for the expected pressure rates to be measured and directly used with common reactive vent sizing methodologies to determine the venting requirements. When the pressure differential across the failure is large, and/or when the resulting relief requirement is close to what is installed, a dynamic evaluation of the failure is recommended to better quantify the impact of the failure.

5 References

[1] CCPS "Guidelines for Initiating Events and Independent Protection Layers in Layer of Protection Analysis," Published February, 2015.

[2] Purple Book, CPR 18E, Publication Series on Dangerous Substances (PGS 3), "Guidelines for quantitative risk assessment," Part One: Establishments, Originally Published 1999, Republished December 2005.

[3] API Standard 521, "Pressure-relieving and Depressuring Systems," American Petroleum Institute, Washington, D.C., Seventh Edition, June, 2020.

[4] Institute of Petroleum "Guidelines for the design and safe operation of shell & tube heat exchangers to withstand the impact of tube failure", Colchester, UK, Aug 2000.

[5] Spouge, J., "New generic leak frequencies for process equipment," Process Safety Progress, 24(4): 249-257, 2005.

[6] Spouge, J., "Leak frequencies from the hydrocarbon release database," IChemE Symposium Series No. 151, 2006.

[7] Harhara, A., "Dynamic modeling of heat exchanger tube rupture," BMC Chemical Engineering, Article 5, 04 March, 2020.

[8] Ennis, C., "Dynamic model for a heat exchanger tube rupture discharging a high-pressure flashing liquid into a low-pressure liquid-filled shell," Journal of Loss Prevention in the Process Safety Industries, Volume 24, Issue 1, pg. 111-121, January 2011.

[9] Kurko, K., "Relief System Sizing for Runaway Chemical Reactions: A Simple Comprehensive Approach," 11th Global Congress on Process Safety, 27-29, April, 2015.

[10] Hui, A. W. and Hamielec, A. E., "Thermal Polymerization of Styrene at High Conversions and Temperatures. An Experimental Study," J. Appl. Polymer Sci. 16, 1972.

[11] Hussain, A. and Hamielec, A. E., "Thermal Polymerization of Styrene," J. Appl. Polymer Sci. 22, 1978.

[12] Askonas, C., "The Versatile VSP2: A Tool for Adiabatic Thermal Analysis and Vent Sizing Applications," 28th North American Thermal Analysis Society, 4-6 October, 2000.

[13] Fauske, H. K., "Emergency Relief Systems (ERS) Design," Chemical Engineering Progress, pp. 53-56 (August, 1985).

[14] Fauske, H. K. and Leung, J. C., "New Experimental Technique for Characterizing Runaway Chemical Reactions," Chemical Engineering Progress, Vol. 81, No. 8, pp. 39-46 (August, 1985).

[15] Leung, J. C., Fauske, H. K., and Fisher, H. G., "Thermal Runaway Reactions in a Low Thermal Inertia Apparatus," Thermochimica Acta, 104, pp. 13-29 (1986).

[16] Leung, J. C. and Fauske, H. K., "Runaway System Characterization and Vent Sizing Based on DIERS Methodology," Plant Operations Progress, Vol. 6, No. 2, pp. 77-83 (April, 1987).

[17] Vechot, L., Buston, J.E.H, Kay, J., Round, G.A., Masharani, S., Tickle, G.A., Rowlands, R., "Experimental Study of the Liquid Phase Hydrolysis Reaction of Titanium Tetrachloride," Symposium Series No. 156, 2011.

[18] Fauske, H. K., "Determine Two-Phase Flows During Releases," Chemical Engineering Progress, pg. 55-58 (February, 1999).

[19] Leung, J. C., "Flashing Two-Phase Flow Including the Effects of Noncondensable Gases," Journal of Heat Transfer, pp. 269-272 (February, 1991).

[20] Leung, J.C., "Vent Sizing for Gassy and Hybrid Systems," Safety of Chemical Batch Reactors and Storage Tanks, 1991.

[21] Leung, J.C. and Epstein, M.A., "A Generalized Correlation for Two-Phase Nonflashing Homogeneous Choked Flow," Transactions of the ASME, Vol. 112, 1990.

[22] FERST powered by CHEMCAD Version 1.0.0.15653, Fauske & Associates, LLC, 2020.

What To Do With Your Maintenance Records for Safeguards?

Angela E. Summers, PhD, PE
SIS-TECH
12621 Featherwood Drive, Suite 120, Houston, TX 77034
asummers@sis-tech.com

Julia V. Bukowski, PhD
Villanova University

Keywords: API 754, reliability, metrics, failure rate, risk

Abstract

Collecting and analyzing operation and maintenance data is essential to process safety. The analysis aims to demonstrate that what is installed in the field achieves the required performance so that adjustments can be made if the performance is not met. Very large organizations routinely have data collection systems and resources for data analysis. Midsized and smaller organizations collect the same data in various ways but often need help to analyze it. This paper presents examples of API 754 Tier 3 and 4 metrics related to safeguards and how to use available plant data to support their analysis.

1 Risk Management and Incident Prevention

Operating a chemical process involves risk. No process can be reasonably designed to be risk-free. The number and kind of risk reduction measures needed are determined by risk assessment, which weighs the mitigated risk against the benefits of further risk reduction measures. Figure 1 illustrates the risk as existing from a level considered unacceptable to an opposite one considered broadly acceptable. The tolerable risk is somewhere in the middle.

The foundation for risk assessment worldwide is to lower the risk to a level that is As Low As Reasonably Practicable (ALARP). ALARP suggests that risk should continue to be reduced until the costs associated with additional risk reduction measures are grossly disproportionate to the benefits of further risk reduction. While a company's risk assessment philosophy may not be directly aligned with ALARP principles, the general consensus is that risk reduction measures should be implemented when practical and these measures should be monitored and managed to ensure that they work as required when required.

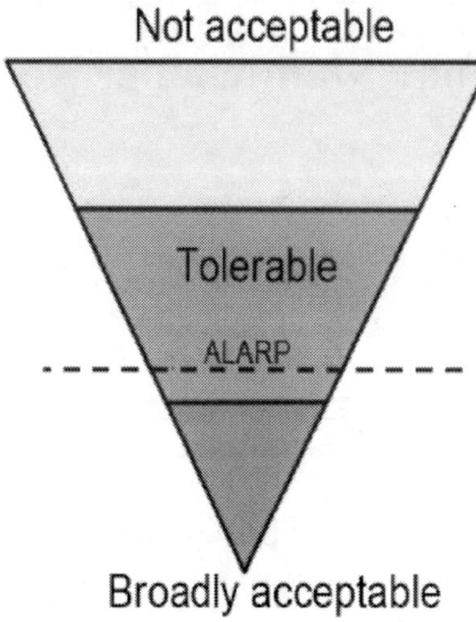

Figure 1. As Low As Reasonably Practicable (ALARP)

The American Petroleum Institute (API) Recommended Practice 754 provides a Tier system for capturing leading and lagging process safety indicators. API 754 [1] takes the ALARP graphic and flips it. Figure 2 delineates incident tiers based on their consequence levels. Tier 1 represents a group of lagging metrics and encompasses the highest impact process safety events. Tier 4 provides leading metrics that can highlight persistent deviations from expected behaviors and conditions. This tiered structure can be likened to the Iceberg graphic, where the visible top represents rare but highly impactful incidents while descending into the iceberg mirrors the increasing frequency of events with diminishing visibility.

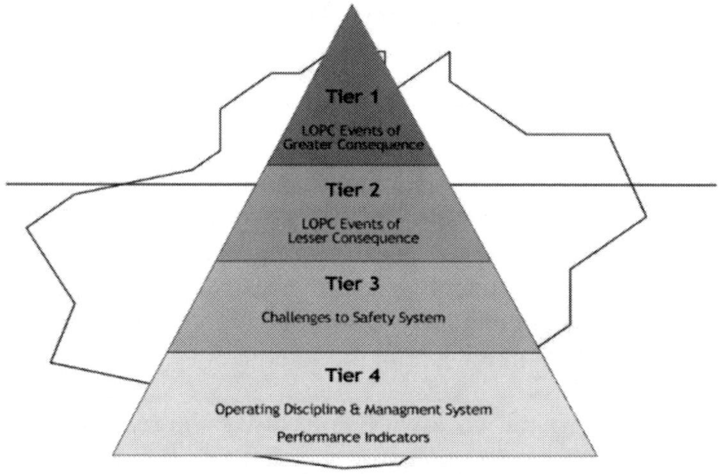

Figure 2. API 754 and the Iceberg

Reactive process safety leverages Tiers 1 and 2 data to initiate intermittent improvement measures. The API 754 framework encourages users to capture data and assess trends associated with Tiers 3 and 4, so that fewer Tiers 1 and 2 events occur. Proactive process safety employs Tiers 3 and 4 data to drive this continuous improvement approach. These tiers provide real-time information about the health of administrative and engineered safeguards. Responding to leading indicators enhances overall process efficiency by minimizing events that could disrupt production and compromise safe operations. Quality data is needed to ensure that gaps become visible.

2 Challenge to Meaningful Risk Assessment

Estimation in any context, whether risk assessment or other analytical process, relies fundamentally on the quality of the underlying data. For estimations to be reliable and meaningful, the data must meet specific criteria, namely being credible, justifiable, applicable, and relevant. Credibility ensures that the data source is reliable and trustworthy. Justifiable requires a sound rationale for data selection and inclusion. Applicability demands that the data is suitable for the specific context or purpose. Relevance emphasizes the importance of the data aligning with the particular needs of the estimation process.

Failure to adhere to these criteria can result in a common pitfall known as "garbage in, garbage out." In this scenario, the accuracy and validity of the estimation process are compromised because the initial data inputs are flawed, of poor quality, excessively filtered, or improperly aggregated [2]. Essentially, the output of any estimation is only as good as the data upon which it is based. Therefore, ensuring that data meets the criteria of credible, justifiable, applicable, and relevant is imperative to prevent the collapse of the estimation process into a state where the output is misleading or fraudulent. This underscores the critical importance of data quality to estimate validity.

3 Using Tier 3 and 4 Data for Safeguards

The input data for process safety risk assessments can be classified into two general categories: data regarding the frequency of the failures that lead to initiating events and data regarding the availability of the safeguards. Records offering insight into these categories should be captured and analyzed to support the Tier 3 and 4 metrics.

Tier 4 Data and Metrics for Safeguards

Monitoring work orders, bypasses, and test deferrals provides valuable insights into a site's Operation and Maintenance culture.

<u>Work Orders – Diagnostics</u>

Monitor maintenance work orders triggered by diagnostic alarms and notifications. The frequency of detected failures can be determined using quantitative analysis. However, faster feedback can be obtained from investigating repeat failures. For example, a safeguard failing in a particular application more than once yearly signals a deviation from industry expectations. A recurring pattern with a specific technology indicates a problem that needs to be investigated and resolved. The safeguard may be incompatible with the particular application or operating environment. Responding to signals associated with repeat failures and failure patterns fosters continuous improvement.

Bypass Use

Monitor the application and removal of bypasses. Safeguards are unable to operate as intended when bypassed. Tracking bypass records and bypass time enables determining the total time the safeguard spent in bypass per approved bypass and the frequency of bypass occurrence. These metrics help identify patterns or trends related to bypass duration, allowing procedure optimization to minimize the time safeguards spend in this vulnerable state. High bypass frequencies may indicate underlying issues, necessitating a closer examination of the root causes and identification of potential improvements to enhance safeguard availability.

Test Deferrals

Monitor the percent extension and deferral period when proof tests are delayed. The planned intervals for tests are determined based on estimated reliability, and any deviation from these intervals, e.g., the deferral of tests, can have implications for safeguard performance and reliability. The probability of a safeguard failing to operate as specified increases as the time in deferral extends, and the risk of an incident increases. The deferral interval should be tracked and reported as a metric. Deferral risk should be reviewed and approved by management. Repeat deferrals of maintenance, especially deferring online testing or leaving faulted devices out of service, warrant further assessment. Many deferrals are triggered by resource availability, scheduling challenges, or other operational constraints. Addressing these issues may enable maintenance to proceed as planned, eliminating the additional deferral risk.

Tier 3 Data and Metrics for Safeguards

Tier 3 metrics emphasize monitoring the performance of safeguards during abnormal operations and learning from failures identified through testing. The lessons learned can generate rapid safety improvements at minimal cost.

Demands on Safeguards

Safeguards are crucial in maintaining safety when a process deviates from its normal operating range. When abnormal operations occur, it often leads to subpar production quality and necessitates rework. By actively monitoring the utilization of safeguards, a site gains insights into potential issues and can proactively introduce adjustments to process control or procedures. Ultimately, the focus on safeguard use improves process profitability, creating a more efficient and reliable production environment.

Failure on Test

Monitor safeguard failures identified through testing. Safeguards are tested against pass/fail criteria defined by safety requirements. Testing should be performed frequently enough to detect and correct impending wear-out or progressing degradation before the safeguard reaches the point of failure. When a safeguard fails the test, it signals a significant issue: the process had likely been running for some time with a degraded or defeated safeguard, posing a serious safety concern. Swift identification and remediation of such failures are paramount for mitigating safety risks and averting potential hazards. Failures identified during proof testing can be tracked and trended to generate bad actors and repeat failure lists. Proof test records can be used to identify bad actors, repeated failure, patterns of failure, and safeguard failure rates. Analyzing these records contributes to a comprehensive understanding of safeguard availability and facilitates informed decision-making for ongoing safety improvements.

4 Determining the Failure Rate from Test Records

Users can collect and analyze test records to verify that the designed-for risk reduction is achieved in the field. The analysis needs credible, justified, applicable, and relevant failure data. Determining the applicability and relevancy requires the classification of the failure data. First, does the failure prevent the equipment under test from performing its safety function? If so, this is a dangerous failure that needs further analysis. Second, is the root cause of the failure a functional failure of the safeguard or a human error? Functional failure should be analyzed quantitatively, while human error should trigger an examination of administrative controls. The analysis also needs to be mindful of some underlying principles, such as confidence, homogeneity, data exclusion, and data aggregation, that are essential for the results to be credible and justified.

Confidence and Homogeneity

The dangerous failure rate, λ_D, should be estimated not as a point estimate but as the upper bound of a confidence interval of no less than 70%. Typically, analysts will use a one-sided confidence interval. (More detailed information about confidence intervals and their interpretations can be found in [3] Appendix A). The appropriate formula for the upper bound of a one-sided 100x% confidence interval for λ_D (in FITS = 1 failure/10^9 hrs) is

$$\lambda_D = 10^9 \text{ chiinv}(1\text{-}x, 2(n+1))/ (2\tau) \tag{1}$$

where chiinv is the Excel function for $\chi 2$ inverse, n is the number of dangerous failures, and τ is the total operating time for the data set.

Additionally, there is an essential statistical *requirement* on the data set being analyzed for the output of Eqn. 1 to be valid, viz., **all equipment contributing to the data set must have the same constant failure rate.** This latter condition is also known as homogeneity of the data. Lastly, τ should be at least 1,000,000 operating hours.

Truly homogeneous data sets are difficult, but not impossible, to find. They are produced at a single site from data collected from identical or nearly identical equipment (manufacturer and model/series) operating in the same environment and subject to the same safety and maintenance culture. However, when homogeneous data is unavailable, it is often possible to identify nearly homogeneous data sets. This data is produced by non-identical equipment of similar technology, operating in similar applications and environments, and maintained under similar safety cultures. While this collection of equipment may not have identical failure rates, the similarities are sufficient to support the estimation of the failure rates for the site.

Some examples may clarify the concept of nearly homogeneous. Similar technologies include designs, design parameters, and material construction. For example, ball valves and gate valves have inherently different designs. Actuated pressure relief valves differ from spring-operated pressure relief valves (SOPRV). Even within a data set consisting entirely of SOPRV, different failure rates prevail between soft-seated and hard-seated SOPRV, even for otherwise identical designs.

A similar application means that all the equipment in the data set functions in essentially the same way. For example, all equipment operates in control functions (where mechanical parts are subject to frequent motion), or all equipment operates in safety functions (where mechanical parts are stationary for extended periods of time). Another application example is equipment that opens on-trip vs. equipment that closes on-trip, even if both are in safety functions.

Operating environments include ambient temperatures and pressures, vibration, interference, and exposure to process fluids. Thus, cabinet-mounted temperature-controlled equipment will likely have a different failure rate than the same equipment deployed outdoors. Similarly, equipment exposed to corrosive fluids will likely have a different failure rate than the same equipment exposed to noble gases.

The Operations and Maintenance culture can impact the failure rate. So, data from a single site over a time period during which it is reasonable to assume the safety/maintenance culture was relatively constant supports a claim that the data are nearly homogeneous. It is possible to extend the data aggregation to an entire company when it is demonstrated that the company's sites share similar operations and maintenance practices.

Finally, the data should be relevant to the analysis being performed. The intent of this analysis is to determine the safeguard's achieved performance, so the data set should include all failures that cannot be eliminated by a change in the specification. Failures due to the properties of the process, the operating environment, and the Operation and Maintenance culture are part of the real world, and their analysis can elevate the need to find new solutions to eliminate the failures or to control them better, lowering their likelihood of reoccurrence.

Analyzing Proof Test Records

Good proof test data analysis begins with sorting the proof test records into homogeneous or nearly homogeneous data groups. End users who have collected significant amounts of proof test data (as required by the standards) but who do not routinely analyze that data may need more expertise to perform this first crucial step well. They might consider contracting an experienced third party to assist in this endeavor until in-house expertise is developed. ISO-14224 [4], ISO-TR12489 [5,6], and ISA-TR84.00.02 [7] provide guidance on data classification and analysis.

Next, the operating time for each equipment record needs to be established. If the proof test record (or other maintenance record) records the date of the last test and the date of the current test and the equipment is known to be operational for the entire time between tests, then establishing operating time is trivial. However, if the equipment was only in operation part of the time between known test dates, then a reasonable guestimate of operation time must be made. Because this operating time increases τ and decreases λ_D, it is essential that a guestimate be conservative (meaning smaller rather than larger). τ is the sum of the operating times for each proof test record.

Finally, the number of dangerous failures, n, needs to be counted. All identified dangerous failures discovered by proof test must be presumed to have been present in the field and, therefore, have contributed to the achieved risk reduction factor (RRF). These must be included in the count for n. It is possible that in some cases, the records clearly show a failure of the proof test but with insufficient documentation to determine if the failure was safe or dangerous. The authors suggest that all such failures be classified as dangerous to obtain a maximum upper bound for λ_D.

With these three steps accomplished, λ_D can be calculated from Eqn. 1.

Examples of Good and Bad Proof Test Data Analyses

Consider the data in Table 1. This data was previously published in Ref. 8 Table 4 and Ref. 9 Table 1. In these cases, the data sets were already established to be homogeneous or nearly homogeneous.

Table 1. Summary of homogeneous and nearly homogeneous data sets used in the examples.

Row	Site	# Dangerous Failures, n	Total Operating Time, τ, hrs	Notes
1	A	1	3,726,288	Identical soft-seat SOPRV in clean service
2	A	9	4,468,301	Nearly identical hard-seat SOPRV in clean service
3	B	1	1,418,304	Identical hard-seat Actuated PRV in clean service
4	C	0	1,051,200	Identical gate valves in severe service

Table 2 shows the good (valid) and bad (invalid) proof test data analyses. Because each row in Table 1 is homogeneous or nearly homogeneous, Eqn. 1 can be applied directly, yielding valid results for the 70% upper bound confidence interval for λ_D. The results are summarized in the first four rows of Table 2. In those rows, note the relatively wide range (655-2548 FITS) of valid values for the 70% upper bound on λ_D for the various valves.

Table 2. Summary of good and bad analyses of proof test data from Table 1.

Data Source rows in Table 1	n	τ, hrs	Upper Bound 70% λ_D, FITS	Notes
1	1	3,726,288	655	Analysis valid
2	9	4,468,301	2548	Analysis valid
3	1	1,418,304	1720	Analysis valid
4	0	1,051,200	1145	Analysis valid
1+2+3+4	11	10,664,093	1270	**Analysis invalid – data set non-homogeneous**
1+2+3+4 with some failures neglected	4	10,664,093	552	**Analysis invalid – data set non-homogeneous *and* 7 field failures not included in *n***

The last two rows of Table 2 show the invalid results from the proof test data analysis. In the first invalid row (row 5 of Table 2), the data from the four homogeneous and nearly homogeneous data sets are aggregated into a super data set, which is inherently non-homogeneous. Because of the non-homogeneous nature of the aggregated data set, Eqn. 1 cannot be validly applied. However, this does not deter some analysts from plugging numbers into Eqn. 1 and obtaining the invalid estimate of λ_D equal to 1270 FITS – a number that does not accurately characterize the safety performance of any of the four valves in question.

Lastly, the final row aggregates the valid data into an invalid non-homogeneous data set and removes seven of the total eleven failures on the basis that they were the user's fault not a random failure of the equipment. This final invalid analysis produces an upper bound estimate for λ_D of 552 FITS, significantly less than the estimates produced by valid analyses for the four valves in question. An analysis of these removed failures revealed that they were largely due to the operating environment and the application. These failures should not be removed from the analysis as the failure mechanism cannot be eliminated. This type of invalid analysis is especially dangerous because it does not measure the safety achieved in the field. Instead, it provides a false sense of safety based on what might be (but is not) achieved [2].

Potential Consequences of Using Invalid λ_D in Safety Calculations

Estimates of λ_D from field failure data have several uses. One is to determine the actual RRF achieved in the field so that it may be compared to the target RRF to determine if any changes in the safety system's testing and maintenance schedule/practices are required. Another is to support prior use (or proven-in-use) failure rates to be used in future designs.

Assume that a reasonable approximation of PFDavg is given by

$$PFDavg \cong \lambda_D \, TI / 2 \qquad (2)$$

where TI is the proof test interval. Then RRF, as a function of TI, is approximated by

$$RRF = 1/PFDavg = 2/(\lambda_D \, TI). \qquad (3)$$

Given the calculations for λ_D as shown in Table 2, Table 3 provides the corresponding RRF computed for various proof test intervals.

Compare the RRF results for a single example of using both valid and invalid λ_D values. Consider the nearly identical hard-seated SOPRV valves (from row two of Table 1) with a TI of three years. Assume the designed-for RRF was 40. The valid analysis shows that the RRF actually achieved in the field is 30 and therefore, the site needs to reevaluate the safety provided by this valve. The proof test schedule is an obvious place to start. Reducing TI from three years to two years would, based on the site's data, make the achieved RRF 45 under current O&M practices and thus greater than the designed-for RRF. This is an appropriate conclusion based on the available field failure data.

On the other hand, supposed that either of the invalid λ_D values (rows five and six of Tables 2 and 3) were used to determine the RRF actually achieved in the field. In these cases, the RRFs at TI equal to three years would be 60 and 138, respectively, suggesting not only that the designed-for RRF of 40 is being met but also that the site can justify increasing its proof test interval to at least four years and perhaps five years. These invalid conclusions provide a false sense of safety.

The reader can deduce other examples of the consequences from use of invalid λ_D values from Table 3. However, whether the estimated λ_D is used to verify the achieved RRF or as a prior use failure rate for future designs, λ_D values from invalid analyses will almost certainly lead to conclusions that safety is greater than that actually achieved.

Table 3. Summary of RRF for various proof test intervals based on the good and bad analyses of proof test data from Table 2.

Data Source rows in Table 1	*n*	τ, hrs	Upper Bound 70% λ_D, FITS	RRF as a function of TI in years				
				TI = 1 yr	TI = 2 yrs	TI = 3 yrs	TI = 4 yrs	TI = 5 yrs
1	1	3,726,288	655	349	174	116	87	70
2	9	4,468,301	2548	90	45	30	22	18
3	1	1,418,304	1720	133	66	44	33	27
4	0	1,051,200	1145	199	100	66	50	40
1+2+3+4	11	10,664,093	1270	180	90	60	45	36
1+2+3+4 with some failures neglected	4	10,664,093	552	413	207	138	103	83

5 Conclusion

Monitoring API 754 Tier 3 and 4 data is valuable for understanding how well safeguards are managed. Tier 4 records encompass diagnostic work orders, bypass records, and test deferral time, offering detailed insights into site culture and attitude toward safeguards. Tier 3 records monitor the use of safeguards and how well they perform under test. Learning from Tier 3 and 4 data benefits the risk management in five critical ways:

1. Representation of Actual Conditions: API Tier 3 and 4 data accurately reflects the variability and complexity of actual conditions in each application. Using this data allows systems to be designed and tested under conditions representative of the environment in which the safeguard is used.
2. Identification of Variability: API Tier 3 and 4 data helps identify the natural variability occurring with safeguard performance. This variability may arise from different sources, such as environmental factors, material variations, or human factors. Understanding and accounting for this variability is crucial for establishing a realistic process safety strategy.
3. Adaptability to Changes: Industries and processes are dynamic, and they can change over time due to advancements in technology, changes in regulations, or shifts in purchasing preferences. API Tier 3 and 4 data allows the site personnel to adapt to these changes by providing the latest and most relevant information.

4. Detection of Unforeseen Issues: API Tier 3 and 4 data can identify unforeseen issues or anomalies that may not be apparent to the manufacturer or in other applications. These issues can include unexpected interactions between safeguards, variations in operating conditions, or other factors that may impact the safeguard's operation.
5. Performance Evaluation: API Tier 3 and 4 data is crucial for evaluating the actual performance of a safeguard. It helps assess whether operating and maintenance practices are sufficient to ensure that safeguards meet performance expectations under realistic conditions.

Regular data collection and analysis provide a defensible basis for refining risk assessments and implementing necessary administrative and engineering safeguards adjustments. In proof test data analysis, ensuring homogeneity is key to valid results. Examples illustrate the pitfalls of aggregating data into heterogeneous data sets or selectively neglecting failures, emphasizing the importance of accurate data collection and analysis to ensure that "garbage in-garbage out" is avoided. Identified gaps in the safeguards trigger revised risk assessments, fostering a continuous improvement cycle that adapts to evolving risks and ensures the effectiveness and sustainability of the safety measures in place.

6 References

[1] API 754, "Process Safety Performance Indicators for the Refining and Petrochemical Industries," 3rd edition, 2021.

[2] Summers, A.E., Bukowski, J.V., Goble, W.M., Limaye, R., and Gutierrez, J., "Summary of the Panel Discussion-Using Credible Failure Rates for SIS Ensures Process Safety," Process Safety Progress, 42(4): 603-611, December 2023.

[3] Bukowski, J.V. and A.E. Summers, "Statistical concepts and data requirements for valid failure rate estimation," Process Safety Progress, 10.1002/prs.12468, 42(3): 516-522, 2023.

[4] ISO 14224, "Petroleum, Petrochemical and Natural Gas Industries: Collection and exchange of reliability and maintenance data for equipment," 2016.

[5] ISO TR12489, "Petroleum, Petrochemical And Natural Gas Industries: Reliability Modeling and Calculation Of Safety Systems," 2013.

[6] Estecahandy, M, J.T. Selvik, J. Signoret, A.E. Summers, S.L. Isaksen, F. Brissaud, N. Clave, "Industry Application and Benefits of ISO/TR 12489 for Reliability Modelling and Calculation of Safety Systems," OTC-32468-MS, Offshore Technology Conference, Houston, TX May 1-4, 2023.

[7] ISA, ISATR84.00.02-2022, Safety Integrity Level (SIL) Verification of Safety Instrumented Functions, Research Triangle Park, NC, 2022.

[8] Bukowski, J.V., W.M. Goble, R.E. Gross, S.P. Harris, "Analysis of spring operated pressure relief valve proof test data: findings and implications," Process Safety Progress, 10.1002/prs.12006, 37(4): 467-477, 2018.

[9] J. V. Bukowski, W.M. Goble, S.P. Harris, "Comparison Of Valve Failure Rates Estimated From Field Failure Data To Those Predicted By Calibrated FMEDA," Proceedings Of The ASME 2021 Pressure Vessels And Piping Division Conference, July 2021.

The Need for Proper Planning
Case Studies from Challenging DHAs

Michelle R. Murphy
Mica, LLC
murphy@micadhm.com

Keywords: combustible dust, dust hazard analysis, DHA, case studies, combustible dust characterization, dust testing

Abstract

Dust hazard analyses (DHAs) have been conducted for over 10 years in many industries. The approach and format of these analyses has evolved due to changes in the National Fire Protection Association code as well as recent guidelines from the Center for Chemical Process Safety. The author has reviewed many different DHAs conducted throughout the US and the world and personally conducted over 120 studies. Structured in the format of case studies, the presenter shares hard-earned lessons from her own experience to shed light on some of the pitfalls that can be avoided with proper planning. These lessons learned touch on some of the challenging decisions that are made during the preparation phase of the DHA. Topics to be addressed include material combustibility, application of worst-case test data, mechanical sparks as an ignition source, inconsistent risk ranking, and prescriptive vs. performance-based approach approach. The author is sharing these lessons learned in hopes of helping others better plan their DHAs for improved (i.e., actionable) results, with less time and expense.

1 Introduction

Dust hazard analyses (DHAs) are required by relevant the National Fire Protection Association (NFPA) codes for all industries handling combustible dust [1], [2], [3], [4], [5]. NFPA codes are considered recognized and generally accepted good engineering practices (RAGAGEP) and are referenced by the Occupational Health and Safety Administration (OSHA) when citing facilities under their general duty clause [6]. Similar assessments are required by ATEX [7] in Europe and by DSEAR [8] in the UK. Other areas, such as South America, Asia and the Middle East follow one or more of these standards.

DHAs have been conducted for over 10 years in many industries. The approach and format of these analyses has evolved through the years due to changes in NFPA code as well as recent guidelines from the Center for Chemical Process Safety [9]. The author has reviewed many different DHAs conducted throughout the US and globally and personally conducted over 120 studies. Structured in the format of case studies, this paper shares hard-earned lessons from her own experience to shed light on some of the pitfalls that can be avoided with proper planning. As one of the principles of lean management, a "check" step is important to evaluate a management system with the goal of taking action to correct or improve the process as shown in Figure 1. These shared case studies are part of a check step where actions can be taken to improve the overall DHA process. It is hoped that others will learn from these mistakes to perform better studies, with less resources, saving both time and expense.

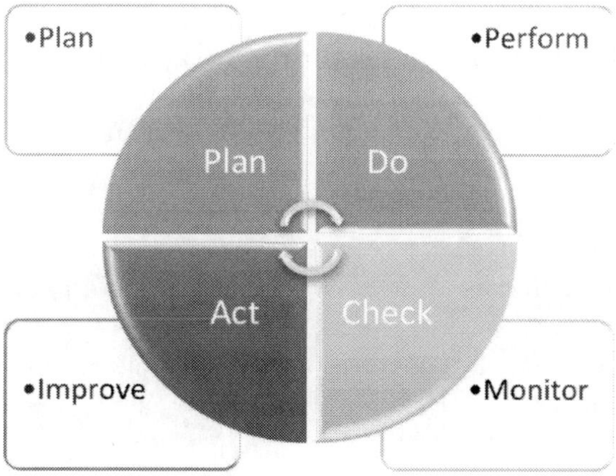

Figure 1. Plan-do-check-act cycle [10]

2 Dust Hazard Analysis Process

Similar to many analyses, DHAs are split into steps of preparation, on-site analysis, and follow-up. In the preparation step, documents are provided and reviewed to determine the scope of the study, the team is formed, pre-analysis meetings are conducted, and the team is prepared for their responsibilities in the analysis. The on-site step begins with an opening meeting, typically a walk-through of the process, followed by team meetings to complete the DHA. In the follow-up step any open items are resolved and a report of the findings is written. Once the site receives the report, it then drives improvements by developing and implementing action plans to address the recommendations from the analysis. For more detail on the DHA approach see Murphy and Borene [11] and the CCPS Book, *Guidelines for Dust Hazard Analysis* [9].

All phases of the DHA are important to its success, but careful preparation sets the critical path. Five case studies are presented here detailing failures in preparation that derailed the analysis, increased the time and cost of the DHA, and in one case, resulted in loss of credibility in the DHA process. Decisions made during the preparation phase, such as determining combustibility of materials handled and which combustible dust test data to apply will have a substantial impact on the success of the DHA. These five case studies help to reinforce the need to properly prepare for the analysis, including:

1. Fully understanding the basic material characteristics (i.e., chemical composition, particle size range, and moisture content) of the dust in each of the pieces of equipment and building spaces to be evaluated
2. Conducting testing of samples that represent the materials handled in the process
3. Completing the appropriate test for the application and considering the impact of operating parameters on the test results (e.g., test results may need to be adjusted to reflect operating parameters)

4. Determining the DHA approach in advance and applying the same approach in each analysis within the company (e.g., same risk criteria to allow for comparison between plants)
5. Considering the benefits of a risk-based team approach vs. a prescriptive audit approach

While it is common to say, "We will apply a conservative approach to process safety," it is unrealistic to do in every single case. Financial resources are limited, and they need to be spent where they will make a significant reduction in risk. It is better to apply "a reasonably conservative approach that considers the material and the process to identify the true hazard." Better preparation provides for a better analysis which leads to real risk reduction.

3 Case Studies

The lessons learned are presented as a series of case studies from actual DHA experiences. The case studies presented are:

1. Material combustibility
2. Application of worst-case test data
3. Mechanical sparks as an ignition source
4. Inconsistent risk ranking
5. Prescriptive vs. performance-based approach

2.1 Case Study 1: Material combustibility

Facility – Production of plastic food containers

Operation – Regrind off-spec material, transport through a pneumatic conveyor

Difficulty – Regrind material was larger than the typical powder (>> 500 micron) with some fines. Site personnel questioned whether the material was explosible as handled in the process. During the DHA, much time was wasted in debating if the material in the regrind process was "explosible." To determine if the material transported in the pneumatic conveyor was truly explosible and measure the minimum ignition energy (MIE), a process sample was collected. This sample included the full particle size range of the process (up to 1" in diameter, with fines). This sample was subject to a series of tests, maintaining the full particle size range of the material in the process, including explosion severity (K_{St}) and MIE.

The sample measured as explosible.

After the testing, the DHA had to be revisited to accurately answer the questions regarding the presence of hazards, existing safeguards, and, where needed, make recommendations for additional safeguards.

Improvement – Understand the particle size of material in the process equipment. Conduct combustible dust testing of samples that represent the materials in the process before beginning the DHA.

Benefit – Time saved, avoidance of unnecessary explosion protection systems or implementation of necessary explosion protection systems, significant cost savings or reduced risk

2.2 Case Study 2: Worst-case test data

Facility – Production of plastic food containers

Operation – Pneumatic transport of pellets and fines

Difficulty – Corporate engineer wanted to reduce testing costs. Samples were gathered from 5 locations in the process and sent for particle size analysis. The finest sample was submitted for additional dust characterization testing. This sample had been collected from a dust collector.

The corporate engineer intended to evaluate all the equipment in the process using the dust characterization data of the dust collector sample.

Applying this worst-case data to all equipment in the process resulted in numerous recommendations for explosion protection. The facility people began to question if this was the right approach as the material in the process was much larger than the material tested in the worst-case data. Much time was wasted debating the validity of this data as "representative of the process". Facility people were frustrated with the approach and began to question the validity of the DHA.

Improvement –Test multiple representative samples before beginning the DHA to assure representative data can be applied in the analysis.

Benefit – Goodwill, credibility, time and money saved

2.3 Case Study 3: Incendive mechanical sparks

Facility – Production of food intermediates

Operation – Mechanical conveying

Difficulty – Site wanted to evaluate the potential for mechanically generated sparks to provide an incendive spark in the process. This approach can be found in the International Section for Machine and System Safety document, *Dust Explosion Prevention and Protection for Machines and Equipment Basic Principles* [12]. It was also published earlier by Eckhoff [13].

The approach involves assigning an equivalent electrical energy E_Q to the mechanically generated spark based on the minimum ignition temperature of the dust cloud, MIT_{cloud}. Once known, the E_Q is compared with measured minimum ignition energy (MIE). The client had measured MIE data for the materials in their process.

The initial evaluation completed by site personnel applied a standard MIE value, measured at room temperature. This value did not represent the hazard in the process.

Improvement – The E_Q approach is valid for evaluation of mechanically generated sparks when the correct data is used in the evaluation.

1. It is important that the MIE data used in this type of analysis is measured with an induction test (to represent ignition energy generated from an impact spark rather than ignition energy generated from an electrostatic spark)

2. If the process is at elevated temperature, the measured MIE value needs to be corrected for temperature

Benefit – Avoidance of unnecessary explosion protection equipment, significant cost savings

2.4 Case Study 4: Inconsistent Risk Ranking

Facility – Production of finished food

Operation – Supersack unloading and pneumatic transfer

Difficulty – Application of a different risk matrix. Site wanted to use a local company to conduct their DHA. The risk matrix applied was different from the one used at other sites in the corporation. This led to very different risk rankings for similar findings.

Improvement – Adopt a corporate risk matrix that is communicated to all facilities for application.

Benefit – Consistent key performance indicators (KPIs) across all sites

2.5 Case Study 5: Prescriptive vs. Performance-Based Approach

Facility – Production of food intermediates

Operation – Full process

Difficulty – Concerned about the looming deadline to complete their DHAs, the client hired a firm to conduct all their DHAs using a prescriptive analysis. This type of analysis applies all of the prescriptive requirements in the NFPA codes to equipment design and operation. As a result, any recommendations for explosion protection were developed. The client company spent significant money on improvements which did not reduce the risk of their operations.

Improvement – Conduct a performance-based DHA, where the facility personnel are involved in each step. Conducting a performance (or risk)-based DHA is described in references [9] and [11], including application of a risk matrix to the defined scenarios. In these approaches the assignment of consequence and likelihood is based on the team's experience with the equipment in their operation. For a more quantitative approach, Zhao and Weber [14] present an equation for estimating scenario likelihood as a function of initial event, probability of ignition, probability of an ignitable dust cloud, and the estimated time at risk. In a risk-based DHA, site personnel are involved in the discussions and help generate the recommendations leading to a more unified solution. They partake in development of the scenario with a detailed analysis of the causes and consequences. They can explain and defend the recommendations up the management chain.

Benefit – Capital funding resulting in true risk reduction. In one different example, an additional success: an operations person was so impressed by what was learned during the DHA that he himself said he would share the learnings with the other people on the floor. Large increase in combustible dust understanding and reduction in hazard level. Risk reduction dollars spent where true risk reduction is realized.

4 Conclusions

As these examples demonstrate, it is imperative to take a thoughtful approach to DHAs, including:

1. Understand the materials in the process, gather representative samples, and conduct the necessary tests to be able to define the hazard during the DHA

2. Assure that ignition source avoidance strategies reflect the materials in the process and the design and operation of the process equipment and systems

3. Define a risk ranking approach that can be applied to all DHA activities, throughout the company, to allow comparison of consistent KPIs

4. Apply a performance-based, team approach to DHAs to assure impactful risk-reduction that facility personnel believe in and generates a higher-level understanding of combustible dust hazards, risk and mitigations

Significant time and money are spent preparing and completing DHAs. Orders of magnitude more time and money are spent in responding to recommendations and taking actions to add additional safeguards. Thoughtful planning and preparation will ensure that all these resources are used to define and implement impactful risk reduction actions.

5 References

[1] NFPA 652, *Standard on the Fundamentals of Combustible Dust*, 2019 Edition, National Fire Protection Association, Quincy, MA, May 4, 2018.

[2] NFPA 654, *Prevention of fire and Dust Explosions from the Manufacturing, Processing, and Handling of Combustible Particulate Solids*, 2020 Edition, National Fire Protection Association, Quincy, MA, August 2019.

[3] NFPA 61, *Standard for the Prevention of Fires and Dust Explosions in Agricultural and Food Processing Facilities*, 2020 Edition, National Fire Protection Association, Quincy, MA, November 2019.

[4] NFPA 484, *Standard for Combustible Metals*, 2019 Edition, National Fire Protection Association, Quincy, MA, May 2018.

[5] NFPA 664, *Standard on the Prevention of Fires and Explosions in Wood Processing and Woodworking Facilities*, 2020 Edition, National Fire Protection Association, Quincy, MA, November 4, 2019.

[6] OSHA Act of 1970, Section 5(a)(1)

[7] ATEX Directive 2014/34/EU, April 20, 2016.

[8] Dangerous Substances and Explosive Atmospheres Regulations 2002 (DSEAR) (SI 2002 No.2776). Health and Safety Executive, 2002

[9] CCPS Guidelines for Dust Hazard Analysis, Wiley, May 2017.

[10] WHAT IS THE PLAN-DO-CHECK-ACT (PDCA) CYCLE? https://asq.org/quality-resources/pdca-cycle, accessed January 27, 2023.

[11] Murphy, M. and Borene, M. How to Conduct a Dust Hazards Analysis. *Process Safety Progress.* 2018, October 2018.

[12] International Section for Machine and System Safety, *Dust Explosion Prevention and Protection for Machines and Equipment*, ISSA Prevention Series No. 2033 (E), International Section of the ISSA for Machine and System Safety Dynamostr. 7-11, D-68165 Mannheim, Germany, 2004.

[13] Eckhoff, Rolf K., *Dust Explosions in the Process Industries*, Third Edition, Gulf Professional Publishing, Amsterdam, 2003.

[14] Zhao, Fuman, Weber, Robert, *A Risk-based Dust Hazard Analysis (DHA)*, Mary Kay O'Connor Process Safety Center, 21st Annual International Symposium, October 2018.

SPRING24
+20TH GCPS
A Joint AIChE and CCPS Meeting

March 24–28, 2024
New Orleans Ernest N. Morial
Convention Center
New Orleans, LA

SPRING24+20TH GCPS
A Joint AIChE and CCPS Meeting

March 24–28, 2024
New Orleans Ernest N. Morial
Convention Center,
New Orleans, LA

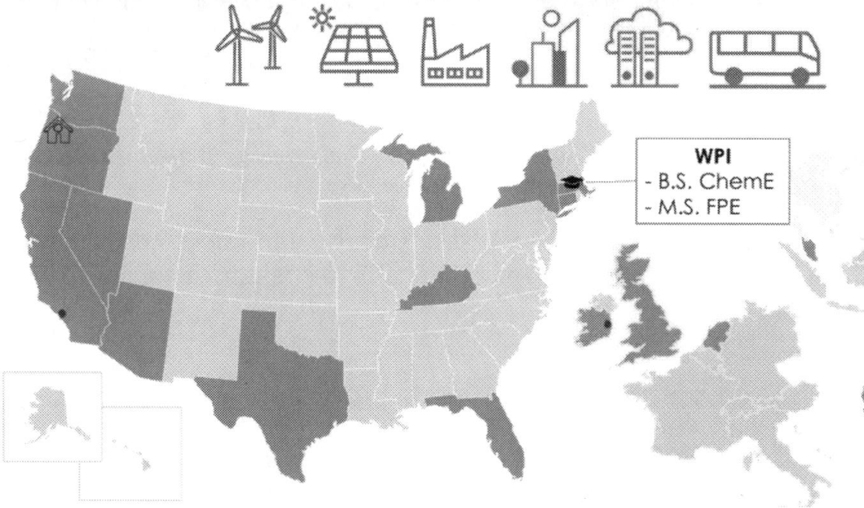

Veronica Goldsmith, PE
Arup US, Inc.

THE ENERGY TRANSITION

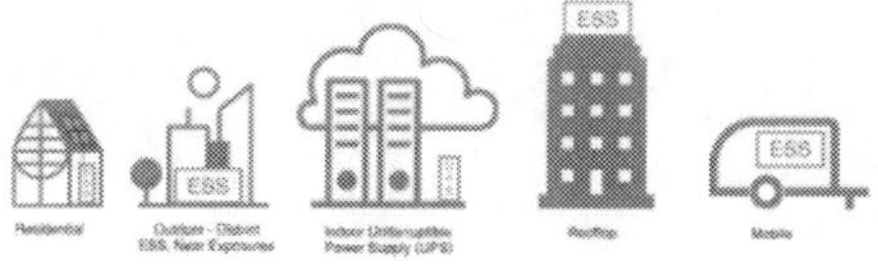

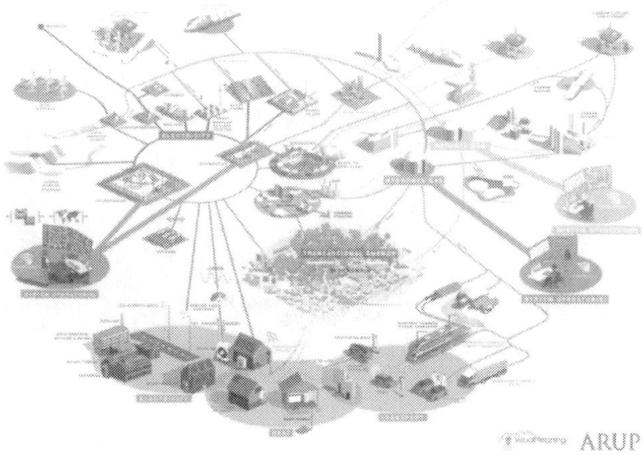

ARUP

SPRING24+20TH GCPS
A Joint AIChE and CCPS Meeting
March 24–28, 2024
New Orleans Ernest N. Morial Convention Center,
New Orleans, LA

Scaling up to from Cell to BESS

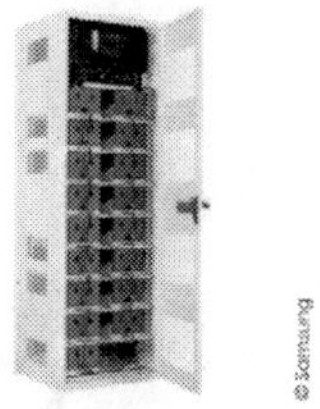

Cell

- Basic unit

Module

- Assembly architecture
- Sensors

Cabinet/Indoor Unit

- Thermal management
- Battery management system

System/Outdoor Unit

- Fire/gas detection
- Explosion control/venting
- Energy management system

SPRING24+20TH GCPS
A Joint AIChE and CCPS Meeting

March 24–28, 2024
New Orleans Ernest N. Morial Convention Center,
New Orleans, LA

Codes & Standards

Hazard Mitigation Analysis

- Evaluating the consequences of failure modes (e.g. thermal runaway, mechanical failure, BMS, protection systems)
- Needs to show that the fire is contained within the room for the minimum duration of the fire resistance rating, fire/products of combustion do not prevent occupants from evacuating, and the deflagration hazard is addressed.

SPRING24+20TH GCPS
A Joint AIChE and CCPS Meeting

March 24–28, 2024
New Orleans Ernest N. Morial Convention Center,
New Orleans, LA

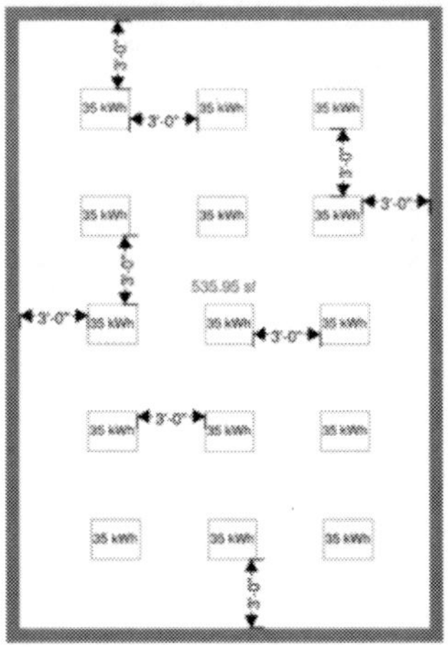

Size and Separation Requirements

Full-scale fire testing becomes necessary for most designs

Up to 50 kWh groupings must be spaced 3 feet apart. For a 525 kWh system (less than the 600kWh threshold) this becomes very space-onerous.

Prescriptive Spacing per IFC/NFPA 855 of a room with LIB cabinets

UL9540A

Full-scale thermal runaway testing

Cell

Understand the propensity of a cell to go into thermal runaway.

If thermal runaway is induced, or cell vent gas is flammable in air, tests proceed to module.

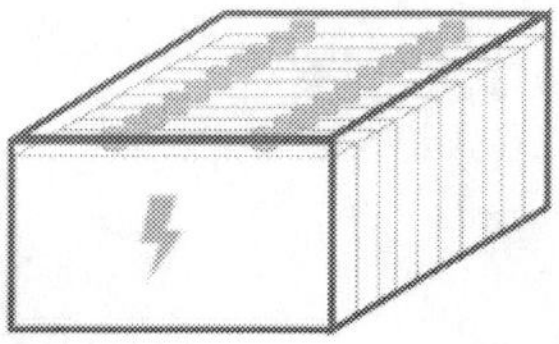

Module

Understand the ability for system architecture to contain thermal runaway.

As of the 4th edition, if the cell vent gas is flammable, tests proceed to unit level.

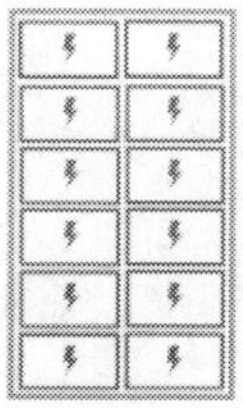

Unit

Understand the ability for the larger unit to contain thermal runaway.

Testing of indoor floor-mounted and outdoor-only units varies.

SPRING24+20TH GCPS
A Joint AIChE and CCPS Meeting
March 24–28, 2024
New Orleans Ernest N. Morial Convention Center, New Orleans, LA

Indoor vs. Outdoor Unit Level Testing

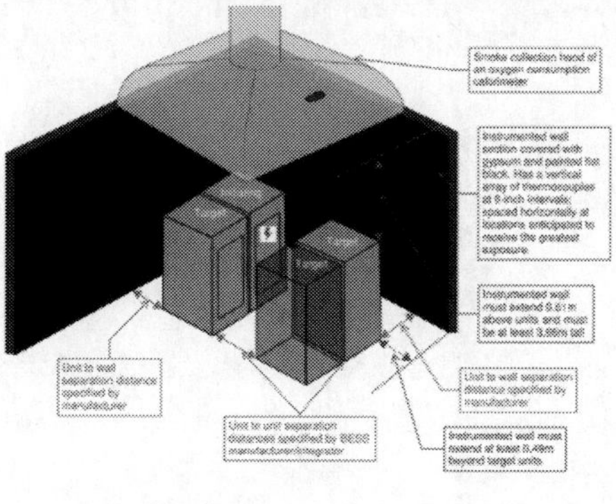

Pass Criteria: No Flaming Outside Initiating Unit, Surface temperatures < T_{vent} or 97°C, no deflagration

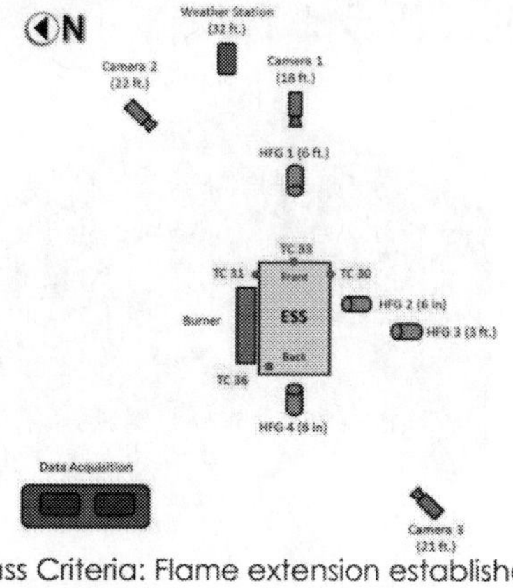

Pass Criteria: Flame extension establishes separation distance, Surface temperatures < T_{vent} or 97°C, no deflagration

SPRING24+20TH GCPS
A Joint AIChE and CCPS Meeting

March 24–28, 2024
New Orleans Ernest N. Morial Convention Center,
New Orleans, LA

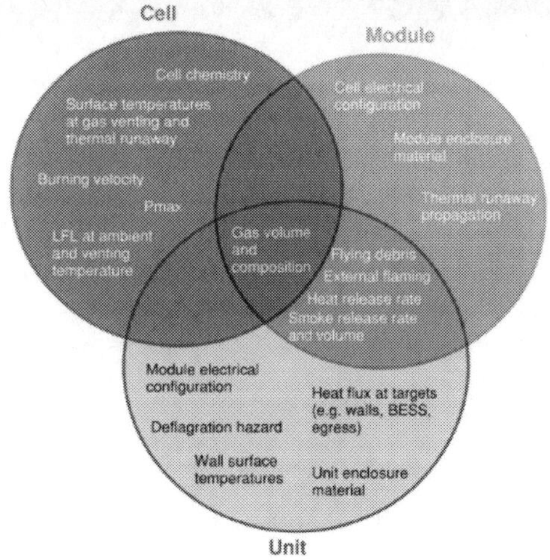

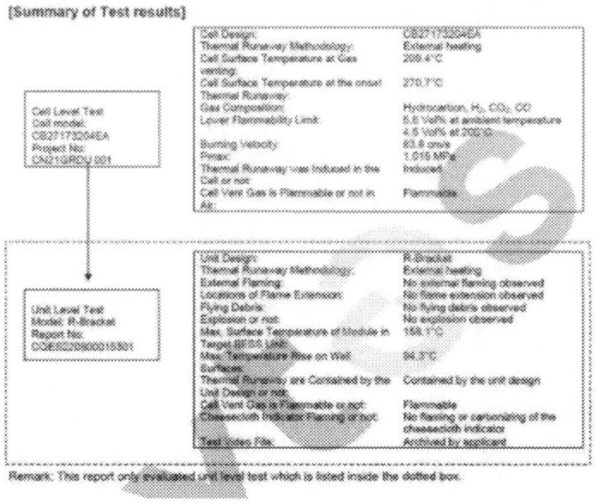

[Summary of Test results]

Cell Level Test
Cell model:
CB27173204EA
Project No:
CN21GRDU 001

Cell Design:	CB27173204EA
Thermal Runaway Methodology:	External heating
Cell Surface Temperature at Gas venting:	209.4°C
Cell Surface Temperature at the onset Thermal Runaway:	270.7°C
Gas Composition:	Hydrocarbon, H_2, CO_2, CO
Lower Flammability Limit:	5.8 Vol% at ambient temperature 4.5 Vol% at 200°C
Burning Velocity:	83.8 cm/s
Pmax:	1.015 MPa
Thermal Runaway was Induced in the Cell or not:	Induced
Cell Vent Gas is Flammable or not in Air:	Flammable

Unit Level Test
Model: R-Bracket
Report No:
CQES220900015301

Unit Design:	R-Bracket
Thermal Runaway Methodology:	External heating
External Flaming:	No external flaming observed
Locations of Flame Extension:	No flame extension observed
Flying Debris:	No flying debris observed
Explosion or not:	No explosion observed
Max. Surface Temperature of Module in Target BESS Unit:	158.1°C
Max. Temperature Rise on Wall Surfaces:	94.3°C
Thermal Runaway are Contained by the Unit Design or not:	Contained by the unit design
Cell Vent Gas is Flammable or not:	Flammable
Cheesecloth Indicator Flaming or not:	No flaming or carbonizing of the cheesecloth indicator
Test Video File:	Archived by applicant

Remark: This report only evaluated unit level test which is listed inside the dotted box.

SPRING24+20TH GCPS
A Joint AIChE and CCPS Meeting

March 24–28, 2024
New Orleans Ernest N. Morial Convention Center,
New Orleans, LA

Case Study 1

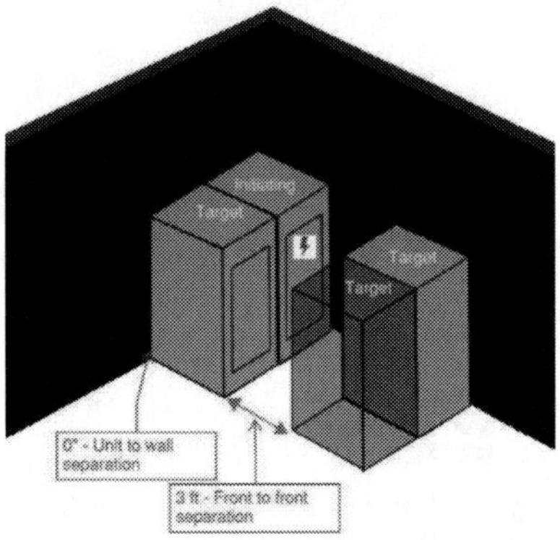

Parameter	Cell Level	Module Level	Unit Level - Indoor
Vent Temperature	132°C	N/A	N/A
Thermal Runaway Temperature	198°C	N/A	N/A
Pmax	91 psig	N/A	N/A
LFL at 25°C	8.5%	N/A	N/A
Gas composition	Carbon Monoxide – 36.2%, Carbon Dioxide – 22.1%, Hydrogen – 31.7%, Hydrocarbons – 10%	Carbon monoxide – 35% Carbon dioxide – 25% Hydrogen – 10% Hydrocarbons – 30%	Carbon monoxide – 8.9% Carbon dioxide – 2.2% Hydrocarbons - 88.9%
Gas volume	70 L	700 L	225 L
Smoke Release Rate	N/A	10 m^2/s	1.0 m^2/s
Flaming Jets	N/A	3-4ft from vent	None
Flying Debris	N/A	Sparks/embers ejected	None
Re-ignitions	N/A	No re-ignitions	No re-ignitions
Separation Distances	N/A	N/A	0" side to side 0" to wall 36" front to front
Temperatures at Targets	N/A	N/A	40°C at Target 23°C at Wall
Heat Flux at Targets	N/A	N/A	10 kW/m^2 at target
Peak Heat Release Rate	N/A	370 kW	N/A – No flaming observed

SPRING24+20TH GCPS
A Joint AIChE and CCPS Meeting

March 24–28, 2024
New Orleans Ernest N. Morial Convention Center, New Orleans, LA

Case Study 1

374 ft^3 room volume

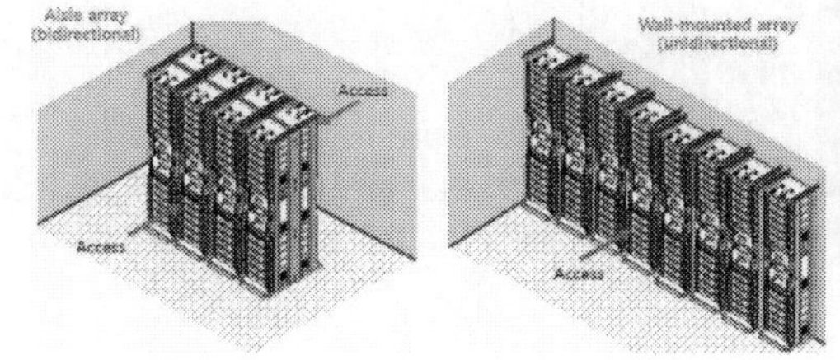

Parameter	Cell Level	Module Level	Unit Level - Indoor
Vent Temperature	132°C	N/A	N/A
Thermal Runaway Temperature	198°C	N/A	N/A
Pmax	91 psig	N/A	N/A
LFL at 25°C	8.5%	N/A	N/A
Gas composition	Carbon Monoxide – 36.2%, Carbon Dioxide – 22.1%, Hydrogen – 31.7%, Hydrocarbons – 10%	Carbon monoxide – 35% Carbon dioxide – 25% Hydrogen – 10% Hydrocarbons – 30%	Carbon monoxide – 8.9% Carbon dioxide - 2.2% Hydrocarbons - 88.9%
Gas volume	70 L	700 L	225 L
Smoke Release Rate	N/A	10 m^3/s	1.0 m^3/s
Flaming Jets	N/A	3-4ft from vent	None
Flying Debris	N/A	Sparks/embers ejected	None
Re-ignitions	N/A	No re-ignitions	No re-ignitions
Separation Distances	N/A	N/A	0" side to side 0" to wall 36" front to front
Temperatures at Targets	N/A	N/A	40°C at Target 23°C at Wall
Heat Flux at Targets	N/A	N/A	10 kW/m² at target
Peak Heat Release Rate	N/A	370 kW	N/A – No flaming observed

SPRING24+20TH GCPS
A Joint AIChE and CCPS Meeting

March 24–28, 2024
New Orleans Ernest N. Morial Convention Center, New Orleans, LA

Case Study 1

- Adapting for Vent Temperature – NFPA 86

$$LFL_{venting\ temp} = LFL_{25°C}[1 - 0.000784\,(T_{venting} - 25°C)]$$

- What about the other gas data? – Le Chatelier's

$$LFL_{mix} = \frac{1}{\sum \frac{x_i}{LFL_i}}$$

Parameter	Cell Level	Module Level	Unit Level - Indoor
Vent Temperature	132°C	N/A	N/A
Thermal Runaway Temperature	198°C	N/A	N/A
Pmax	91 psig	N/A	N/A
LFL at 25°C	8.5%	N/A	N/A
Gas composition	Carbon Monoxide – 36.2%, Carbon Dioxide – 22.1%, Hydrogen – 31.7%, Hydrocarbons – 10%	Carbon monoxide – 35% Carbon dioxide – 25% Hydrogen – 10% Hydrocarbons – 30%	Carbon monoxide – 8.9% Carbon dioxide - 2.2% Hydrocarbons - 88.9%
Gas volume	70 L	700 L	225 L
Smoke Release Rate	N/A	10 m^3/s	1.0 m^3/s
Flaming Jets	N/A	3-4ft from vent	None
Flying Debris	N/A	Sparks/embers ejected	None
Re-ignitions	N/A	No re-ignitions	No re-ignitions
Separation Distances	N/A	N/A	0" side to side 0" to wall 36" front to front
Temperatures at Targets	N/A	N/A	40°C at Target 23°C at Wall
Heat Flux at Targets	N/A	N/A	10 kW/m² at target
Peak Heat Release Rate	N/A	370 kW	N/A – No flaming observed

SPRING24+20TH GCPS
A Joint AIChE and CCPS Meeting

March 24–28, 2024
New Orleans Ernest N. Morial Convention Center,
New Orleans, LA

- Think critically about the numbers provided
- Adjust for vent temperature
- Consider failure-scenarios and room congestion

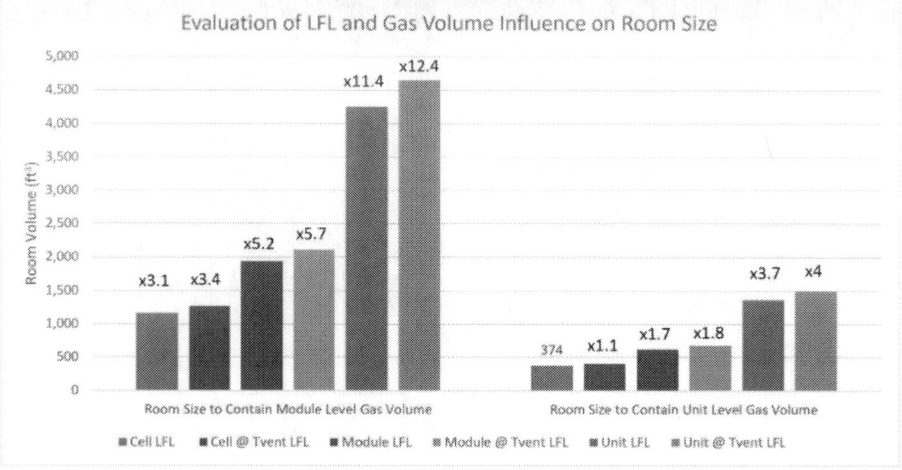

SPRING24+20TH GCPS
A Joint AIChE and CCPS Meeting

March 24–28, 2024
New Orleans Ernest N. Morial Convention Center,
New Orleans, LA

- Get ALL the data
- Work with the mechanical engineers and architects on the room design
- Specify gas detection based on the test

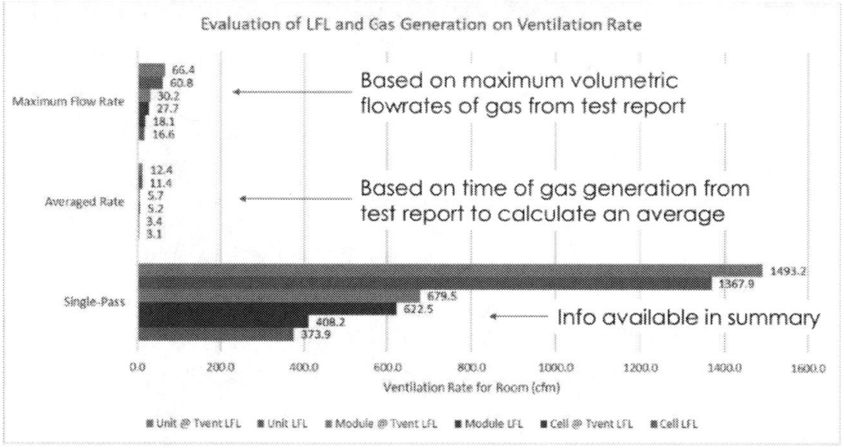

Case Study 2

Parameter	Unit Level Test - Outdoor
Gas composition	N/A
Gas volume	N/A
Smoke Release Rate	N/A
Flaming Jets	10 ft from vent on roof 5 ft from door
Flying Debris	None
Re-ignitions	No re-ignitions
Separation Distances	6" back-to-back 6" side to side
Temperatures at Targets	80°C at Target
Heat Flux at Targets	$15 kW/m^2$ at 3 ft to side $40 kW/m^2$ at 8 ft from front
Peak Heat Release Rate	N/A

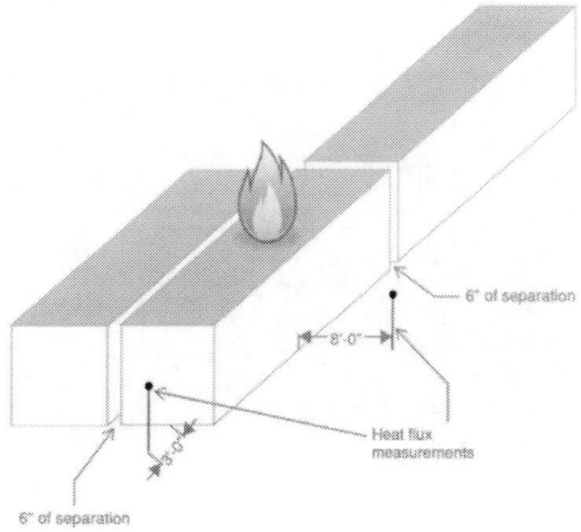

SPRING24+20TH GCPS
A Joint AIChE and CCPS Meeting
March 24–28, 2024
New Orleans Ernest N. Morial Convention Center,
New Orleans, LA

Victoria Big Battery Fire – July 2021

Lessons Learned from Victoria Big Battery Fire

Layout was per UL9540A test results, however, 35mph winds tilted flames impacting the roof of an adjacent unit. Such wind conditions are not considered in the UL testing.

The fire department had to use significant amounts of water to cool exposures(e.g. transformers) to avoid further spread. Had greater separation distances to transformers been used, cooling efforts could have been avoided.

Image Left: Tesla Megapack thermal runaway event during commissioning.

Case Study 2

Parameter	Unit Level Test - Outdoor
Gas composition	N/A
Gas volume	N/A
Smoke Release Rate	N/A
Flaming Jets	10 ft from vent on roof 5 ft from door
Flying Debris	None
Re-ignitions	No re-ignitions
Separation Distances	6" back-to-back 6" side to side
Temperatures at Targets	80°C at Target
Heat Flux at Targets	15kW/m² at 3 ft to side 40kW/m² at 8 ft from front
Peak Heat Release Rate	N/A

Methods for Back-Calculating Heat Release Rates:

- Module level data
- Module data + Multiplication factor
- Flame height data
- Heat flux data

Understanding Wind

Are the UL9540A separations (6 inches) enough? What winds conditions are common in the area?

UL9540A testing requires a maximum wind speed of no more than 12 mph.

Where flame tilt is expected, closer collaboration with BESS integrators/manufacturers is needed to understand the enclosure and its weakest components.

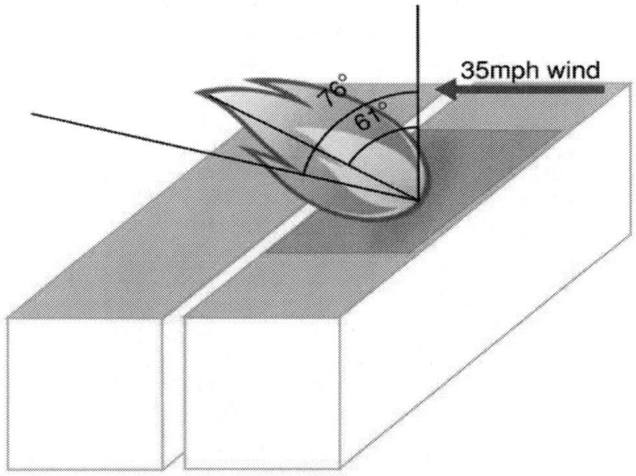

SPRING24+20TH GCPS
A Joint AIChE and CCPS Meeting

March 24–28, 2024
New Orleans Hilton Riverside
New Orleans Convention Center

Transformer Separation

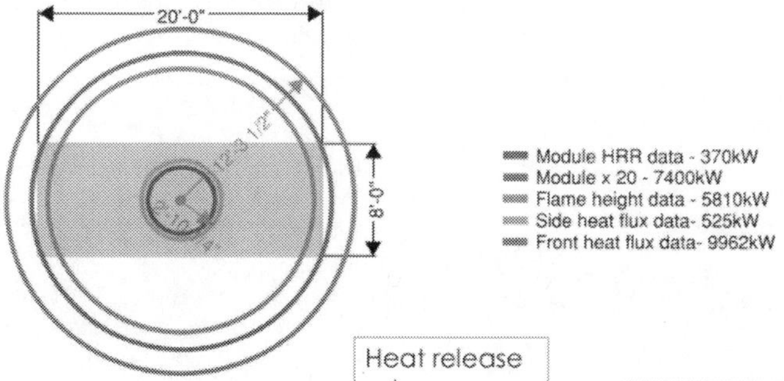

With an acceptable heat flux of 17 kW/m² for the tanks of oil-filled transformers; additional, separation distances may be warranted to mitigate spread or minimize the need for fire department cooling efforts on nearby exposures.

Heat release rate

Radiative Fraction (0.3)

Separation distance

$$R = \sqrt{\frac{\dot{Q} \times X_r}{4 \times \pi \times q_r}}$$

Acceptable heat flux

SPRING24+20TH GCPS
A Joint AIChE and CCPS Meeting

March 24–28, 2024
New Orleans Hilton Riverside
New Orleans Convention Center

Separation to People

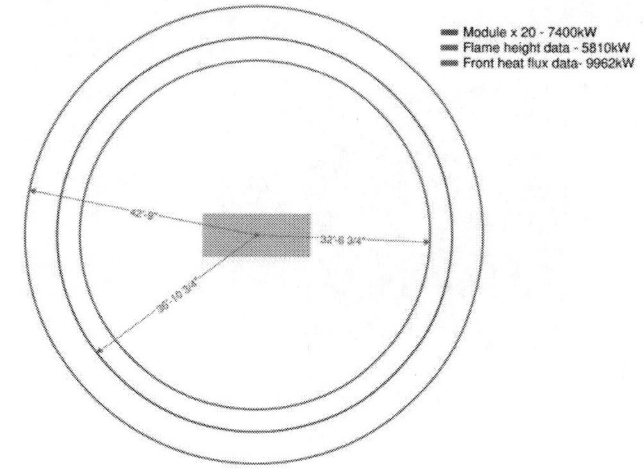

People can withstand heat fluxes up to 1.4 kW/m² without pain.

Such analyses are important to consider for fire department pre-planning and especially as BESS units are integrated in urban areas. Where such distances cannot be met, alternative protection features like fire walls or spray systems may be considered.

Collaboration and data sharing is essential

Design, permitting, and emergency response approaches need to be coordinated. Information cannot be locked behind NDAs or muddled for sales pitches

Think critically and understand the test method

Data in the test reports can be misleading or incomplete. Results must be adapted based on sound engineer principles and consider the site-specific conditions.

Full-scale testing will evolve

Future editions need to further standardize test setup and reporting. Engineers need additional data to answer owner, insurers, and fire department questions.

SPRING24+20TH GCPS
A Joint AIChE and CCPS Meeting
March 24–28, 2024
New Orleans Ernest N. Morial Convention Center, New Orleans, LA

Thank you. Questions?

Veronica Goldsmith, PE
Arup US, Inc.

veronica.goldsmith@arup.com

Carbon Dioxide (CO_2) Pipeline Dispersion Modeling Comparison

Cynthia Spitzenberger
DNV Energy USA Inc.
1400 Ravello Drive, Katy, TX 77449
cynthia.spitzenberger@dnv.com

Kejia Wu, kejia.wu@dnv.com
Gabriele Ferrara, gabriele.ferrara@dnv.com
Gean Rocha, gean.rocha@dnv.com
DNV Services UK Limited

Tatiana Norman, tatiana.norman@dnv.com
DNV GL USA, Inc.

Keywords: Carbon Dioxide, CO2, Pipeline, Dispersion, Dense vapor

Abstract

Carbon dioxide (CO_2) capture, use, and sequestration is important to achieve stated climate goals. An essential element of carbon capture and sequestration (CCS) programs is the pipeline transport of CO_2. Public concern has focused on the importance of dispersion modeling related to potential pipeline releases. There are choices of methodology for dispersion modeling depending on various factors such as analysis time, required detail of analysis, and local terrain complexity.

This paper will explore a case study to compare integral dispersion modeling with computational fluid dynamics (CFD) modeling for locations representative of simple terrain versus complex terrain. The example cases will provide guidance for when integral (simplified) modeling is sufficient versus when consideration of terrain and complex modeling is necessary and can provide value for stakeholder engagement.

1 Overview

Carbon dioxide (CO_2) is a nonflammable, colorless and odorless gas that is about 1.5 times heavier than air (at standard temperature and pressure conditions). Depending on the pressure and temperature of the system, CO_2 can exist as a gas, liquid, solid, or supercritical fluid. It is worth noting that under normal atmospheric pressures, carbon dioxide can only exist as a gas or solid; CO_2 cannot exist as a liquid under atmospheric conditions.

Carbon dioxide transportation via pipelines have operated for decades within the United States for various purposes. Carbon dioxide capture, transportation, injection, and geological storage (CCS) requires the linking and integration of separate process activities to capture CO_2 at an emitter facility, transport it to designated storage sites, and then inject it deep underground for permanent geological storage. CCS operations may range in complexity from relatively simple point-to-point chains with a single capture facility, storage site, and interconnecting pipeline, to multiple capture plants feeding CO_2 into a network of pipelines from which the CO_2 is supplied to multiple storage sites.

With the development of CCS systems, where pipelines transport large inventories of CO_2, operators and stakeholders along pipeline routes want to understand the potential hazard zones related to a pipeline release of CO_2. Dispersion modeling of the release can assist in predicting the potential hazard zone of a release, however as with all modeling there are important limitations to be aware of and consider when performing the evaluation.

2 Dispersion Models

Dispersion models are mathematical tools that simulate the behavior of a vapor in the atmosphere after a release from a source, such as an industrial plant or a storage site. Dispersion models can be used to estimate the spatial and temporal distribution of vapor concentrations, as well as the potential exposure and health risks for the surrounding area.

Dispersion models can account for various factors that affect the movement and concentration of the vapor, such as atmospheric conditions (wind speed, atmospheric stability, temperature, and humidity), surface roughness, and the release source conditions [1]. The location and height of the release is also an important input detail applied in the modeling. Dispersion models can also be used to evaluate the effectiveness of mitigation measures, such as emergency response plans or monitoring systems.

Complex discharge phenomena exist at the immediate vicinity of the release point. Therefore, an equivalent release source at a distance slightly downstream from the release point is needed to avoid the problematic region. This equivalent release source is obtained from a source term model that estimates the discharge rate and momentum conditions of the release based on the material properties, storage conditions (pressure, temperature) and orifice size is required as input to the dispersion model. Inventory and volume amounts contained within the release source are important as well for determining the discharge duration and depressurization rate over time. In some models the source term and dispersion calculation are integrated together, but it is important to recognize that they are separate calculations.

There are different types of dispersion models, and these are described briefly in the following sections.

2.1 Integral models

Integral or free-field dispersion models are based on empirical or semi-empirical formulas that describe the behavior of the vapor plume as a function of the distance from the source and the direction of the wind. There are many different integral models that have been developed to represent various dispersion conditions and the modeling of different vapor cloud behaviors. A selection of the different types is discussed.

- Neutrally buoyant gas plumes or puffs are often represented by Gaussian models which describe a gas release at height in the wind direction at the wind speed. Gaussian models assume that the vapor concentration follows a normal distribution along the horizontal and vertical directions, with the mean value being the centerline of the plume. The standard deviation of the concentration distribution depends on the atmospheric stability, the wind speed, and the distance from the source. The predicted concentrations are also time averages.
- Dense gas models are meant to represent dispersion of releases that are heavier than the ambient air. Dense gas dispersion has significantly different mechanisms than neutrally buoyant cases especially with respect to mixing with air. Liquid or supercritical CO_2 releases from pipelines are expected to be dense gas releases.
- Lagrangian and Eulerian models are also available, but these topics are beyond the scope of this discussion.

Integral models apply mathematical formulas developed and validated based on experiments and large-scale test cases. As they are correlations, they are not meant to model complex geometries, terrains or obstacles that are in the path of the dispersing cloud. The developed integral models are often part of commercially available software packages that can execute the modeling calculations with simple setup and within minutes.

For this study the DNV dispersion model PHAST v9.0 [2] was applied. PHAST (Process Hazard Analysis Software Tool) is a comprehensive consequence modeling software tool which is applicable to all stages of design and operation across a wide range of process industries.

There are two main modules that PHAST uses to model atmospheric dispersion. The first one (*ATEX* module) generates the source term i.e., the outflow conditions at expanded conditions starting from the process conditions: release rate, velocity, temperature, and phase distribution. PHAST generates a suitable and realistic source term also for releases from buried pipelines. In this case, the model will predict the outflow rate following the depressurization in the pipeline and the size of the crater formed by the release. The second module (*Unified Dispersion Model* module) is the core model in PHAST for the atmospheric dispersion modeling. The UDM can model a wide range of scenarios covering unpressurised or pressurized releases, steady-state or time-varying releases, buoyancy driven or heavy-gas-dispersion.

2.2 Computational Fluid Dynamics

Computational fluid dynamics (CFD) is a numerical method that solves the governing equations of fluid flow and mass transfer, including the Navier-Stokes equations and the species transport equations. CFD can simulate complex scenarios involving turbulence, multiphase flow, chemical reactions, heat transfer, and other physical phenomena. The methodology is capable of capturing flow-structure interaction through appropriately set boundary conditions.

The spatial geometry of the problem is discretized into cells in a computational grid. The fluid flow equations are solved simultaneously in each cell. One-equation or two-equation turbulence approximation models can be used to resolve the turbulence which is an important factor in any fluid flow problem. More advanced approaches to resolve turbulence are also available in CFD, for example Large Eddy Simulation (LES) or Direct Numerical Simulation (DNS). These advanced technologies try to resolve the turbulence to the smallest scale possible instead of using approximation models.

CFD can provide detailed information on the spatial and temporal distribution of the flow properties, for example, velocity, temperature, and pressure fields. The methodology has also been validated against experimental data or simplified models to ensure its accuracy and reliability.

CFD modeling requires computational resources and expertise in setting up the appropriate boundary conditions, grid resolution, and turbulence models.

Conducting the dispersion modeling in CFD is considered the most realistic approach to estimate the gas cloud sizes in conditions where geometric details play an important role (such as releases in confined spaces, presence of obstructions, complex terrain etc.). In particular, and relevant to the present study, a release of heavy gas will tend to follow the profile of the terrain and will be considerably affected by the topography.

This work used ANSYS CFX that is a general-purpose CFD code [3]. CFX is considered one of the most advanced Computational Fluid Dynamics (CFD) codes and is widely used for industrial applications around the world.

3 Methodology

The study approach, further discussed in the following, consisted of relying on PHAST for the evaluation of the source term. A correct and realistic prediction of the source term is key for the next step of the atmospheric dispersion modeling. PHAST has been thoroughly validated in this regard and is considered a valid choice for the task. Once the source term is established, PHAST is also deemed a reliable tool to predict the atmospheric dispersion on flat terrain and in the absence of important obstacles. For the case of complex terrain where CFD is necessary, as mentioned above, the best strategy was considered the one to use the source term generated by PHAST and the 3D capabilities offered by CFX. This combination was considered the merger of the best contributions from both tools.

4 Release Scenarios

3.1 Scenarios

A range of release scenarios were defined for the evaluation to reflect the range of possible releases from a CO_2 pipeline for comparison across the different dispersion models. Table 1 outlines the four release scenarios that were defined. All are modeled as pipeline releases from a 20-inch (508 mm) diameter pipeline of 100-mile length. For simplification of the modeling and to be conservative, no isolation valves are included in the modeling to divide the pipeline into sub-segments. Normally isolation valves would be placed at regular intervals (maximum of 15 or 20 miles depending on location [4]) along the pipeline to create smaller segments that would allow isolation and separation of the pipeline inventory. Also, for simplification no pumped flow is considered in the model.

The released material is assumed 100% CO_2. Liquid and vapor phase operation is considered along with two different release sizes – 2-inch diameter hole and a full-bore rupture (FBR). Note that the FBR case is assumed to be the entire 20-inch diameter, and no reduction in the orifice diameter is given for wall thickness. The FBR case is represented as a breach in the pipeline such that there is flow from 20-inch diameter sections from both the upstream and downstream pipeline sections of the breach.

Table 1. Defined Release Scenarios

Scenario ID	Operating Phase	Operating Pressure, gauge bar [psi]	Operating Temperature degC [degF]	Release Size
1	Liquid	151.7 [2200]	20 [68]	2" diameter hole
2	Liquid	151.7 [2200]	20 [68]	Full bore rupture
3	Vapor	40 [580]	20 [68]	2" diameter hole
4	Vapor	40 [580]	20 [68]	Full bore rupture

3.2 Weather Conditions

Weather conditions are extremely important in the consideration of dispersion modeling. Atmospheric stability is a measure of the degree of turbulent mixing in the atmosphere; the more stable the conditions, the less mixing occurs and less dilution of the cloud concentrations, while the more unstable the conditions, the greater the degree of mixing [1]. The stability is often defined by the Pasquill stability classes (also referred to as Pasquill-Gifford-Turner stability classes), ranging from A (extremely unstable) to F (moderately stable).

Wind speed is also an important factor in dispersion modeling as the wind can act to push the cloud downwind but also promotes mixing with air and dilution of the cloud.

For a given location it is important to evaluate the potential weather conditions, including turbulence and wind speed, and select a range of representative combinations to use for the dispersion evaluation. Conservative conditions should be evaluated to understand the potential worst-case scenarios for emergency planning.

For the study, two weather conditions defined in terms of wind speeds and their associated atmospheric stability conditions were considered to assess the impact of atmospheric condition on the CO_2 gas dispersion pattern. The stability class is a measure of turbulence in the atmosphere:

- F stability (moderate), 1.5 m/s (3.4 mph) wind speed, and
- D stability (neutral), 5 m/s (11 mph) wind speed.

The case study assumed atmospheric temperature of 20°C (68°F), with a humidity of 70%. Each of the release scenarios was evaluated for the two weather conditions.

3.3 Terrain

As noted earlier, free-field models do not account for the specific terrain or geometry obstacles in the path of the dispersing cloud. These models do often have an entry for a "surface roughness" parameter or length, which is meant to reflect a generalized description of the terrain. More frequent and large obstacles increase the turbulence in the wind flow and thus have an impact on the cloud dispersion. The case study applied a surface roughness length of 10 cm, indicative of low crops, occasional obstacles [2].

For CFD models two terrain conditions were applied. The first was a simple model with no terrain input (flat terrain) with the same surface roughness length applied as the free-field model. This model setup is directly comparable to the free-field modeling. The second setup in CFD included an example complex terrain to demonstrate the difference in dispersion between a flat and a complex terrain under the same release condition. A relatively mountainous terrain was chosen for the assessment to clearly demonstrate the impact of local topography on gas dispersion. The chosen terrain has an elevation difference of approximately 100 m in the vicinity of the release location.

5 Model Results

This section discusses the results from the discharge and dispersion calculations for the defined release cases. The discharge results provide the source term definition that is applied within the different dispersion models applied in the study.

4.1 Discharge Calculation

As the focus of the study was the comparison of the dispersion models, the discharge modeling for all cases was performed with DNV's PHAST v9.0 [2] long pipeline discharge calculation and used to provide the source term input for the dispersion models.

The pipelines are assumed buried at depth of 4 ft. Different durations of interest may be specified based on the user interest, but the study applied the default value of 3,600 s (1 hr); thus, the calculations will stop at 1 hr even if the release inventory has not fully depressurized. The discharge rate is predicted for time steps up to 3,600s. For the dispersion modeling, the discharge is represented with two standard approaches. The first approach is to calculate the average discharge rate over the first 20s of the release, and assume this rate is constant for the duration of the release till the inventory is exhausted; note this is the default calculation within PHAST. The second approach is to calculate the multiple representative discharge segments (PHAST allows up to 10) along the duration of the release. Both approaches to the discharge calculation are applied for the comparison evaluation, with results presented in Table 2. Figure 1 and Figure 2 present a comparison of the FBR discharge rates for the liquid and vapor FBR pipeline release cases, respectively.

The velocity term represents the velocity after depressurization to atmospheric conditions. As the buried long pipeline model is applied, there are calculations performed to reflect the crater modeling and the impact on the discharge definition – these are reflected in pre-dilution air rates that are not presented here for brevity.

Note that based on the results for the liquid FBR case, the average release rate over the first 20s is 5,336 kg/s which will de-inventorize the example pipeline in 775 s (~13 min); whereas the time-varying discharge calculation demonstrates the discharge duration of at least 3,600 s (1 hr) with varying discharge rates as the pipeline depressurizes.

Table 2. Scenario Discharge Rates

ID	Operating Phase	Release Size	Avg 20s			10 Time Segments		
			Release time s	Mass flow rate kg/s	Final Velocity m/s	Release time s	Mass flow rate kg/s	Final Velocity m/s
1	Liquid	2" diameter hole	0 3,600	188 188	58 58	-	-	-
2	Liquid	Full bore rupture	0 775	5,336 5,336	16 16	0 152 410 739 1,128 1,565 2,030 2,520 3,041 3,600	17,341 2,128 1,572 1,274 1,112 1,015 964 910 853 792	53.1 6.4 4.8 3.9 3.4 3.1 2.9 2.8 2.6 2.4
3	Vapor	2" diameter hole	0 3600	23.5 23.5	62 62	-	-	-
4	Vapor	Full bore rupture	0 594	2,439 2,439	23 23	0 129 374 691 1,067 1,493 1,964 2,475 3,020 3,600	6,786 805 566 462 400 358 327 305 287 268	60.9 7.9 5.6 4.6 4.0 3.5 3.2 3.0 2.8 2.7

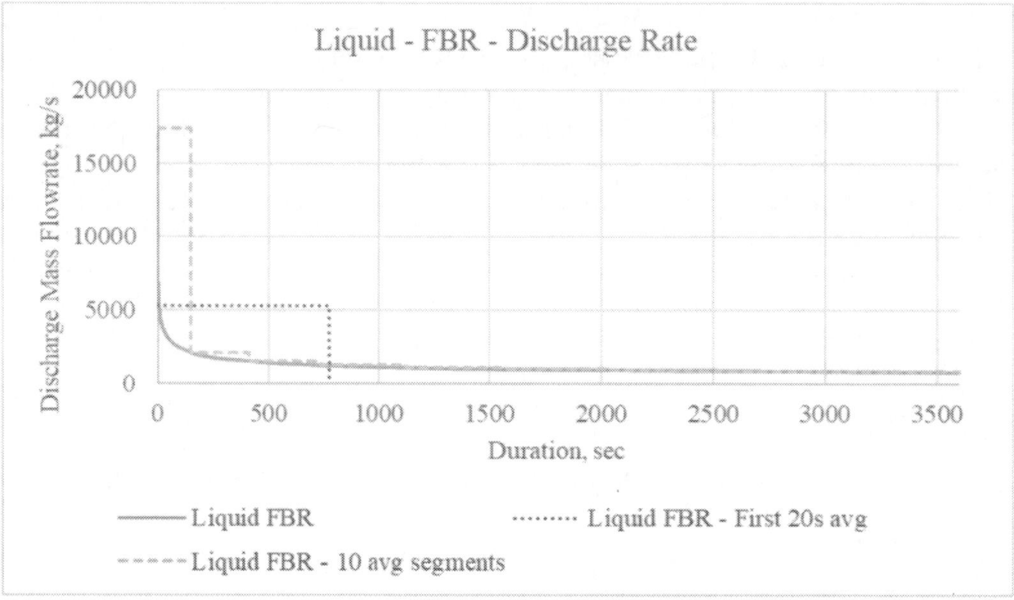

Figure 1. Liquid CO_2, FBR Pipeline, Discharge Rate

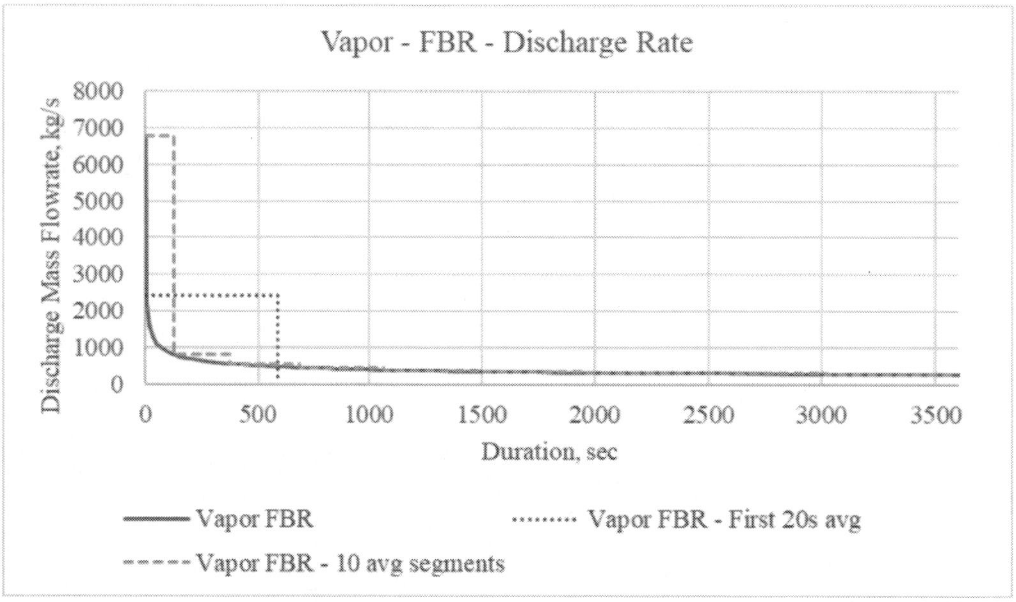

Figure 2. Vapor CO_2, FBR Pipeline, Discharge Rate

The 2-inch hole size releases maintain a constant discharge rate through the modeled 3,600s as the inventory is sufficient to prevent significant depressurization and thus decay of the discharge rate over time. The FBR releases demonstrate rapid depressurization of and decay of the discharge rate over the first 100s of the release.

4.2 Dispersion Endpoints

A key element of dispersion modeling is the endpoint concentration used to evaluate the dispersion extent. Asphyxiation, displacement of oxygen, is a concern for releases of large quantities of CO_2. However, CO_2 will cause toxic physical responses and harm at concentrations lower than needed to cause asphyxiation. The following dispersion endpoints are modeled for carbon dioxide, following the API CO_2 Tactical Guidance [5]:

- 30,000 ppm – American Conference of Governmental Industrial Hygienists (ACGIH) threshold limit value (TLV) for short-term explosion limit (STEL) [6]
- 40,000 ppm – National Institute for Occupational Safety and Health (NIOSH) immediately dangerous to life or health (IDLH) after 30 minutes exposure time [7]. This limit is used as a conservative threshold for modelling purposes and should not be considered as a maximum emergency limit.

Additional concentration endpoints could be evaluated as needed, but for simplicity of the comparison study, just these two concentrations were evaluated.

4.3 Integral / Free-field Dispersion

The free-field dispersion modeling is performed with PHAST v9.0 [2]. Table 3 presents the downwind dispersion distances to the concentrations of interest, depending on the model scenario and weather condition. The distances are reported at a height of 1 m above the ground.

Table 3. Scenario Dispersion Downwind Distances - PHAST Modeling

ID	Operating Phase	Release Size	Discharge model	Weather	Downwind Distance to Concentration at 1 m height	
					STEL 30,000ppm	IDLH 40,000ppm
1A-F1.5	Liquid	2" diameter hole	Avg 20s	F1.5	271	50
1A-D5	Liquid	2" diameter hole	Avg 20s	D5	1	1
2A-F1.5	**Liquid**	**Full bore rupture**	**Avg 20s**	**F1.5**	**2,554**	**2,260**
2A-D5	**Liquid**	**Full bore rupture**	**Avg 20s**	**D5**	**2,107**	**1,696**
2B-F1.5	Liquid	Full bore rupture	10 segments	F1.5	1,999	1,698
2B-D5	Liquid	Full bore rupture	10 segments	D5	1,526	1,352
3A-F1.5	Vapor	2" diameter hole	Avg 20s	F1.5	13	11
3A-D5	Vapor	2" diameter hole	Avg 20s	D5	1	1
4A-F1.5	**Vapor**	**Full bore rupture**	**Avg 20s**	**F1.5**	**1,590**	**1,397**
4A-D5	**Vapor**	**Full bore rupture**	**Avg 20s**	**D5**	**660**	**581**
4B-F1.5	Vapor	Full bore rupture	10 segments	F1.5	1,158	976
4B-D5	Vapor	Full bore rupture	10 segments	D5	944	838

The F1.5 weather condition generally results in greater downwind distances compared to the D5 weather results. The simplified discharge model with the average over the first 20s predicts greater distances versus the time-varying 10 segment approach. The hole size release cases produce much small hazard distances than the FBR cases, given the significantly smaller discharge rate between the cases. Figure 3 and Figure 4 present the maximum cloud footprints for the FBR cases for the F1.5 and D5 weather conditions, respectively. Figure 5 presents the dispersion side view of the liquid FBR release case, based on the first 20s average discharge rate.

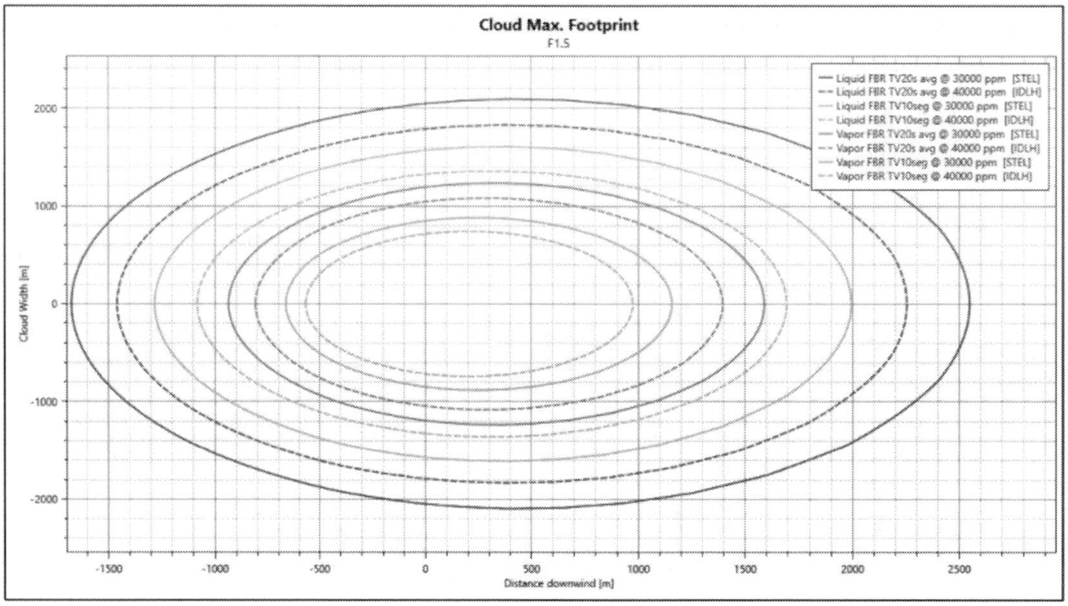

Figure 3. PHAST Cloud Maximum Footprint, FBR Pipeline Cases, F1.5 Weather, 1m elevation above ground, Release Location at (0,0).

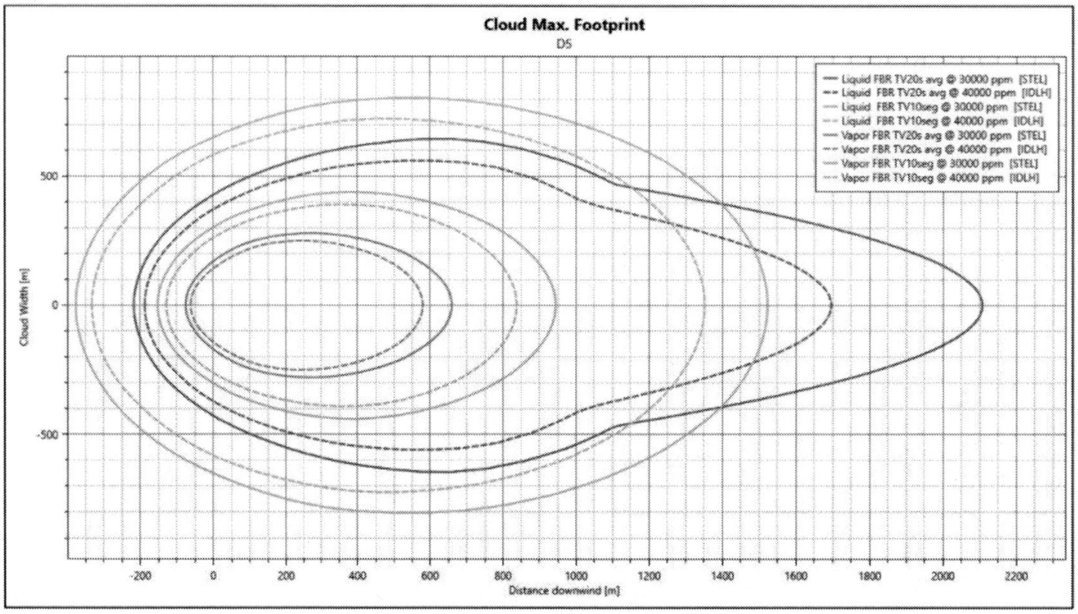

Figure 4. PHAST Cloud Maximum Footprint, FBR Pipeline Cases, D5 Weather, 1m elevation above ground, Release Location at (0,0).

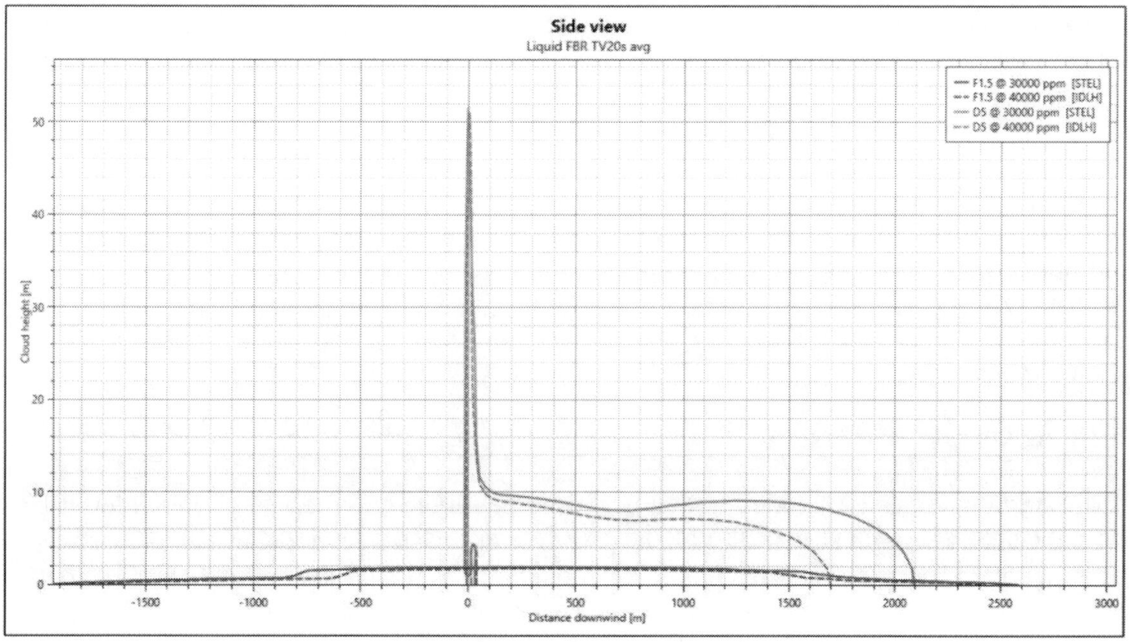

Figure 5. PHAST Side View, Liquid FBR Case, Discharge Rate 20s Average, Release Location at (0,0).

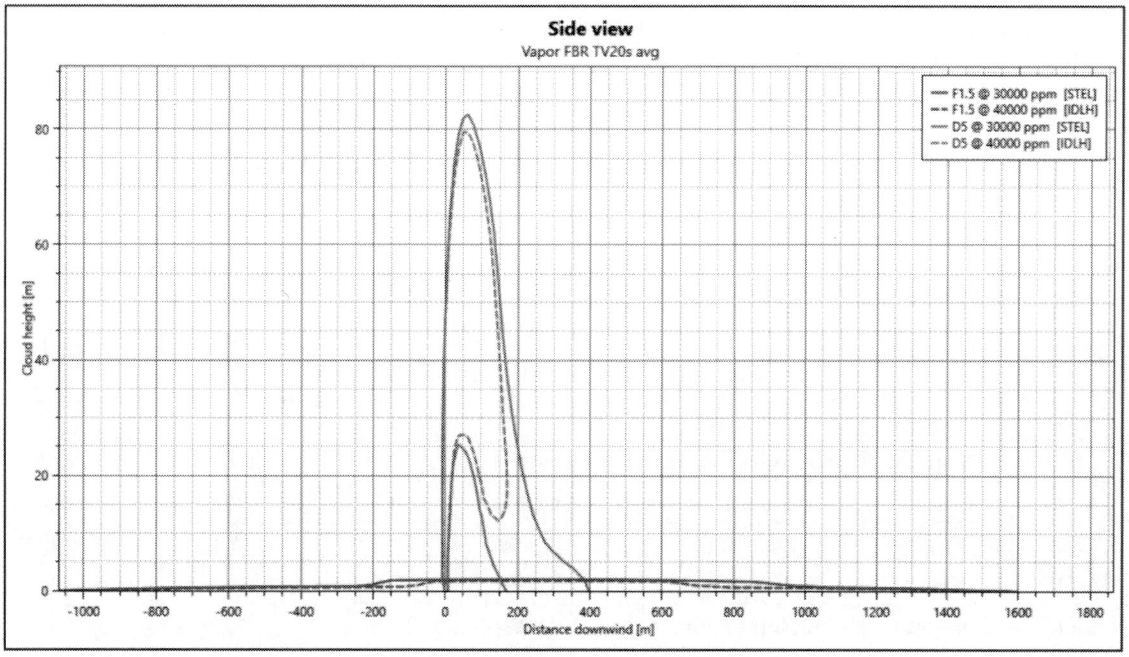

Figure 6. PHAST Side View, Vapor FBR Case, Discharge Rate 20s Average, Release Location at (0,0).

As demonstrated by the footprints, the F1.5 weather has significant upwind components to the dispersion cloud in addition to the downwind dispersion. This is illustrative of the slumping of the heavy gas dispersion cloud and gravity spread from the release location that has been indicated by large scale testing [8]. The test indicated that plumes generated from CO_2 releases (ruptures) from buried pipelines in liquid phase have tendency to collapse under their own weight and disperse in all directions forming a "blanket-type", heavy-gas envelope at the ground level. It is important to be able to predict this type of behavior as considerable amount of gas can travel upwind under the effect of the gravity spreading. In addition, when heavy-gas dispersion conditions occur, the topography of the terrain may become the leading factor that affects the gas behavior. For the D5 weather condition this effect is not as demonstrated from the modeling.

4.4 CFD Dispersion

4.4.1 Flat Terrain Dispersion

The flat terrain dispersion modeling was performed with the commercial CFD modeling package ANSYS CFX. The eight scenarios with the 20 s average discharge model in Table 3 were assessed. The atmospheric and the release conditions considered in the CFD simulations were identical to the input data used in the free-field modeling. A simple flat terrain was used with an appropriate surface roughness value matching the free-field modeling setup. Atmospheric wind and turbulence profiles appropriate for the weather conditions were set as the domain inlet boundary condition. Steady-state CFD simulations using k-ε turbulence model were carried out for all eight cases.

Table 4 presents the downwind dispersion distances to the concentrations of interest, for the modeled scenarios and weather conditions considered in the study. The distances are reported at a height of 1 m above the ground.

Table 4. Scenario Dispersion Downwind Distances - CFD Modeling Flat Terrain

ID	Operating Phase	Release Size	Discharge model	Weather	Downwind Distance to Concentration at 1 m height	
					STEL 30,000ppm	IDLH 40,000ppm
1A-F1.5	Liquid	2" diameter hole	Avg 20s	F1.5	95	73
1A-D5	Liquid	2" diameter hole	Avg 20s	D5	1	1
2A-F1.5	**Liquid**	**Full bore rupture**	**Avg 20s**	**F1.5**	**1,237**	**881**
2A-D5	**Liquid**	**Full bore rupture**	**Avg 20s**	**D5**	**930**	**744**
3A-F1.5	Vapor	2" diameter hole	Avg 20s	F1.5	0.4	0.4
3A-D5	Vapor	2" diameter hole	Avg 20s	D5	0.4	0.4
4A-F1.5	**Vapor**	**Full bore rupture**	**Avg 20s**	**F1.5**	**772**	**593**
4A-D5	**Vapor**	**Full bore rupture**	**Avg 20s**	**D5**	**522**	**406**

The F1.5 weather condition generally results in greater downwind distances compared to the D5 weather results. Similar to the free-field modeling results, the hole size release cases produce much smaller hazard distances than the FBR cases, given the significantly smaller discharge rate between the cases. Figure 7 and Figure 8 present the maximum cloud footprints for the FBR cases for the F1.5 and D5 weather conditions respectively. Dispersion side view of the liquid and vapor FBR release cases are presented in Figure 9 and Figure 10 respectively.

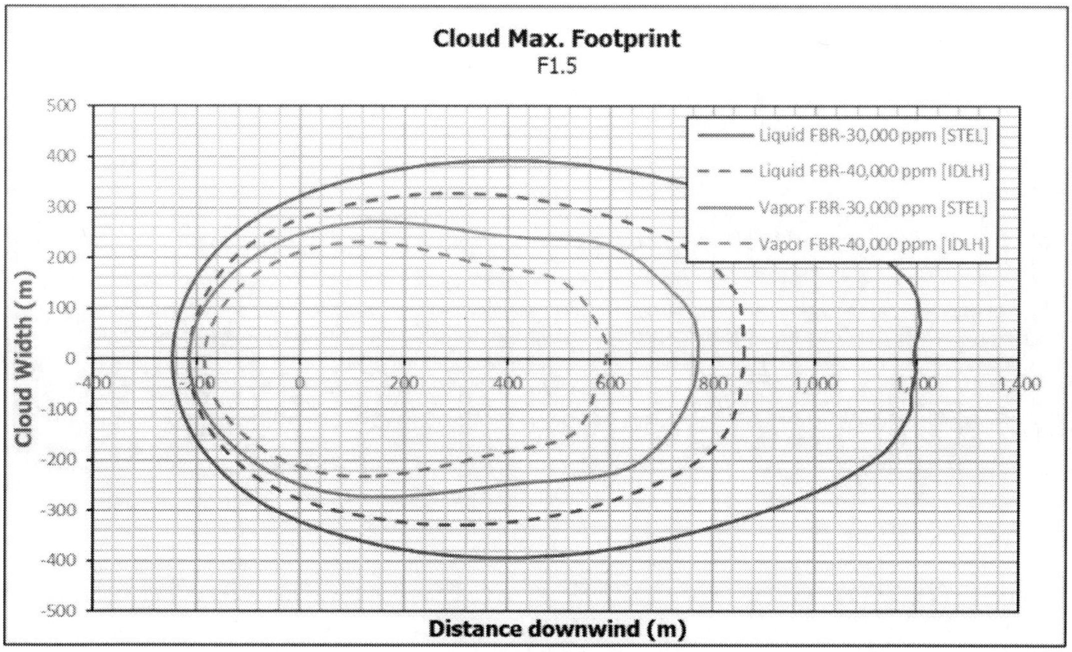

Figure 7. CFD Cloud Maximum Footprint, FBR Pipeline Cases, F1.5 Weather, 1m elevation above ground. Release Location at (0,0), Flat Terrain.

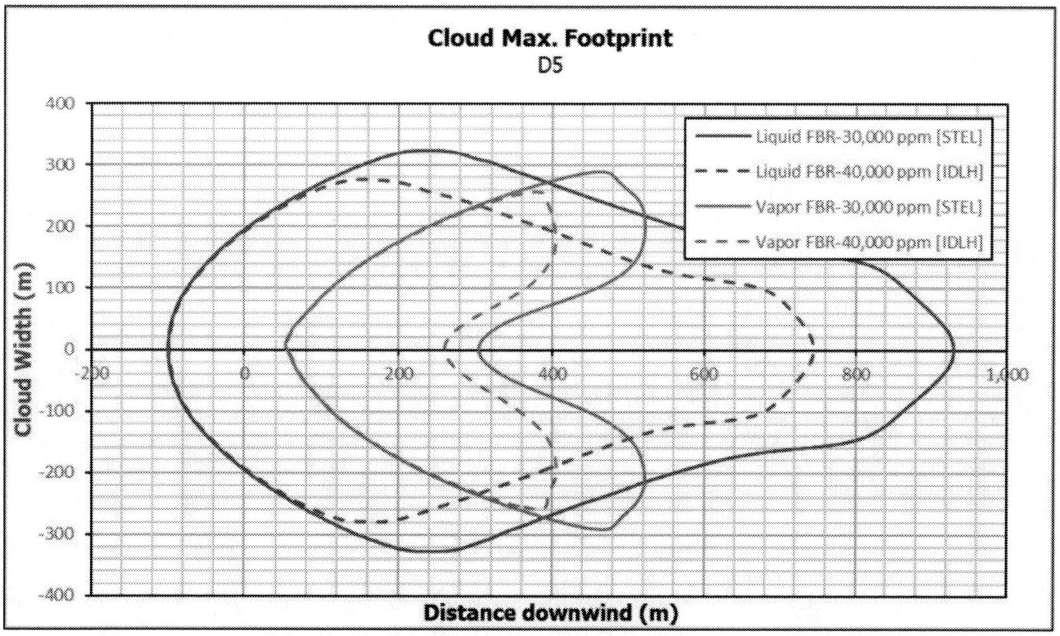

Figure 8. CFD Cloud Maximum Footprint, FBR Pipeline Cases, D5 Weather, 1m elevation above ground. Release Location at (0,0), Flat Terrain.

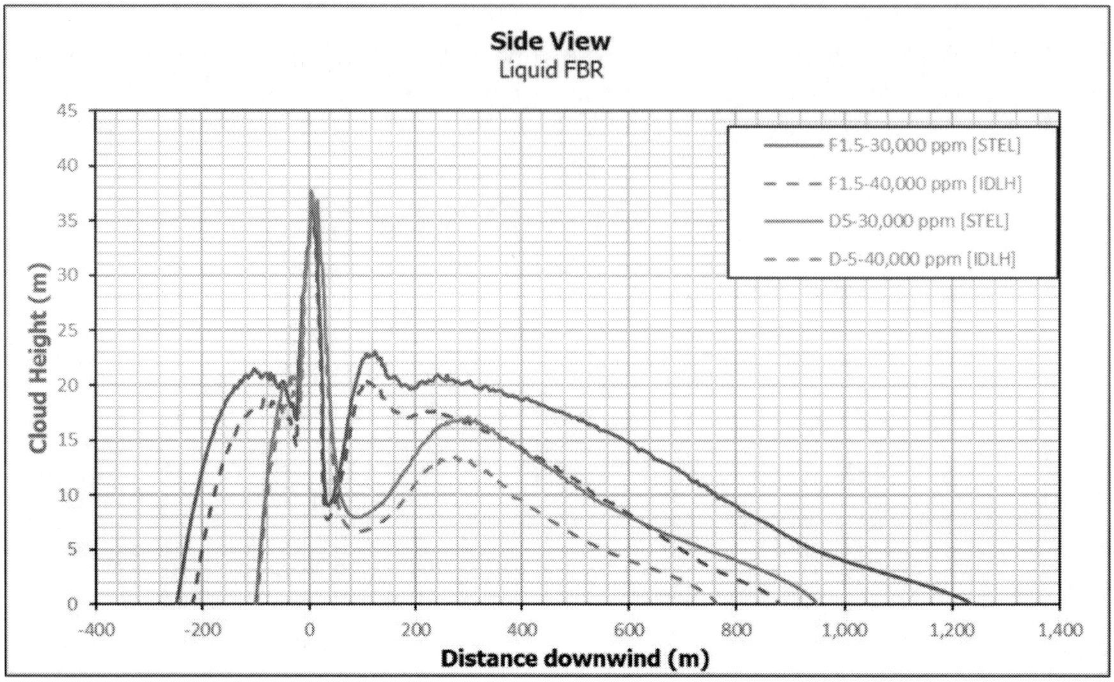

Figure 9. CFD Side View, Liquid FBR Case, Discharge Rate 20s Average. Release Location at (0,0), Flat Terrain

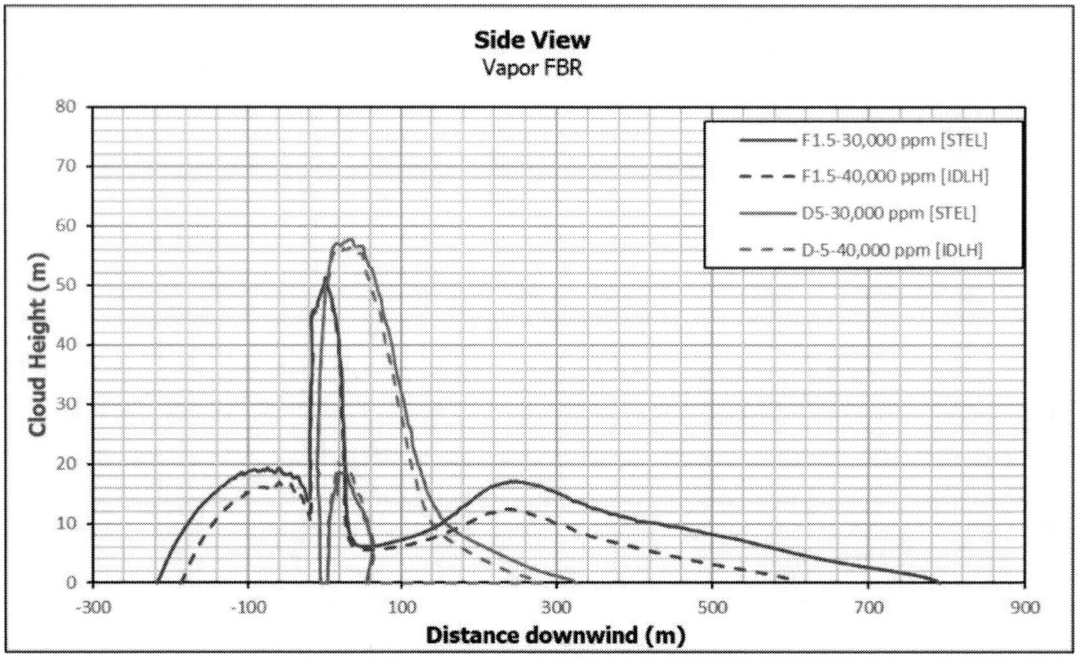

Figure 10. CFD Side View, Vapor FBR Case, Discharge Rate 20s Average. Release Location at (0,0), Flat Terrain

As demonstrated by the footprints, the F1.5 weather has a more significant upwind components to the dispersion cloud in addition to the downwind dispersion compared to D5 weather. The lower wind speed allows the released gas to spread more radially on its own momentum whereas gas dispersion is more dominated by the wind speed under the higher D5 weather.

4.4.1 Complex Terrain Dispersion

Additional CFD simulations were also carried out over a non-flat complex terrain topography to compare the dispersion results with the corresponding flat terrain case. The liquid FBR case with 20 s average release rate under D5 weather was selected for the comparison (Case 2A-D5). A generic hilly topography was modeled for this purpose as shown in Figure 11. Two different release locations and the associated wind directions were assessed for the terrain topography. One release location and wind direction direct the gas plume in a downhill direction (Case $2C_\alpha$-D5) while the other case directs the gas plume in an uphill direction (Case $2C_\beta$-D5) as indicated in Figure 11. For both scenarios, a vertical release direction was considered in alignment with the flat terrain dispersion and free-field modeling.

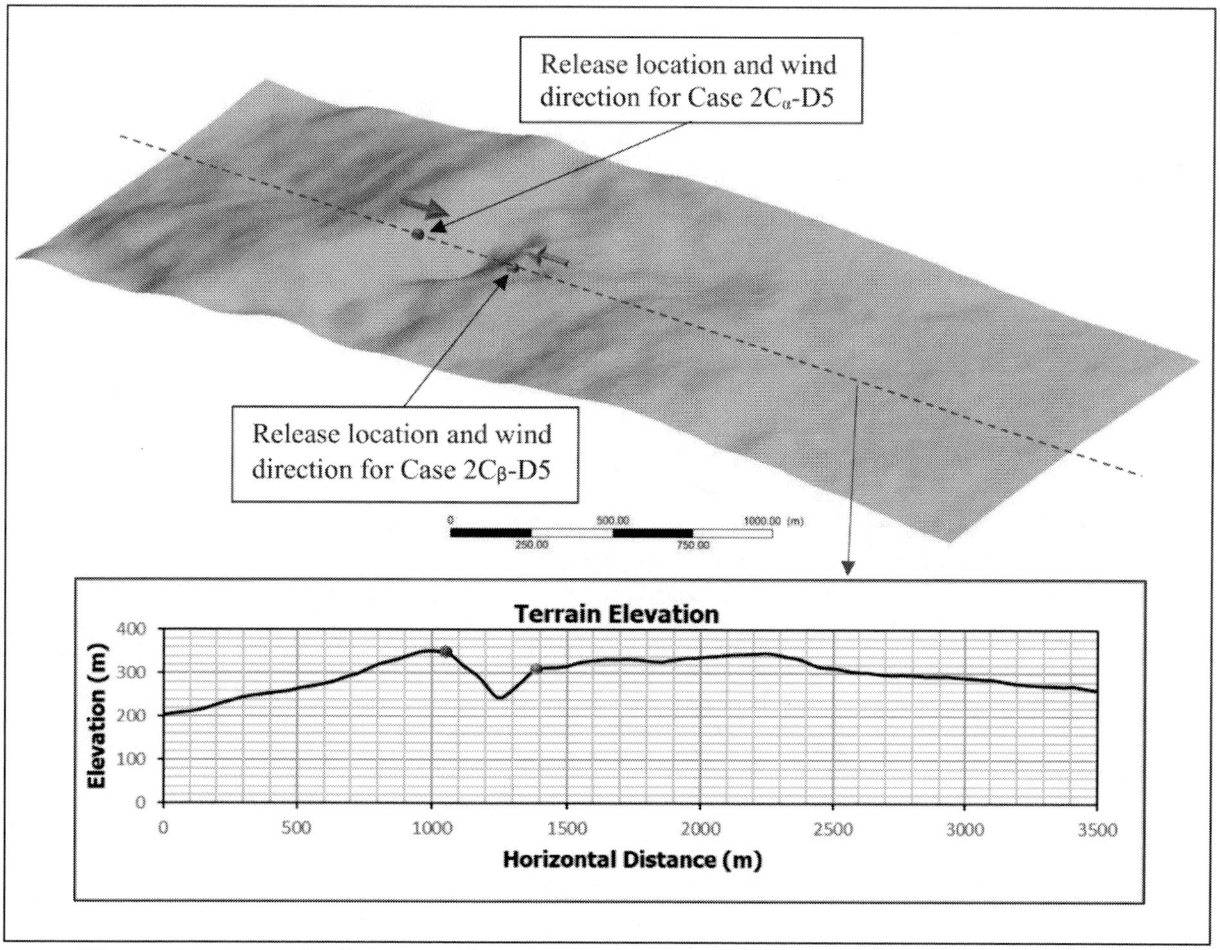

Figure 11. Terrain Topography and Release Locations Considered in the Study

Figure 5 presents the downwind dispersion distances to the concentrations of interest, for the model scenario and weather conditions considered in the study. The distances are reported at a height of 1 m above the ground. The downwind distance is defined as the length between the maximum extent of the gas plume and the release location projected in the x-axis.

Table 5. Scenario Dispersion Downwind Distances - CFD Modeling Complex Terrain

ID	Operating Phase	Release Size	Discharge model	Weather	Downwind Distance to Concentration at 1 m height	
					STEL 30,000ppm	IDLH 40,000ppm
$2C_{\alpha}$-D5	Liquid	Full bore rupture	Avg 20s	D5	852	710
$2C_{\beta}$-D5	Liquid	Full bore rupture	Avg 20s	D5	682	494

Top and perspective views of the gas plume at the concentrations of interest for the 2 complex terrain cases as well as the corresponding flat terrain case are shown in Figure 12. Figure 13 presents the maximum cloud footprints for the complex terrain cases together with the corresponding flat terrain case while the dispersion side views are presented in Figure 14.

In both complex terrain cases, the gas plume dispersion patterns are significantly affected by the topography when compared to the flat terrain dispersion. In the downhill dispersion case ($2C_{\alpha}$-D5), the gas plume reaches the valley floor and with sufficient momentum remaining, climbs up the relatively shallow valley wall on the other side to continue the spread downwind. The resultant downwind distance is similar but slightly shorter compared to the corresponding flat terrain case. Some of the gas plume also follows the valley floor and spreads in crosswind directions along the valley.

However, in the uphill dispersion case ($2C_{\beta}$-D5), the gas plume loses all its momentum before reaching the top of the hill. The gas plume changes its direction almost by 90 degrees and with the help of the wind tries to navigate around the hill. This resulted in a maximum distance in the crosswind direction instead of the expected directly downwind direction.

In both complex terrain cases, very limited upwind dispersion was observed compared to the flat terrain case. This is caused by the interaction of the gas plume and the atmospheric wind with the local topography.

Case No.	Top View	Perspective View
2A-D5	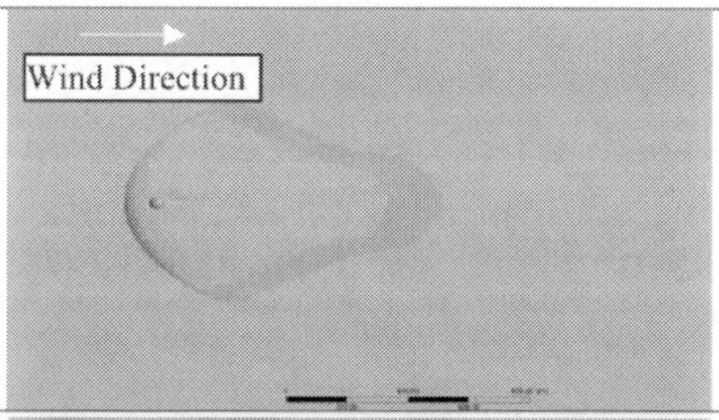	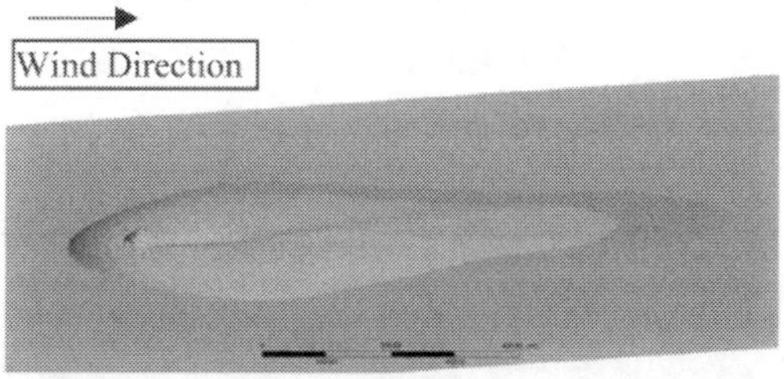
$2C_{\alpha}$-D5	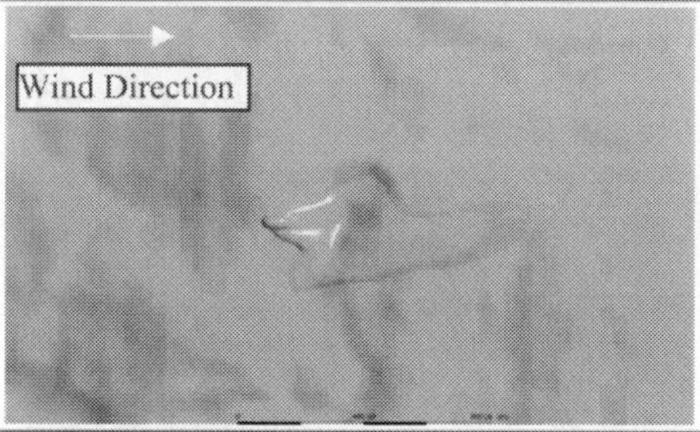	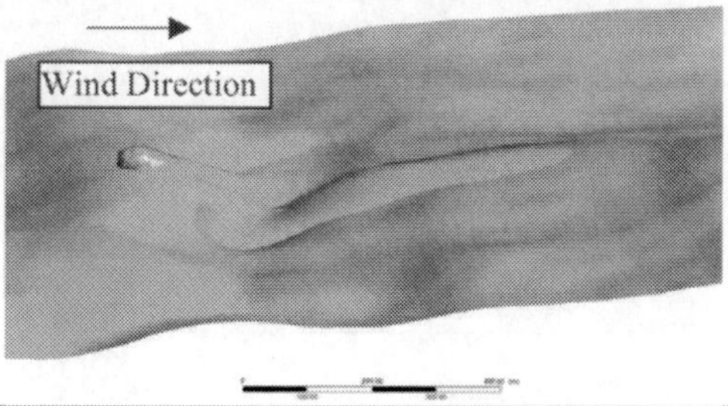
$2C_{\beta}$-D5	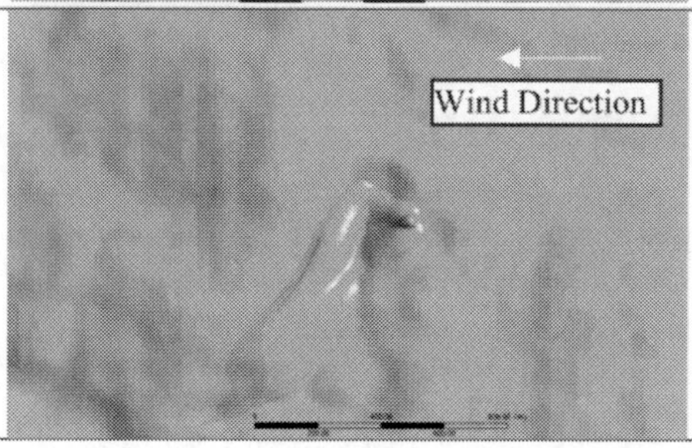	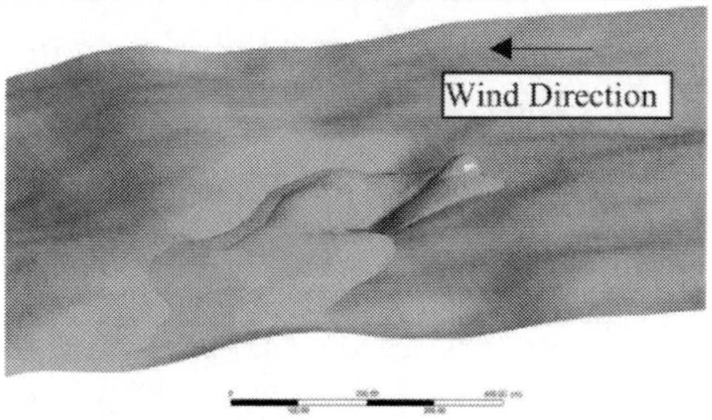
30,000 ppm CO_2 (STEL) 40,000 ppm CO_2 (IDLH)		

Figure 12. CFD Top and Perspective Views, Liquid FBR Case, D5 Weather, Complex Terrain

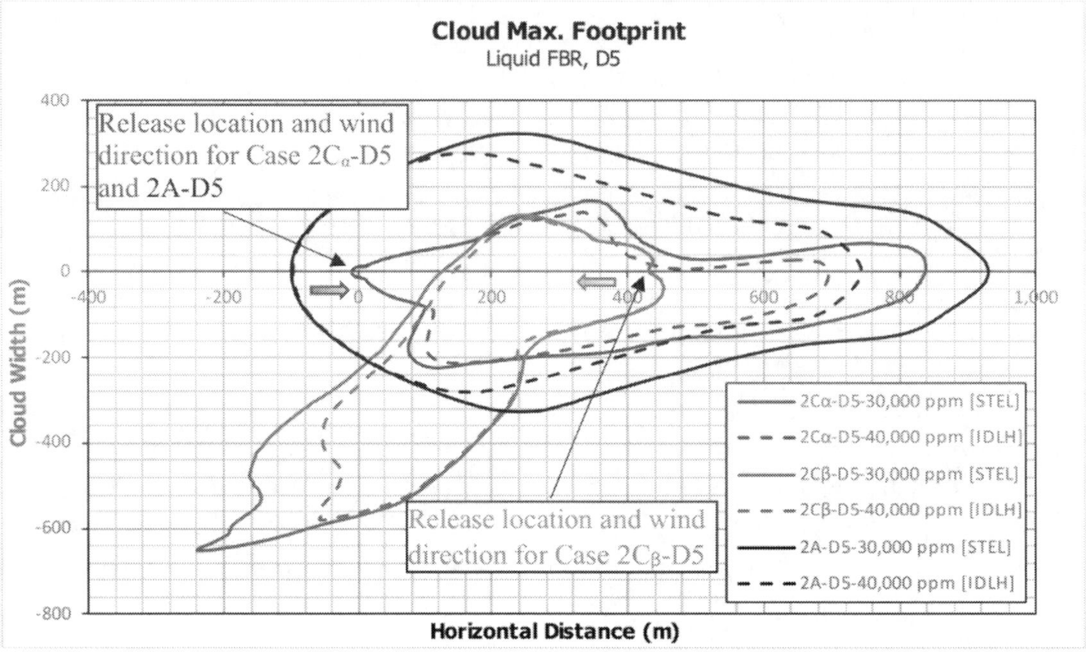

Figure 13. CFD Cloud Maximum Footprint, Liquid FBR Cases, D5 Weather, 1m elevation above ground, Complex Terrain

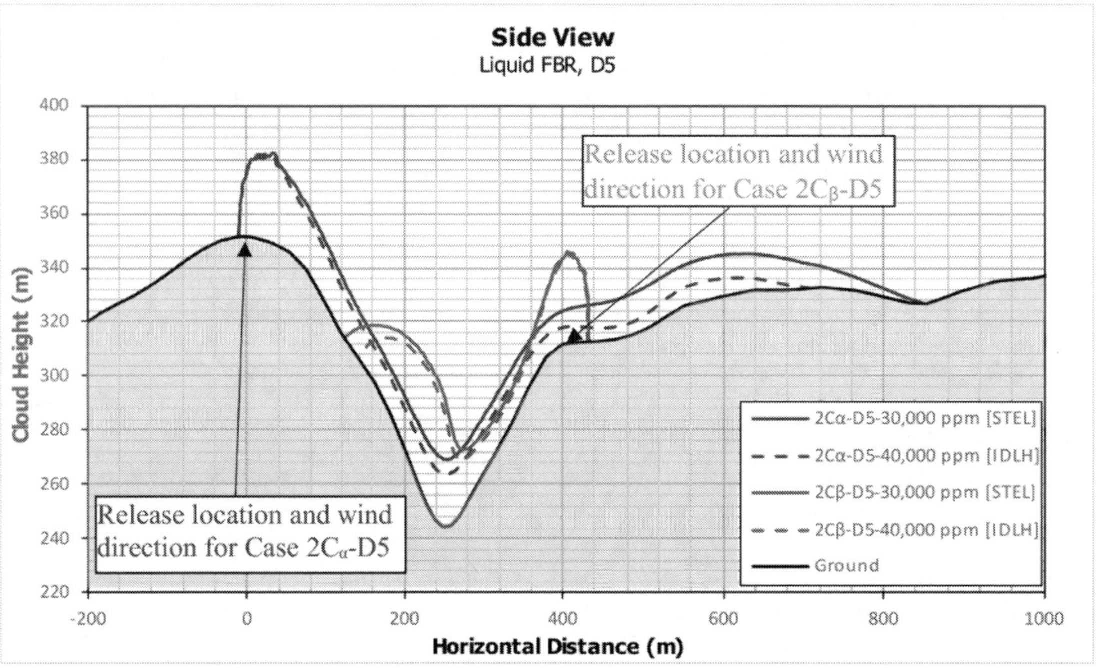

Figure 14. CFD Side View, Vapor FBR Case, Liquid FBR Cases, D5 Weather, Complex Terrain

6 Comparison

For the majority of the cases considered in the study, the gas plumes between the free-field and the CFD modeling show very similar dispersion patterns. This is particularly evident in the liquid FBR case under D5 weather condition. For this release case, the gas plumes from both modeling methodologies show the initial radial dispersion near the leak location where the gas plume gains the initial momentum from the collapsing of the gas plume to the ground under its own weight. When the release momentum dissipates further downstream from the leak location, the high atmospheric wind starts to dominate in the dispersion process and elongates the gas plume. Both the free-field and CFD modeling capture this behavior. At low wind speed of F1.5 weather condition, both modeling methodologies also predict the elliptical gas plume shape radiating out from the release location.

For small 2" hole release cases, both modeling methodologies were able to predict the slumping behavior of the gas plume at low wind condition (F1.5) and the small upward gas plume that dilutes quickly and does not fall back to ground under high wind condition (D5).

The only case that shows different dispersion behavior is the vapor FBR release case under D5 weather condition. This case has a relatively high release velocity with a relatively lighter density of the vapor phase release. This results in an initial vertical high momentum jet fountain above the release location. The atmospheric wind then brings the gas plume back towards ground when the initial release momentum has been dissipated. This gas plume dispersion pattern is predicted in both the free-field and CFD modeling methodologies. However, the free-field modeling was not able to capture the interaction of the gas plume with geometry, in this case the ground plane, due to the nature of the modeling approach. This interaction was captured by CFD modeling and can be seen by the branching of the gas plume after reattaching to the ground downwind from the release fountain.

Although the general gas plume behavior shows similarities between the two modeling methodologies, differences are shown in the predicted downwind dispersion distances as illustrated in Figure 15. In general, except the cases where the plume does not fall back to ground, the free-field modeling predicts longer dispersion distances than the corresponding CFD modeling cases. The PHAST free-field modeling uses empirical models and assumes perfect idealized conditions while CFD modeling solves all the governing flow equations explicitly. Therefore, the free-field modeling results are in general more conservative compared to CFD results for similar release scenarios. CFD modeling was also able to capture the interaction between the atmospheric wind and the obstacles created by the released gas itself. This changes the flow and gas dispersion patterns locally and downwind from the release. This would not be possible to achieve with free-field modeling.

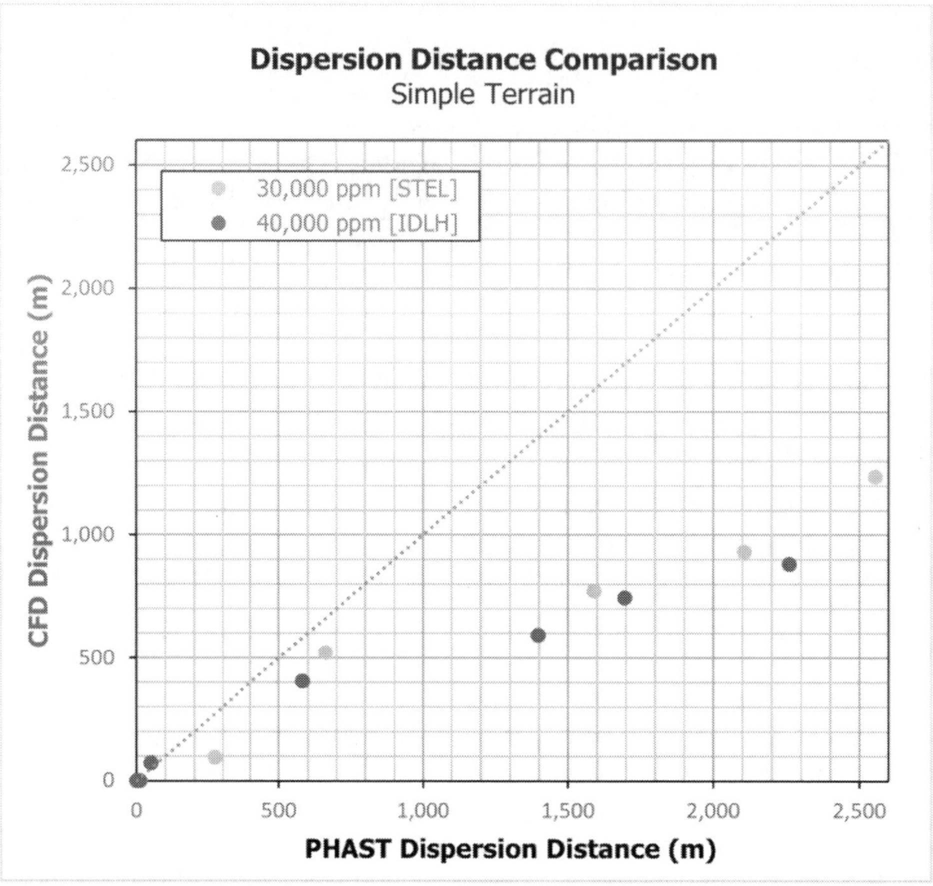

Figure 15. Dispersion Distance Comparison Between PHAST and CFD Modeling, Simple Terrain

7 Conclusion

A comparison study was carried out to assess CO_2 gas dispersion modeling capabilities between the integral free-field modeling approach and the CFD modeling approach. Both approaches are capable of capturing the important phenomena in a CO_2 gas dispersion event with the integral model generally providing a more conservative result in terms of dispersion distances compared to the CFD model.

On the other hand, CFD methodology is a more detailed modeling approach and is capable of solving the governing flow equations explicitly. This removes some of the conservatism present in the integral model. However, CFD modeling is more demanding in terms of computational resources and expertise in setting up the appropriate boundary conditions, computational mesh, and turbulence models.

In light of the advantages and disadvantages associated with each modeling approach, each would have its appropriate uses in an assessment investigating CO_2 dispersion risks. Free-field models can be used at the start of a study to screen out critical scenarios that might have high risk impacts on the surrounding environment. CFD modeling can then be employed to investigate these identified scenarios further and provide a more detail result in terms of impact zone size. CFD modeling is also very crucial for releases over complex terrains especially in mountainous areas or river crossings as the local topography will not only affect the dispersion distance but also significantly affect the gas plume shape creating unexpected impact zone shapes.

The current study briefly compares CO_2 dispersion modeling between an integral model and CFD modeling approaches. Future work involving more extensive comparisons would be needed to better understand the result differences between the two modeling approaches. This would also facilitate the possibly of creating a more comprehensive ruleset for CO_2 modeling approaches.

It has to be noted, the current study was carried out using steady-state CFD modeling. Future investigation using transient CFD simulation and release profile will enable the incorporation of the temporal element into the solution. This will allow the assessment of dosage, a combination of both concentration and exposure time, and therefore, offer a better understanding of the risk associated with an accidental CO_2 release.

8 References

1. **Center for Chemical Process Safety (CCPS).** *Guidelines for Chemical Process Quantitative Risk Analysis, 2nd Edition.* New York : American Institute of Chemical Engineers, 2000.

2. **DNV.** Phast. 2024. Vol. v9.0.

3. **Ansys.** CFX. *https://www.ansys.com/en-gb/products/fluids/ansys-cfx.* 2024.

4. **Code of Federal Regulations, Title 49 - Transportation.** *Valves: Location.* 2023. 49 CFR 195.260.

5. **American Petroleum Institute.** *Carbon Dioxide (CO2) Emergency Response Tactical Guidance Document.* Washington DC : API Publishing Services, November 2023.

6. **American Conference of Governmental Industrial Hygienists (ACGIH).** Carbon dioxide. [Online] 2024. https://www.acgih.org/carbon-dioxide/.

7. **National Institute for Occupational Safety and Health (NIOSH).** NIOSH Pocket Guide to Chemical Hazards. *Carbon dioxide.* [Online] https://www.cdc.gov/niosh/npg/npgd0103.html.

8. *COSHER joint industry project: Large scale pipeline rupture tests to study CO2 release and dispersion.* **Ahmad, et al.** s.l. : International Journal of Greenhouse Gas Control, 2015, Vol. 37, pp. 340-353.

Expect the Unexpected: A Process Safety Story Involving Dangerous Diborane

Ralph Zhao
Merck & Co., Inc.
126 E. Lincoln Ave, Rahway NJ 07065, USA
Ralph.zhao@merck.com

Brittany Armstrong, Taylor Behre & Nadine Kuhl
Merck & Co., Inc.
126 E. Lincoln Ave, Rahway NJ 07065, USA

Keywords: diborane, calorimetry, thermal hazard, process hazard analysis

Abstract

During process development to produce the penultimate amino alcohol TsOH salt intermediate to Nemtabrutinib, diborane was unexpectedly discovered to be the in-situ generated reducing reagent. The reaction used boron trifluoride etherate as a Lewis acid and triethylsilane as the reducing agent. Diborane is highly flammable and acutely toxic and poses many process safety hazards, however, the process is required to be conducted in a closed vessel to meet the quality parameters. The reaction headspace was studied using React IR and GC analysis, while the reaction stream was studied using heat flow and adiabatic calorimetry to ensure this process could be safely scaled up. Herein, we describe how it was discovered, how process safety analysis was conducted and how we handled it safely at our pilot plant.

1 Introduction

Diborane is a highly reactive, flammable and acutely toxic gas.[1-3] Some literature sources including the commonly used Bretherick's Handbook of Reactive Chemical Hazards reference it as pyrophoric with an autoignition temperature of 38 °C.[1-4] Its lower explosive limit (LEL) is 0.8% (v/v) in the air. The US health exposure limit from the National Institute for Occupational Safety and Health (NIOSH) lists a permissible exposure limit (PEL) of 0.1 ppm and an immediately dangerous to life and health (IDLH) limit of 15 ppm.[1] More recently, a paper was published[5] stating pure diborane is not pyrophoric, and the autoignition temperature of pure diborane is 136-139 °C, much higher than previously reported but still reachable within a typical maximum jacket temperature. With this in mind, one should avoid using any reagents or combination of reagents that potentially generates diborane. However, elimination and substitution are not always feasible when considering other aspects of process development.

The reaction called acetal ring opening in Scheme 1 is used to produce an intermediate called the amino alcohol tosylate (TsOH) salt. This intermediate is further processed to manufacture an active pharmaceutical ingredient (API) called nemtabrutinib, an oral small molecule reversible inhibitor of Bruton's tyrosine kinase that is currently being developed for the treatment of chronic lymphocytic leukemia (CLL). Originally, sodium borohydride ($NaBH_4$) was proposed as the reducing reagent. It is known that using sodium borohydride would generate diborane.[4,6] Thus, triethylsilane was selected as the reducing agent and boron trifluoride diethyl etherate was selected to activate the acetal ring to ideally avoid the production of hazardous diborane as a side product.

Scheme 1. Acetal Ring Opening Step

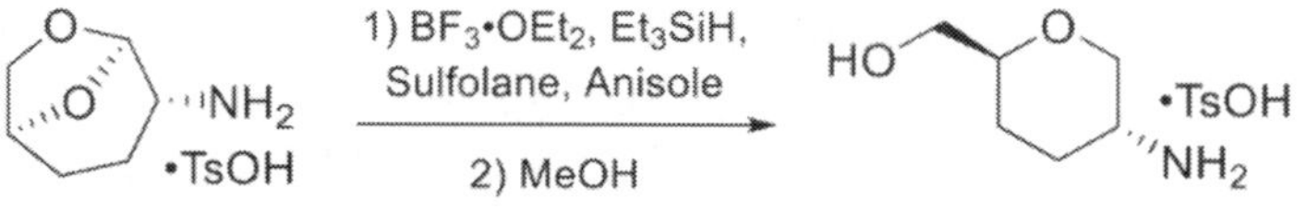

1 2

The detailed process development and optimization including all process safety testing for this particular process has been published[7] in the Organic Process Research and Development (OPRD) journal in 2023. Only the key process safety findings are discussed here.

2 Results and Discussions

2.1 Discovery of Diborane in the new process

The Environmental and Process Safety Engineering (EPSE) group at Merck & Co., Inc., Rahway, NJ, USA carries out thermal hazard characterization for all processes going into our pilot plant and manufacturing facilities. Before the process parameters are finalized for formal safety testing, a preliminary headspace screening analysis was conducted using single injection gas-phase infrared spectroscopy (IR) of the reaction headspace after 30 minutes and at the end of reaction (EOR) since gases, if any, were expected at either stage of approximately 18-20 hours reaction. Diborane has characteristic peaks in the 2450–2640 cm^{-1} region[8] that correspond to symmetric and asymmetric stretches of -BH2. No diborane was present at these specific time points.

However, detailed mechanistic studies[9] by nuclear magnetic resonance (NMR) spectroscopy of samples taken throughout the reaction revealed that the reagent combination of boron trifluoride diethyl etherate and triethylsilane in sulfolane results in the in-situ formation of diborane and triethylsilylfluoride (Scheme 2), a phenomenon that was previously unknown in literature. These NMR studies did not contradict the preliminary headspace screening since diborane was not detected by NMR after 30 minutes or at EOR. Three other observations from the NMR data revealed a key mechanistic detail: 1) the formation of diborane is delayed until after addition of the final reagent of BF_3-OEt_2 and heat up to 40 °C, which explains why diborane was not observed in the 30-minute headspace IR sample; 2) there is no conversion to the product **2** until after formation of diborane; and 3) the diborane concentration in solution decreases throughout the reaction and is not detected at EOR. These findings led us to conclude that diborane acts as the active reducing agent in this synthesis.

Scheme 2. Reduction of BF_3-OEt_2 by Triethylsilane to Form Diborane

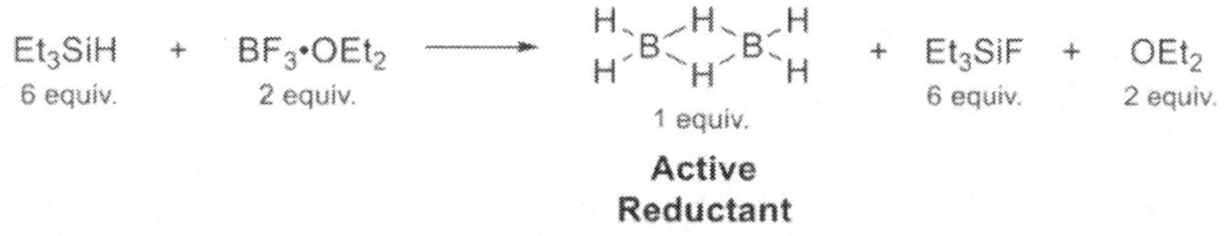

If only trace amount of diborane is detected in the headspace, it would not be a concern for the scale-up facility. However, the peak diborane concentration under typical reaction conditions was found to be around 9.6% (v/v), significantly above its LEL (0.8% v/v), PEL (0.5 ppm), and IDLH (15 ppm). Correspondingly, Figure 1 from a standard reaction run in a Mettler Toledo EasyMax vessel connected to a gas phase IR probe showed no diborane was detected in the reaction headspace prior to heat up to 40 °C (Figure 1a, green trace). However, after heating up to 40 °C diborane began to form in solution, partitioned into the headspace, and continued to increase in concentration until its peak around 2 hours (Figure 1a, red trace). After 2 hours, the diborane concentration decreased as it was either consumed by the reaction or swept away. At EOR the diborane was no longer detected (Figure 1a, purple trace). Figure 1b shows the profile of diborane over the course of the reaction. Diborane also exhibits a very strong peak in the 1610–1540 cm^{-1}, associated with the asymmetric stretch of the bridging hydrogen atoms, however, the calibration curve/measurement was built by integrating the diborane peak around 2512-2450 cm^{-1}due to much less interference in this region from other compounds.

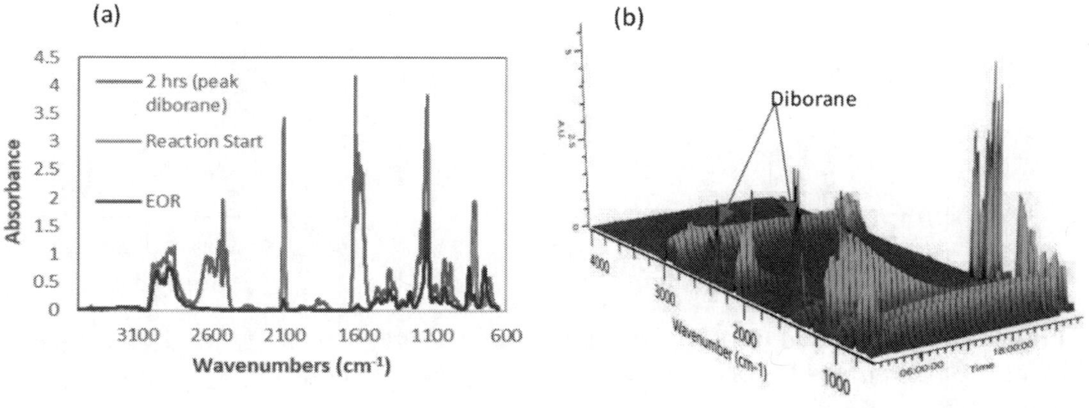

Figure 1. (a) IR spectra of the reaction at key timepoints, and (b) IR surface of reaction over time, with diborane peaks at ca. 2500 cm^{-1} and 1600 cm^{-1}.

In addition to diborane, the reduction reaction also generates a large amount of hydrogen. Headspace hydrogen levels were analyzed by GC through a series of reactions run in vials with pressure release caps and found to be around 75% (v/v) in the headspace 30 minutes into the reaction and increased to nearly 100% by EOR, indicating that hydrogen is produced at a rate that purges the original nitrogen out of the headspace through the pressure release cap. The presence of these high amounts of flammable (hydrogen and diborane) and toxic (diborane) gases in the reaction headspace means appropriate venting of headspace gases is of utmost importance for the safe operation of the reduction reaction at large scale.

2.2 Open vessel process vs closed vessel process

From a safety point of view, any process which generates a flammable gas should be vented directly to the roof through a pipe from the vessel. If it is a toxic gas or a gas with an environmental emission limit, it should go through an appropriate scrubber before being released into the air. In theory, diborane can be easily scrubbed since it reacts with water. However, its LEL is very low. In addition, a huge amount of hydrogen is also generated in the process, and it is not possible to dilute it by nitrogen below its LEL before passing through a scrubber which is typically not inerted due to the prohibitive cost. Practically, these gases can be passed through an inerted vessel filled with water to quench any unreacted diborane and then vented directly to the roof. This becomes unnecessary due to subsequent discoveries and process changes.

During process development and optimization, it was discovered that reactions run under a higher nitrogen flow had a decreased reaction conversion (Figure 2a), resulting in a lower yield. One impurity called A is also identified in addition to the product **2**. This impurity cannot be purged through the downstream processes to the API. But interestingly, reactions run under pressure had significantly increased conversion and suppressed this impurity A (Figure 2b).

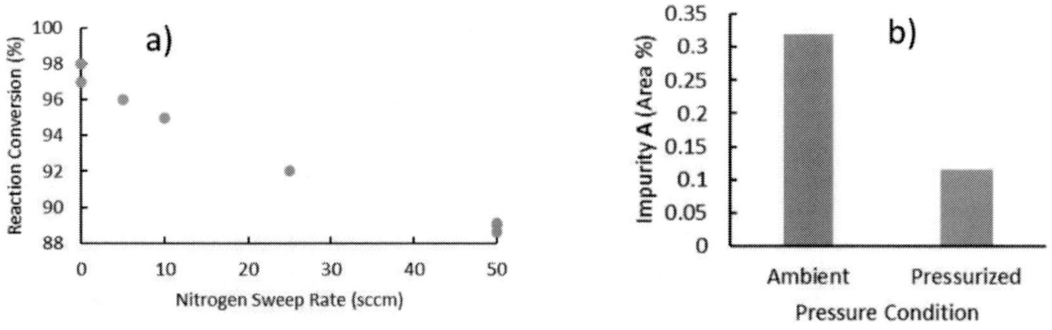

Figure 2. (a) Effect of nitrogen sweep rate on a reaction at 40 °C for 18 hours, and (b) effect of ambient (n=1) versus pressurized (n=4) reaction conditions on impurity levels in isolated **2**.

Considering diborane as the in-situ generated reductive species, it is not unexpected that a higher nitrogen flow would decrease the reaction conversion rate and consequentially the yield since it would sweep away diborane before it fully reacts. The exact mechanism of how the pressurized reaction suppresses Impurity A formation is unclear. But it is clear that operating under pressure would increase the observed diborane concentration within the reactor and maintain more dissolved diborane in the liquid-phase of the batch. As such, there is a process need to conduct this reaction under pressure.

2.3 Thermal and pressure hazard characterization

For any closed vessel process, one should consider if the pressure generated by gas and vapor from both normal operation and a jacket runaway scenario would reach the vessel MAWP (maximum allowable working pressure) since the set pressure of the emergency pressure relief valve is at or near the MAWP. Opening of the emergency pressure relief valve under any circumstance is considered to be a process safety incident, especially with diborane and hydrogen in the vessel. In addition, the closed vessel at elevated pressure would increase the possibility of a leak which poses a risk to operators due to diborane's low PEL and IDLH.

2.3.1 Normal process conditions

Multi-Modular Calorimetry (MMC) is a scanning or adiabatic calorimeter (two modules) which can measure exotherm and pressure generation of a sample during a temperature range. Once all reaction parameters are finalized in the lab for the process, MMC isothermal age was carried out at 45 °C (5 °C above the normal process temperature) for 20 hours with 36% fill level in a 2.8 mL nominal volume test cell to simulate the maximum expected pressure accumulation in a closed vessel at a fill level close to the proposed pilot scale fill. The pressure profile indicated a maximum pressure (PMAX) of 73 psig (5.03 bar, Figure 3, blue trace). The pressure profile showed two distinct slopes: one initial slower pressure increase and then a sudden steeper pressure increase. This corroborates well with the previous headspace analysis of two gas generating reactions: initially, only hydrogen is generated, and then diborane is formed after an apparent induction period following heat up to 40 °C. In addition, a larger scale reaction at 30% fill in a 35 mL nominal volume sealed glass pressure tube with magnetic stir bar and oil bath heating showed similar behavior to this MMC isothermal age, with maximum observed pressure of 71 psig (4.90 bar).

Another MMC isothermal age with 65% fill showed a maximum pressure of 179 psi, indicating that the fill level directly affects pressure accumulation in a closed vessel, as expected. In both MMC tests, no significant exotherm was observed. Both solvents used in this process, anisole and sulfolane, have a very high boiling point, 154 °C and 285 °C respectively. The vapor pressure from the solvents including small amount of diethyl ether from BF_3-Et_2O under the normal process conditions of 45 °C is expected to be negligible. Thus, the pressure observed is assumed to be mainly from gases of diborane and hydrogen.

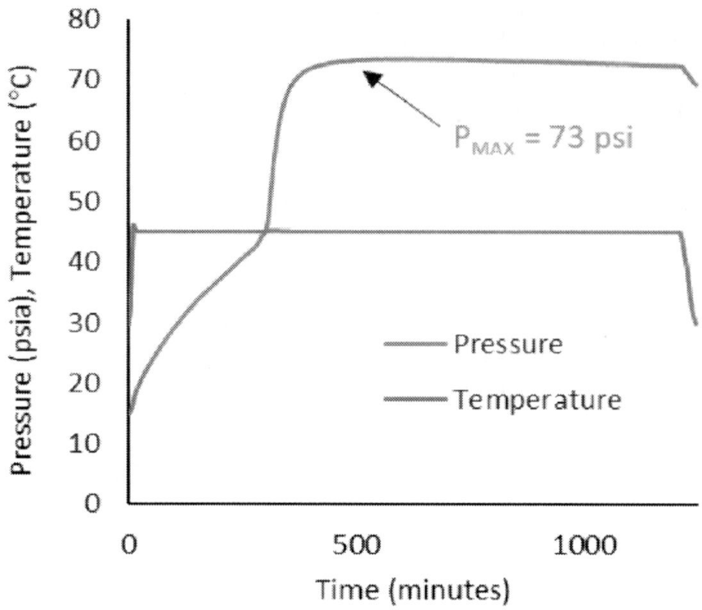

Figure 3. MMC isothermal age of pre-reaction mixture at 45 °C.

The vessel selected for this process scale-up at our pilot plant has MAWP of 120 psig. It is preferred to have a 50% safety margin for maximum accumulated process pressure. Of course, one could select another vessel with a higher MAWP which could allow a higher fill volume.

Another engineering method to control the total accumulated pressure in the selected vessel with a fixed fill volume is to use a pressure control valve (PCV) to maintain the pressure at a specified value at any given time during the whole process. In fact, more detailed studies (Table 1) indicated 50 psig is preferred from both process quality and process safety points of view. Consequently, a 30% target fill volume along with PCV control to 50 psig was chosen for our pilot plant to satisfy overall process objectives and to employ multiple layers of process safety in vessel pressure control.

Table 1. End of reaction purity with varying operating conditions.

Entry	Equipment	Vessel Fill (%)	Pressure (psig)	EOR Impurities[d] (Area %)
1	Glass tube	30	71[a]	2.37
2	EasyMax	30	75[a]	2.28
3	EasyMax	30	50[b]	2.40
4	EasyMax	60	50[b]	2.34
5	EasyMax	30	40[b]	3.13
6	EasyMax	30	0[c]	3.46

[a]Reactions were run in completely sealed vessels with no pressure relief.
[b]Reactions were run with a backpressure regulator to maintain pressure at the desired setpoint.
[c]Reaction was run at ambient pressure with a 5 sccm nitrogen sweep.
[d]Several impurities, including impurity **A**, were not separated in this analytical method.

2.3.2 Jacket runaway scenario

The Global Safety and Environment (GSE) organization standard at Merck & Co., Inc., Rahway, NJ, USA has identified a jacket runaway scenario for a batch process as the single most credible upset scenario for our manufacturing facility. EPSE conducts standard initial tests with a Differential Scanning Calorimeter (DSC) and a MMC up to the maximum jacket temperature to screen for potential secondary exotherm and pressure generation.[10]

DSC scan on the pre-reaction stream (prior to heat-up to 40 °C) was run up to 350 °C at 5 °C/min showed only one insignificant exothermic activity of 27 J/g at 76 °C up to the pilot plant maximum jacket temperature of 145 °C (Figure 4a). MMC scan up to 220 °C at 2 °C/min identified a maximum pressure rise rate of 5.2 psi/min at 96 °C (Figure 4b, blue trace). Similar to the DSC results, the MMC scan indicated a small exotherm (21 J/g) at 72 °C (Figure 4b, red trace). Neither the DSC scan nor the MMC scan indicated significant concern in terms of thermal and pressure hazard up to the standard maximum jacket temperature of 145 °C for the pilot plant. Additionally, when the desired chemistry heat of reaction was studied, an adiabatic temperature rise of 7 °C was observed upon heat-up to 40 °C during lab scale reactions using a EasyMax calorimeter.

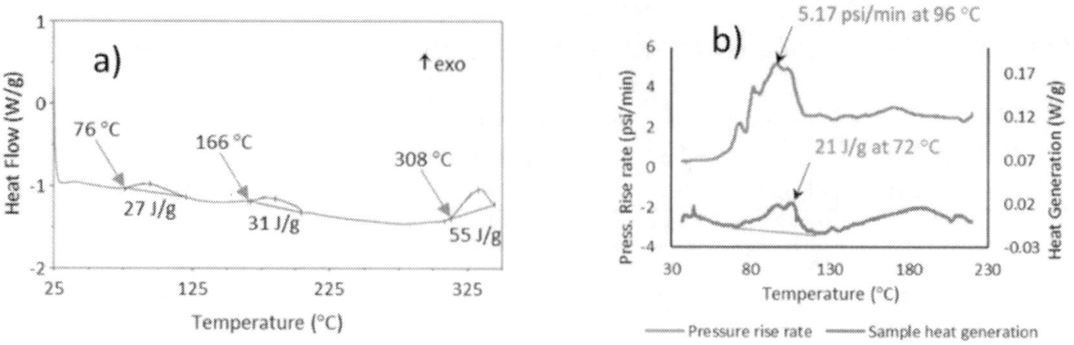

Figure 4. (a) DSC scan of the reaction mixture, and (b) Heat and pressure generation from a MMC scan of the reaction mixture.

However, a significant delayed exotherm was observed upon heat up to 40 °C for a 1.5 kg scale run of the reaction which was carried out under controlled pressure with a pressure regulator in a 38 L Hastelloy reactor with 30% fill volume (Figure 5). This exotherm occurred simultaneously with the second pressure spike, which was associated with the formation of diborane gas.

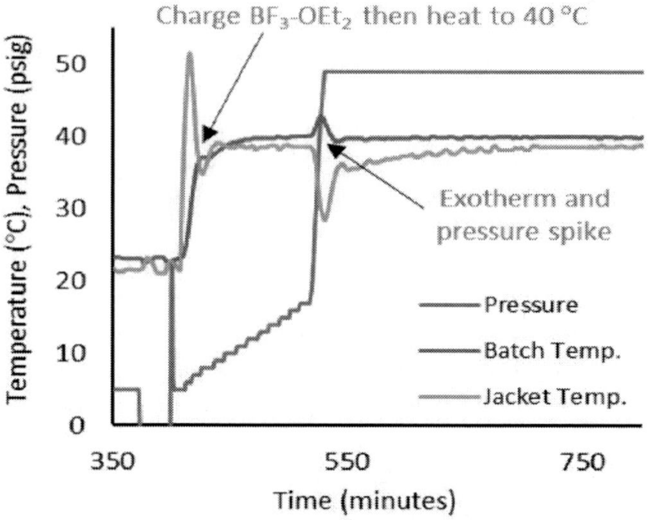

Figure 5. Pressure and temperature profile from the reduction run in 38 L Hastelloy reactor with backpressure regulator set to 50 psig.

This observation led us to believe that the thermal hazards of the desired reaction had not been adequately captured by DSC/MMC small scale screening. The exotherms from DSC and MMC scan of the pre-reaction mixture at 76 °C and 72 °C, respectively, appeared to be the same delayed exotherm that was observed in both the EasyMax run and the larger scale Hastelloy reactor run. The magnitude of this exotherm was probably underestimated due to the scanning mode utilized by DSC and MMC and the more efficient heat transfer in the small-scale EasyMax calorimeter. As such, an Accelerating Rate Calorimeter

(ARC) study of the pre-reaction mixture was carried out to determine if this exothermically driven gas generation would pose a safety hazard to the pilot plant. The ARC instrument is more sensitive to exothermic activity with the ability to detect sample self-heat rates as low as 0.02 °C/min. Adiabatic calorimetry also more accurately mimics conditions found in a pilot plant or manufacturing plant scale due to the increased volume to surface area ratio that limits heat transfer, especially in the case of a thermal runaway.

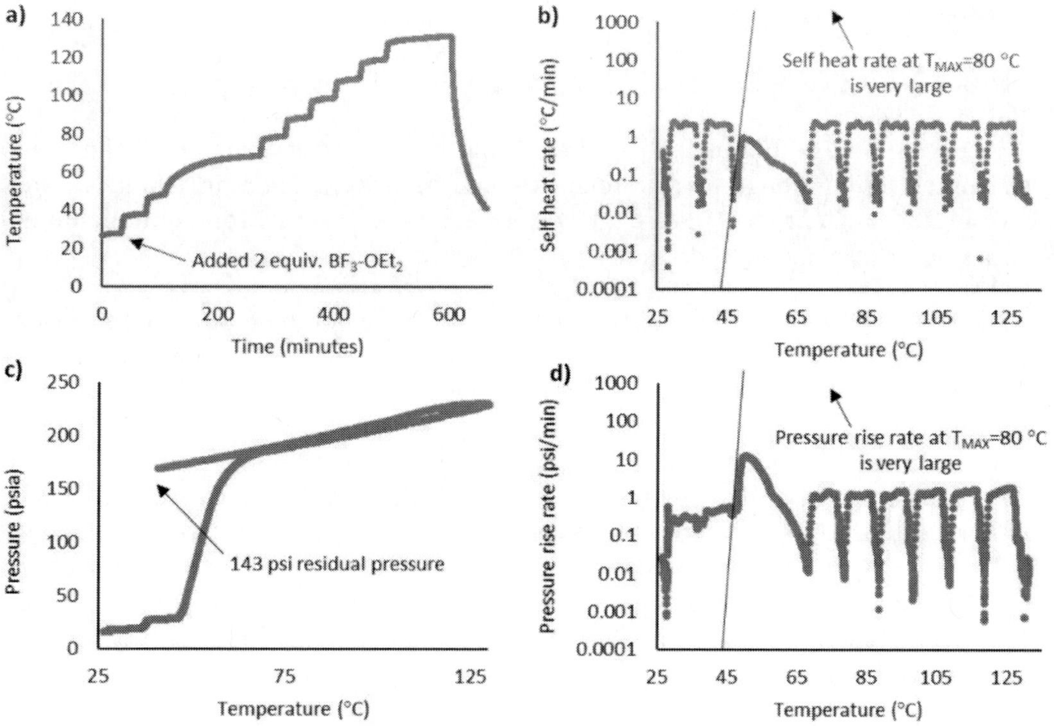

Figure 6. ARC analysis of the reaction mixture showing delayed exotherm upon heat up. (a) Temperature vs. time, (b) self-heat rate, (c) temperature vs. pressure, and (d) pressure rise rate data for ARC HWS of the reaction mixture.

An ARC heat-wait-search (HWS) program was used to identify any sustained exothermic activity in the reaction mixture. ARC analysis of the reaction mixture showed no exotherm upon addition of boron trifluoride diethyl etherate to the reaction mixture (Figure 6a). However, a sustained exotherm was observed with an initiation temperature of 49 °C and an adiabatic temperature rise of 31 °C. Significant residual pressure (143 psi) was observed in conjunction with the exotherm (Figure 6c). This exotherm and pressure rise is from the desired chemistry and no secondary exotherm or pressure spike is observed up to the maximum pilot plant jacket temperature of 145 °C. When corrected for a phi factor of 1.5 from ARC, a maximum temperature (TMAX) the process can reach without cooling from this desired chemistry is 80 °C: TMAX = exotherm onset temperature + adiabatic temperature rise from exotherm.

Since there is no secondary exotherm or pressure spike, the major thermal and pressure hazard of this process is from the desired chemistry. Although it is not an ideal process, the delayed exotherm from the desired chemistry can be controlled by a slower jacket ramping rate and/or fast cooling when the anticipated exotherm initiates in the pilot plant batch. Simulation of the total pressure from the exotherm of the desired chemistry at the TMAX indicated that 79 psi would be generated in the vessel with a 36% fill. This pressure generation did not pose a hazard when compared to the vessel MAWP (120 psig), as such, emergency vent sizing calculations were not necessary if the maximum jacket can be controlled at TMAX. However, at the normal maximum jacket temperature of 145°C, the total accumulated pressure would far exceed the vessel MAWP.

2.4 Process safety controls at the pilot plant

The process involves diborane and hydrogen evolution with a delayed exotherm in a closed vessel. Even though no secondary exotherm or associated pressure rise was observed, a thorough process hazard analysis (PHA) was warranted and conducted at the pilot plant to ensure a safe operation.

The major concern was whether the accumulated pressure in the vessel would trigger the emergency relief valve to open under the worst-case scenario, e.g., if the automatic PCV malfunctioned and/or in a jacket runaway scenario. A total accumulated pressure from the desired chemistry at ARC TMAX is less than the vessel MAWP as mentioned above but it would exceed the vessel MAWP at the maximum jacket temperature of 145 °C if the PCV does not work at the targeted fill volume. When considering the diborane AIT of ~135 °C vs. the ARC TMAX, the maximum jacket temperature was recommended to be limited to 80 °C using a hardwire interlock to ensure that a jacket runaway upset could not raise the batch temperature to the diborane AIT (with 50 °C safety margin) and the pressure accumulated would not reach the vessel MAWP (with 40 psi safety margin) in case of the PCV malfunctioning and/or jacket cooling failure.

Secondarily, reaction pressure was controlled by allowing some venting of the hazardous gases to the atmosphere through an automatic PCV with a 50 psig setpoint. This strategy was designed to keep the vessel pressure below 50% of the vessel MAWP and to allow the batch to be quickly vented down in case of a pressure excursion. Reducing the batch size to ensure 30% fill volume during the reaction was another layer of protection used against over-pressurization. During the periodic venting through the PCV in the process, some diborane would be vented out. The environmental emission assessment was carried out indicating that even if all diborane generated from the batch is released (not reacted at all), it would not exceed its emission limit, so the pilot plant is in compliance without the need to go through a scrubber.

Industrial hygiene (IH) associated with diborane exposure to operators is another major concern due to its significant toxicity. Several controls were put in place: 1) the reaction was run in a closed, isolated area of the pilot plant with diborane detectors installed strategically throughout the production area; 2) operators used personal diborane monitors and wear appropriate PPE, including supplied air respirators, for any operations inside the production area during the reaction and until the quench was complete at which point the diborane was no longer present; 3) a comprehensive emergency response plan was developed and included actions based around low, medium, and high threshold values of diborane levels as observed by the compound-specific detectors related to the PEL and IDLH values. A pressure leak testing of the vessel was mandated prior to the start-up.

Finally, flammability concerns for hydrogen and diborane were addressed by operating in a bay rated for hydrogenation reactions, maintaining inerting in the vessel, and directly venting the vessel headspace to the atmosphere through group B flame arrestors.

Figure 7 shows actual pilot plant data from two batches showing the two-stage pressure increase was also observed consistently at pilot scale. Following the rapid pressure increase associated with diborane generation, the PCV accurately controlled the vessel pressure at 50 psig as desired throughout the remainder of the reduction reaction (Figure 7a). A significant exotherm was observed consistently in conjunction with the pressure spike (Figure 7b). This event occurred within a spread of no more than 10 minutes across both batches. With the prior knowledge of this delayed exotherm from the desired reaction, operators were able to control the batch temperature by manually lowering the jacket temperature in anticipation of the temperature rise. Overall, the reaction was well controlled, consistent, performed with satisfactory purity and without incident.

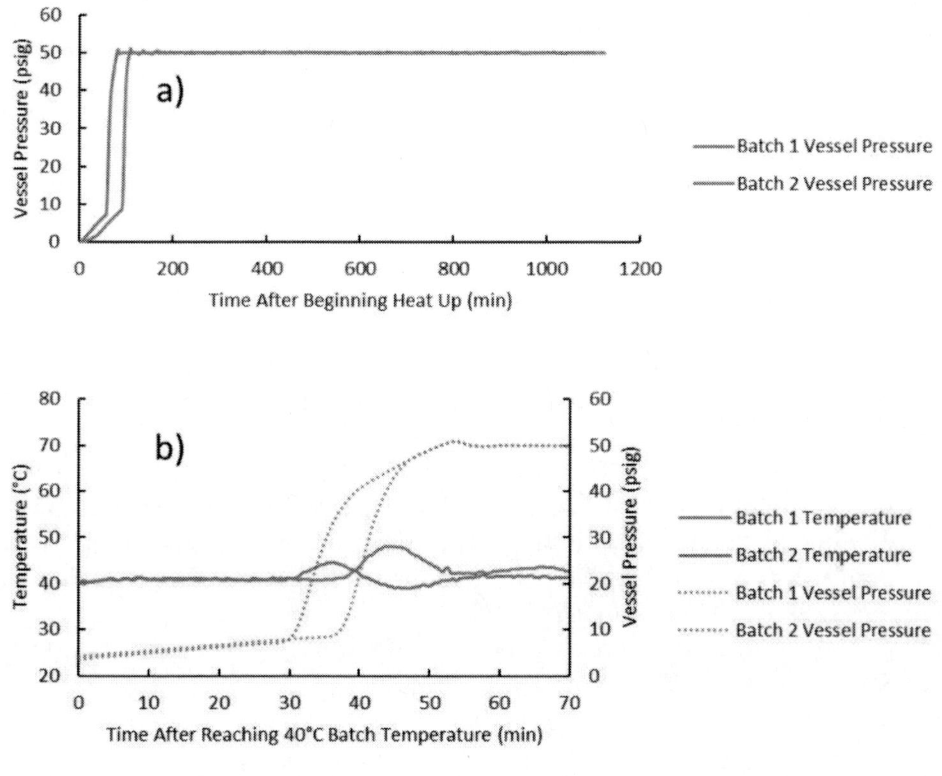

Figure 7. (a) Vessel pressure trends for two batches, and b) Trends showing corresponding pressure spike and exotherm for two batches.

2.5 Conclusion

In conclusion, we have described how we identified and characterized diborane in the synthesis of amino alcohol TsOH salt intermediate in the preparation of Nemtabrutinib. The reaction was a unique process that had an unusual number of high severity hazards. Hydrogen is generated throughout the reaction, and following an induction period, diborane is also generated. The diborane generation is associated with a delayed exotherm with a large self-heat rate and high pressure rise rate. Process improvements, including operating the reaction under pressure, were implemented to control the formation of key dimer impurity and manage the gas-generation. Given the large number of potential upsets that needed to be controlled, a safety basis was established after extensive safety testing and conducting a thorough PHA. Though it is not an ideal process from a process safety point of view, we safely demonstrated this process at our pilot plant scale three times. The team collaborated well through 2020 COVID peak period to address all process safety concerns and has successfully transferred this process to our partner to continue the API clinical supply for the program.

3 References

1) Wikipedia Diborane - Wikipedia
2) Horiguchi, S.; Urano, Y.; Otani, H.; Tokuhashi, K.; Iwasaka, M.; Kondo, S., Explosion hazard of diborane. *Koatsu Gasu* **1988,** *25* (11), 37-43.
3) Krackow, E. H., Toxicity and health hazards of boron hydrides. *AMA Arch. Ind. Hyg. Occup. Med.* **1953,** *8* (4), 335-9.
4) Urben, P. G.; Bretherick's Handbook of Reactive Chemical Hazards, 2017, 8th edition, p51-52.
5) Monteiro, A. M.; Flanagan, R. C., Process Safety Considerations for the Use of 1 M Borane Tetrahydrofuran Complex Under General Purpose Plant Conditions. *Org. Process Res. Dev.* **2017,** *21* (2), 241-246.
6) Kanth, J. V. B.; Brown, H. C., Improved Procedures for the Generation of Diborane from Sodium Borohydride and Boron Trifluoride, *Inorg.Chem.* **2000,** 39 (8), 1795-1802.
7) Armstrong, B. M.; Behre, T.; Turnbull, B. W. H.; Bishara, D.; Hartmanshenn, C.; McCarthy, E.; Whittington, M.; Ji, Y.; Jenks, A.; Desmond, R.; Muzzio, D. J.; Corry, J.; Zhao, R.; Kuhl, N.; Chung, C. K., Integrating Process Development and Safety Analysis for Scale-Up of a Diborane-Generating Reduction Reaction, *Org. Process Res. Dev.* **2023,** 27 (4), 763–774**.**
8) Colthup, N. B., *Introduction to Infrared and Raman Spectroscopy*. Academic Press: San Diego, CA, 1990; Vol. 3rd Edition.
9) Khul, N., Ji, Y., et. al., https://chemrxiv.org/engage/chemrxiv/article-details/63614f7bac45c75ba5a114af.
10) Zhao, R.; Roth, M.; Furman, T., MSD's Process Safety Scale-Up Methodology for Pilot Plant Scale and Beyond, Org. Process Res. Dev. **2021**, 25 (11), 2566–2577.

How to Accurately Predict Thermal Runaway: Experimental Study Involving Lab-Scale and Full-Scale Tests

Scott Davis
Gexcon US
4833 Rugby Ave, Ste 100, Bethesda, MD 20814
Email scott.g.davis@gexcon.com

John Pagliaro, Derek Engel, Thomas DeBold
Gexcon US

Keywords: Thermal runaway, exothermic polymerization, fire and explosion investigation, adiabatic calorimetry

Abstract

Thermal runaway involving chemical reactions can be a complex process, which typically involves an exothermic reaction (i.e., a reaction that gives off heat). This will result in a temperature rise where reaction rates will exponentially increase with temperature. If sufficient cooling is present, the reaction will remain under control; however, if heat generation exceeds the heat loss, then thermal runaway can occur. This paper will present the experimental methods used to predict the runaway condition from the auto-polymerization of 80% grade of divinylbenzene (DVB80) in ISO-containers on July 14, 2012, when an explosion and ensuing fire significantly damaged the container ship MSC Flaminia that was transporting cargo from the United States to Europe. Given DVB80 uses inhibitors to prevent exothermic polymerization, this paper will first demonstrate how appropriate lab-scale calorimetry tests can be used to predict the induction time to auto-polymerization or shelf life. Next, this paper will discuss how certain investigators used incorrect test methods, which on "face" value appeared justified, and erroneously concluded that DVB80 could not violently runaway and that "mild" exotherms were observed. These investigators mistakenly neglected the importance of increased heat loss due to the large surface-area-to-volume ratio inherent in smaller test apparatus. When these tests were repeated with appropriate lab-scale calorimetry tests, violent thermal runaway was predicted. Full-scale runaway tests were also performed with an actual ISO-container and agreed very well with the lab-scale calorimetry tests, demonstrating the importance of taking into account scaling issues when investigating runaway conditions.

1 Introduction

In July of 2012, the MSC Flaminia set forth on an international shipping voyage from the east coast of the United States to Europe. Aboard the ship were 23 crewmembers and two non-crew passengers. Cargo was loaded onboard the MSC Flaminia in New Orleans, LA on July 1, 2012. The ship made port for additional cargo loadings in Mobile AL, Freeport Bahamas, Savannah GA, and lastly, Charleston SC before beginning the voyage across the Atlantic Ocean on July 8, 2012. The ship carried 4,805 containers stowed on, and below deck.

1.1 Incident

On the morning of July 14, 2012, a smoke alarm sounded aboard the Flaminia. The alarm went off at approximately 05:42 and indicated smoke in cargo hold no. 4. Crew reported seeing a small amount of white smoke coming from cargo hold 4, and minutes later at approximately 05:50, the general alarm sounded, alerting the crew of the emergency situation. Several crewmembers prepared water hoses to cool cargo hold no. 4, and other members closed the passive ventilation flaps for the hold. Once the flaps were closed, CO_2 was discharged into the hold. According to alarm records, the first discharge occurred at 06:43 and lasted until approximately 07:05. At 07:07, the second CO_2 discharge took place and lasted until approximately 07:13. The explosion occurred around 08:04 followed by large black smoke cloud. The event was powerful enough to knock containers off the deck on both the port and starboard sides (see Figure 1) and fatally injured three crewmembers. The blue ISO-tank containers seen in the image on the left were stored on deck, above bay 26. Right around the time the explosion occurred, crewmembers initiated the third CO_2 discharge sequence.

Figure 1: Cargo containers thrown from the deck.

The consensus was that white smoke emanated from the area of hold no. 4 before the explosion and thick black smoke billowed out of the same area after the explosion as shown in Figure 2. Passengers also referred to either the general smell onboard the Flaminia, or the white smoke emanating from the hold prior to the explosion as "pungent", "unpleasant", "a bit like a chemical smell", and "similar to that of burning plastic or of burning electrical insulation". The entire event (fire, explosion, and post-explosion fire) damaged or consumed many cargo containers onboard (see upper image in Figure 3) and the majority of the damage occurred in cargo holds no. 4 through

6 (see lower image in Figure 3). The ensuing fire burned for approximately 10 days after the initial explosion.

Figure 2: Post-explosion image showing the smoke from cargo hold no. 4.

Figure 3: The MSC Flaminia after the incident (upper) and post incident image of hold no. 5 (lower)

1.2 Cause of the incident

Investigations have determined that on July 14, 2012, there was an auto-polymerization of 80% grade of divinylbenzene (DVB80) loaded in ISO-containers aboard the MSC Flaminia container ship. The auto-polymerization reaction reached temperatures and pressures that exceeded the 4.4 barg (63.8 psig) set point of pressure relief valves (PRV) on the DVB80 ISO-tank container. This resulted in a violent discharge of DVB80 contents via the PRV into hold 4, followed by a subsequent ignition and explosion, and an ensuing fire that significantly damaged the ship transporting cargo from the United States to Europe. Just under 23 days had passed from the time that the DVB80 was loaded into the ISO-tank containers until the day of the incident (22.7 days). Given there have been no reported incidents of auto-polymerization of DVB80 after only 22.7 days, the goal of our investigation was to determine how DVB80 reached auto-polymerization and thermal runaway after only 22.7 days. This paper will focus on how lab-scale calorimetry tests can be approximately used to predict the induction time to auto-polymerization by comparing lab-scale results to those determined on an exemplar full-scale ISO-tank container test. Prior to discussing the analysis, the temperature exposure of the DVB80 ISO-tank containers is discussed.

1.3 Events prior to the incident

DVB80 was initially loaded into standard 20-ft – 25,000 liter, insulated ISO-tank containers at 9:40 am CDT on June 21, 2012, and the temperature of the DVB80 recorded in these tanks was approximately 6.7°C. The DVB 80 ISO-tank containers were transported by truck to New Orleans, where they remained at the port in New Orleans until July 1, 2012, when they were loaded onboard the Flaminia into hold 4 at approximately 3:00 am CDT. The temperatures at the New Orleans Port are shown in Figure 4.

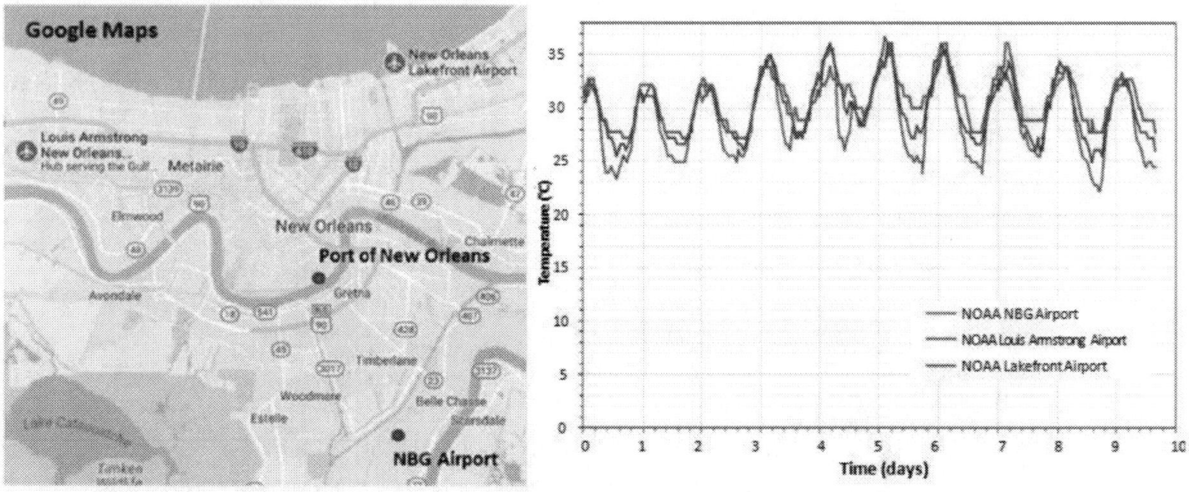

Figure 4: Google maps image showing the location of the Port of New Orleans and the three nearby airports (left image) and the temperature measured at each airport (right image).

Once the DVB80 ISO-tank containers were loaded on the Flaminia in New Orleans, they were exposed to temperatures present in the hold. Hold temperatures were realistically estimated (see green and blue lines in Figure 5) by considering: (1) the mass of contents loaded into the hold; (2) the temperature of the loaded contents; and (3) the external temperature measured on the ship. Figure 5 also shows a conservatively high hold temperature estimate (see red line), which overpredicted the temperature of the hold contents loaded on July 1, 2012

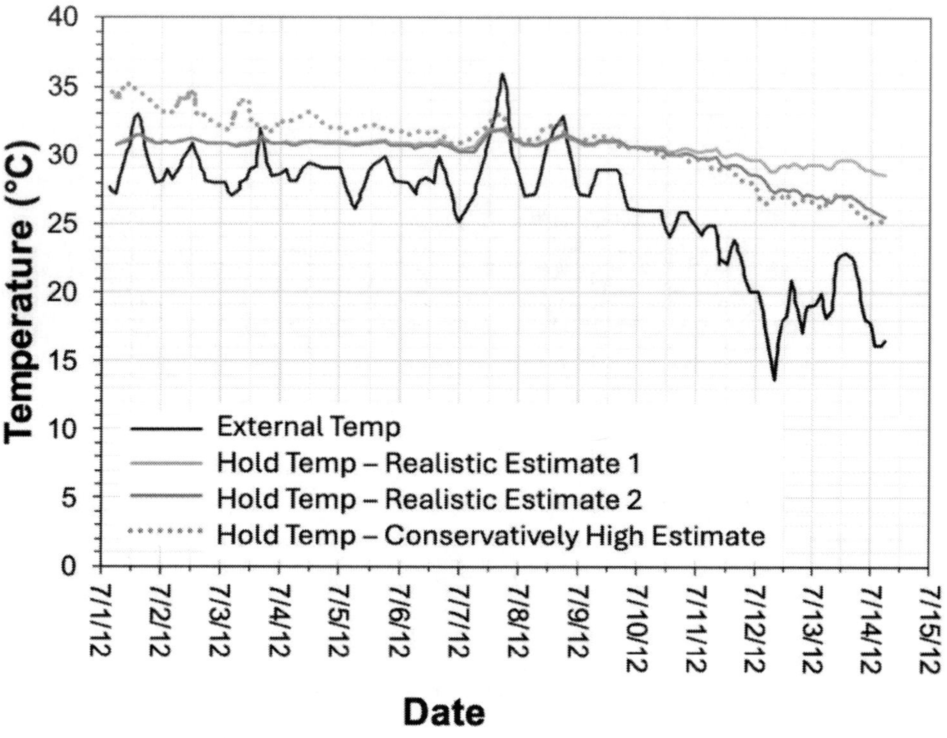

Figure 5: Hold air temperature predicted using the thermal masses for similar vessels.

2 Properties and characteristics of DVB80

Three ISO-tank containers in hold no. 4 of the MSC Flaminia contained DVB80. DVB80 is predominately comprised of divinylbenzene (~80% DVB by weight) and ethylvinylbenzene (~20% EVB by weight). The DVB monomer is a clear liquid at ordinary ambient temperatures, with a boiling point of 200°C (392°F), and a melting point of -88°C (-126.4°F). At 20°C (68°F) and at 7.2°C (45°F), DVB80 has a density of 7.69 lb/gallon (0.92 kg/liter) and 7.84 lb/gallon (0.94 kg/liter), meaning its density is approximately 92-94% that of water. The lower flammability limit (LFL) for DVB80 is 1.1% (vol% in air), whereas its upper flammability limit (UFL) is 6.2% (vol% in air). DVB80 has a closed cup flashpoint of 69°C (156.2°F) and an open cup flashpoint of 76.7 °C (170 °F), with the lowest reported auto-ignition temperature AIT of 470°C (878°F). Table 1 presents a list of physical, thermodynamic and combustion properties of DVB80.

Table 1: Physical, thermodynamic, and combustion properties of DVB80

Parameter	Units	Value
Heat of Vaporization	kJ/kg	405
Boiling Point, at 760 mm Hg	°C	200
Vapor Density (air = 1)	ρ/ρ_{air}	4.5
Liquid Specific Heat Capacity	kJ/(kg*K)	1.76
Liquid Density at 20°C	kg/m³	921
Molecular Weight	kg/kmol	130.19
Heat of Polymerization	kJ/kg	735.7
Flash Point	°C	69
Auto-ignition temperature	°C	470
Upper flammability limit	%Volume	6.2
Lower flammability limit	%Volume	1.1
Vapor-Specific Heat Capacity	kJ/(kg*K)	1.25

Divinylbenzene is a monomer that can undergo heat initiated free radical polymerization depending on the exposure conditions. The DVB monomer has a chemical composition of C10H10, and stably exists in two different forms, or "isomers", known as "meta-DVB", which has the two vinyl groups bonded to the first and third carbons of the 6-carbon aromatic ring (see left image in Figure 6) and "para-DVB" , which has the two vinyl groups bonded to the first and forth carbons of the 6-carbon aromatic ring (see right image in Figure 6).

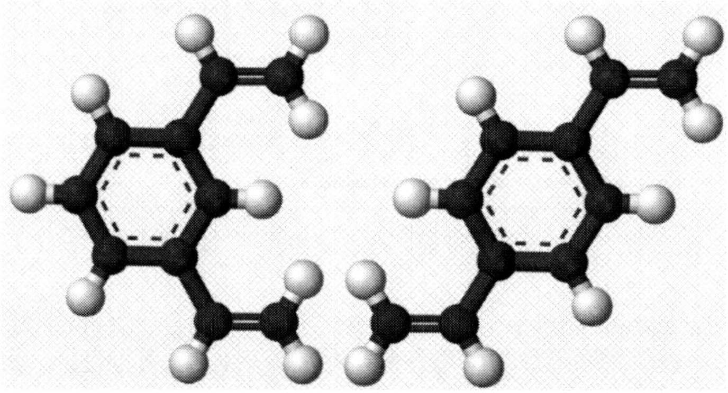

Figure 6: Schematic of meta-DVB (left) and para-DVB (right)

When heated, the monomer molecule becomes unstable and forms a reactive free radical that can react with another DVB molecule to start a polymer chain. Polymerization is the bonding of two or more simple molecules (monomers) to form larger molecules with repeating structural units (polymers). The polymerization reaction is exothermic, which means it releases energy and causes the temperature of the bulk liquid to increase. The process is considered "auto-polymerization" because once started it does not require outside heat to maintain the polymerization reaction. The increase in temperature accelerates the formation of more free radicals and more polymerization, which implies that once the reaction starts the energy generated can rapidly increase the polymerization rate (self-promoting and auto-accelerating).

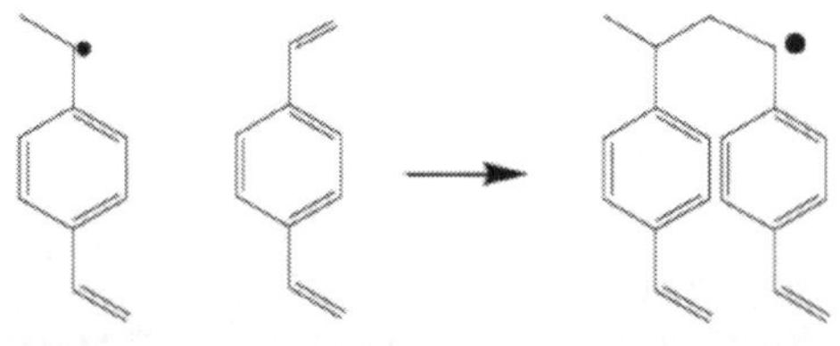

Figure 7: First step in the polymerization of para-DVB

When the heat generated by the exothermic auto-polymerization reaction exceeds the heat lost to the environment, the reaction is said to have reached "thermal runaway", whereby the reaction kinetics will increase exponentially. If the runaway reaction exceeds the capacity of the storage equipment, a pressure relief device is required to prevent damage to the vessel. In the geometry of an ISO-tank container, thermal runaway occurs approximately 36 hours (1.5 days) after the initial exothermic auto-polymerization is detected (discussed later).

Therefore, to avoid the auto-polymerization reaction, inhibitors are added to DVB80 to prevent the polymerization in bulk storage. More specifically, p-tert-butyl catechol (TBC) acts in conjunction with oxygen to inhibit the polymerization reaction. TBC and oxygen create a chain terminating reaction sequence whereby a polymer radical reacts with oxygen to form a polymer peroxy-radical, and then the polymer peroxy-radical reacts with TBC. Together, TBC and oxygen inhibit the DVB80 polymerization reaction. The time to reach the onset of auto-polymerization, referred to as the "induction time" or "shelf life," is the time necessary to deplete the inhibiting materials below a threshold value allowing the auto-polymerization reaction to commence. The induction time depends on the concentration of dissolved oxygen and TBC in the liquid DVB80, in addition to the amount of oxygen in the headspace above the liquid, as well as the temperature of the liquid, which dictates the consumption rate of the inhibiting materials. Given TBC is provided well in excess of what is required, the induction time depends on the concentration of dissolved oxygen and the amount of oxygen in the headspace above the liquid that can diffuse into the liquid.

Extensive testing performed by Gexcon confirmed that the expected oxygen saturation levels in DVB80 that was shipped on the Flaminia was at least 95% when arriving at the New Orleans Terminal. More specifically, testing demonstrated that even worst-case oxygen levels would be very close to fully saturated after standard manufacturing and storage, ISO-container splash filling processes, and delivery to New Orleans Terminal.

3 Lab scale testing

Multiple groups performed lab-scale tests to help evaluate: (1) if DVB80 would reach the initial exothermic auto-polymerization under the current shipment temperatures; (2) if after the initial exothermic auto-polymerization, would DVB80 reach thermal runaway type conditions for the actual ISO-tank container geometry; and (3) would thermal runaway reach critical conditions within the ISO-tank container such that the resulting temperatures and pressures would exceed the containers design criterial resulting in a violent discharge from the pressure relief valve (PRV) set to 4.4 barg (63.8 psig).

3.1 Lab Scale Determinations of the Induction Time of DVB80 Polymerization or Shelf Life

Thermal Activity Monitor ("TAM") tests were performed by Fauske & Associates, LLC (Fauske) on DVB80. The TAM is an isothermal microcalorimeter capable of measuring heat flows from both endothermic and exothermic processes. It is extremely sensitive to even the most minuscule changes in heat flow, capable of measuring heat flows as low as 0.1 μW. Because heated DVB80 will begin to exothermically auto-polymerize (i.e., produce heat) when the reaction constituents for the inhibitor reaction (oxygen) have been depleted below its threshold value, the TAM can be used to determine the shelf-life or induction time to auto-polymerization for DVB80. The shelf-life or induction time to auto-polymerization is the time necessary to deplete the inhibitor, or oxygen in this case, below the threshold value allowing the auto-polymerization reaction to commence and initiate an exotherm. Note induction time (or time to auto-polymerization) is not the time to thermal runaway conditions and venting, which occurs approximately 1.5 days after the onset of auto-polymerization inside a 25,000-liter ISO-tank container (discussed later).

The TAM tests determined the effective lifetime of the constituents for the inhibitor reaction (e.g., oxygen in this case), which prevents auto-polymerization. Two bounding cases were evaluated using the TAM for DVB80. The first case essentially eliminates any possible contribution of oxygen from the headspace by completely filling the sample vessel. As there is no headspace in this set of tests, the induction time shows the conservatively shortest induction time or shelf life for DVB80. On the other end of the spectrum, the second case permits diffusion of oxygen from the headspace by filling the sample vessel to 80%, which is similar to the available headspace volume in the ISO-tank containers aboard the Flaminia. As the test sample size is significantly smaller and has a higher surface to volume ratio as compared to the ISO-tank container, this will provide an upper bound to the induction time (i.e., show the longest potential induction time) as oxygen diffusion from the headspace into the liquid will be greater for the test sample container.

Figure 8 shows the induction time to auto-polymerization as a function of temperature for the no-headspace TAM results (red line) with samples as received (e.g., no additional oxygen was added). As seen in the figure, the induction time required to initiate auto-polymerization was exponentially dependent on the exposure temperature and decreased as temperature increased. From this data, equations known as Arrhenius equations can be applied which describe the mathematical relationship between induction time and temperature for the no-headspace condition, and which therefore can be used to determine the induction time (in days) of the DVB at any given temperature. The equation below can be used to determine the conservatively shortest induction time to auto-polymerization at any given DVB temperature because it is based on testing in which no oxygen was added to the DVB via diffusion from the headspace:

$$t\text{ (days)} = 5.4238 \times 10^{-16}\, e^{11{,}479/T(K)}$$

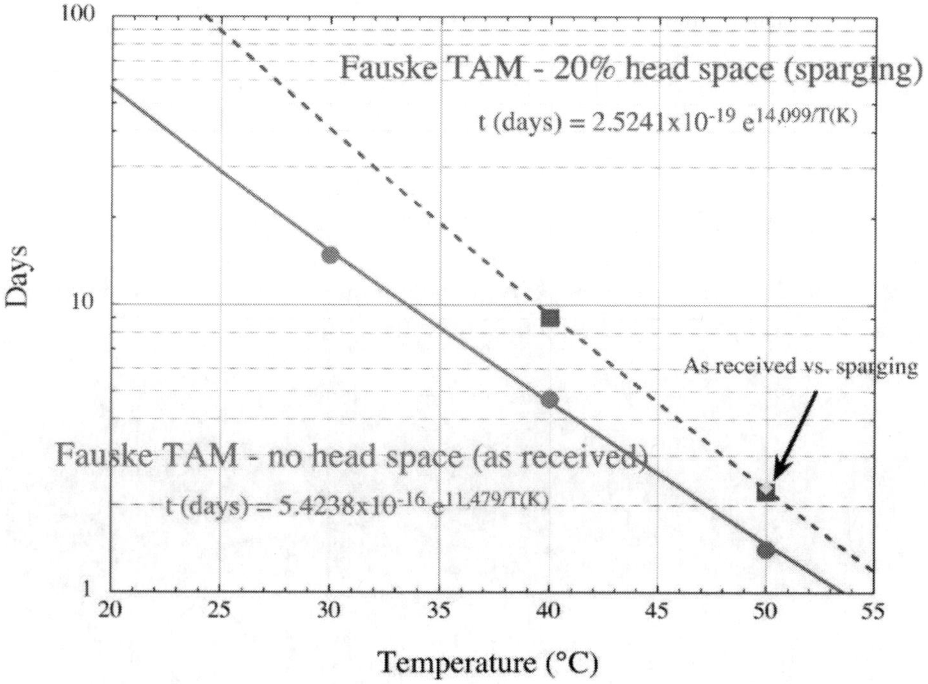

Figure 8: Induction time versus isotherm exposure temperature. The blue symbols and line are Fauske test data and fit for the 20% headspace tests (with sparging) and the red symbols and line are Fauske test data and fit for the no-headspace tests (as received).

In contrast, TAM tests were also performed to determine the limiting case on the other end of the spectrum, in which oxygen from the headspace is able to diffuse through the DVB. Hence, TAM tests were performed with sample containers having approximately 20% headspace and the results are shown in Figure 8 (blue line). These tests were performed with samples that had been sparged with air. While the induction time required to initiate a runaway reaction (auto-polymerization) was exponentially dependent on the exposure temperature, the induction times with 20% headspace condition were longer than those with no headspace, hence confirming the contribution of oxygen diffusing into the DVB80 from the headspace in these tests. The following equation provides the relationship between induction time and temperature for the 20% headspace condition:

$$t\ (\text{days}) = 2.5241 \times 10^{-19}\ e^{14{,}099/T(K)}$$

3.2 Lab Scale Determinations of the DVB80 Auto-Polymerization and Potential for Runaway

Numerous groups analyzed the possibility of DVB80 not only reaching the induction time to auto-polymerization, but also the extent and severity of such an auto-polymerization and whether thermal runaway would be achieved in a 25,000-liter ISO-tank container. In order for thermal runaway to occur, the auto-polymerization would need to reach elevated temperatures and pressures that exceed the threshold pressure of ISO-tank container PRVs set to 4.4 barg (63.8 barg), resulting in a violent discharge of DVB80 from the container.

3.2.1 DVB80 polymerization: small-scale, ISO-thermal analysis

One set of investigators explored the exothermic reactivity of DVB80 by using a two-liter pressure vessel, equipped with a thermowell, pressure gauge, relief valves set to 4.4 bar (63.8 psi), and a manual relief valve for sample collection. The pressure vessel was assembled and was placed in an electrical resistance heater unit with a temperature controller. A thermocouple was placed into the thermowell of the reactor for feedback to the heater control unit. This type of test is called iso-thermal, as the input temperature is kept constant during the tests.

Figure 9: DVB80 sample (jar sample) and bottom part of pressure vessel (right image).

The temperature controller was set to increase the temperature of the pressure vessel to a set point of 50°C, which is 15°C higher than estimated temperatures in hold no. 4. The vessel was held at 50 °C for 139 hours. About 75 hours into heating, the temperature rose to 52 °C, or a net +2°C temperature increase, and the pressure gauge started to drop in the direction of negative pressure. The gauge read negative pressure for the remainder of the heating, indicating that the volume of the polymer was lower than the volume of the liquid DVB80, as expected. The container was opened and a translucent, white colored polymer had formed as shown in Figure 10.

Figure 10: Product from DVB80 polymerization reaction in a 2-liter pressure vessel at 50 °C.

The same test was repeated where the 2-liter pressure vessel of DVB80 was heated to 50 °C for 135 hours, and like the previous test, the temperature rose to 52 °C. Next, the vessel was heated further, with a setpoint of 100 °C and the temperature of the sample continued rising past the setpoint, climbing to about 290 °C, or a net + 190 °C temperature increase. The pressure slowly rose to 50 psi and held steady. The controller was then set to hold the temperature at 300 °C for three hours, and then the vessel was allowed to cool and opened. The vessel contained a slightly yellow-colored solid poly-DVB material as shown in Figure 11.

Figure 11: Product from DVB80 polymerization reaction in a 2-liter pressure vessel at 50 °C, then heated to 100 °C and 300 °C.

The preliminary conclusion based on the lab scale tests performed in the 2-liter pressure vessel was that DVB80 could not result in a runaway reaction, as the exothermic polymerization reaction was too "mild" and yielded insufficient temperatures and pressures even when heated. More specifically, when DVB80 was heated to temperatures of 50 °C, the polymerization reaction had a very mild exotherm with the sample temperature rising only 2 °C and a negligible pressure rise. In fact, even when the sample was heated to 100 °C and above, the resulting reaction yielded a temperature and pressure rise of 190 °C and 50 psi, well below the PRV setpoint of 4.4 barg (63.8 psig). Hence, this iso-thermal testing determined that thermal runaway would not occur and there would be no violent discharge of DVB80.

3.2.2 Scaling and heat loss

While other investigators interpreted the results of the previous testing performed in an iso-thermal pressure vessel to mean that the auto-polymerization of DVB80 cannot result in thermal runaway and hence cannot cause the incident, these groups failed to recognize that such small-scale tests suffer from significant heat loss as compared to the actual full-scale container. Recall, thermal runaway involving chemical reactions can be a complex process. For DVB80, the exothermic polymerization reaction will result in a temperature rise which can increase reaction rates (i.e., reaction rates generally increase exponentially with temperature). If however sufficient cooling is present, the reaction will remain under control. This is due to smaller objects generally having higher surface area to volume ratios and thus are subjected to significant heat loss as compared to the large scales. This was observed in the previously discussed iso-thermal lab scale tests.

When heat generation exceeds the heat loss, thermal runaway can occur. Hence, when conducting tests at the lab-scale to determine exothermic reactivity at the large scale, one must consider heat loss in the experimental apparatus. Thus, to properly evaluate whether the auto-polymerization of DVB80 in ISO-tank containers can result in thermal runaway, lab-scale tests mush be conducted adiabatically. Hence, Gexcon performed adiabatic tests in the VSP2 apparatus, to determine if DVB80 runaway reaction resulting in a violent discharge of DVB80 was likely to occur in ISO-tank containers.

3.2.3 Adiabatic Calorimetry

The VSP2 or (Vent Sizing Package 2) was invented by Dr. Hans Fauske and is a very well-known adiabatic calorimeter and is shown in Figure 12. The test is performed in a small test cell (normally of stainless-steel construction) containing a magnetic stirrer bar (or mechanical stirrer) and surrounded by a heater element. The test cell and heater are enclosed by additional guard heater elements (which are in turn enclosed by thermal insulation material) and placed in a containment vessel. The test cell heater is used to heat the sample to its launch temperature and the outside guard heaters are used to control the adiabatic environment. When heat output is detected from the sample, these heaters maintain the adiabatic environment by accurately matching the sample temperature. A very accurate control system is used to detect very small changes in reaction mass temperature and thus the adiabatic control is very efficient. The test cell is housed inside a containment vessel capable of withstanding rupture of the test cell. During testing, the containment vessel pressure is controlled to balance the test cell pressure. This enables the use of low-pressure test cells (with low phi factor) up to high pressures (typically 100 barg).

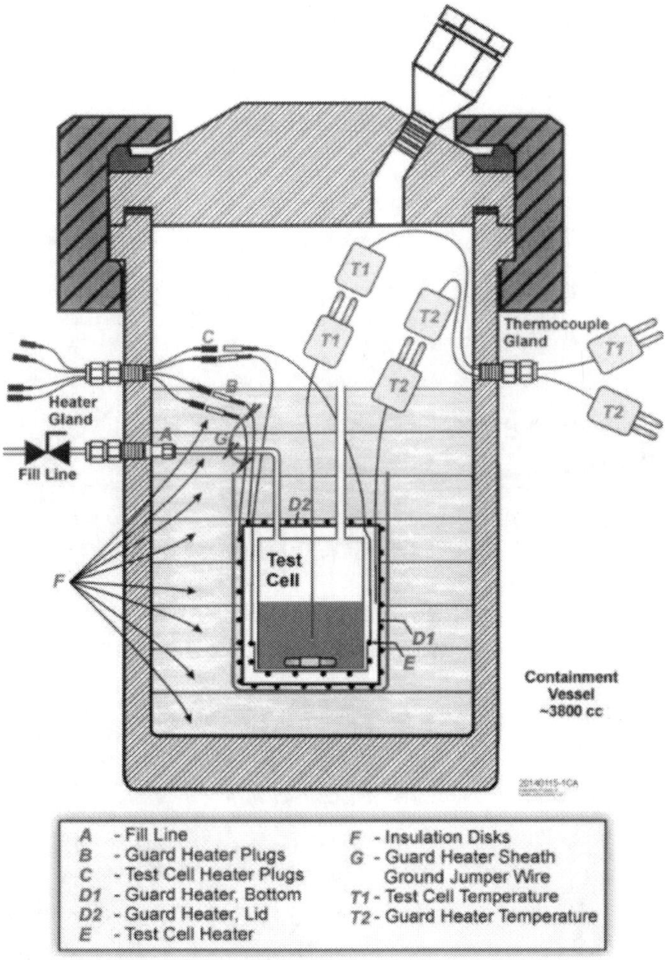

Figure 12: VSP2 testing diagram.

The test was first performed by heating the BVD80 sample to 50 °C to first react the inhibitors (TBC and oxygen) and then by slightly increasing the temperature to 55 °C to continue the inhibitor reaction, and show onset of auto-polymerization. Unlike the iso-thermal tests, when an exothermic reaction and associated temperature rise is detected in the VSP2, it tracks and maintains a similar external temperature to minimize heat loss to the surroundings, similar to what would occur in large-scale tests. Hence, tests performed in the adiabatic VSP2 resulted in thermal runway with a maximum measured temperature and pressure of 428.2 °C and 5.85 barg (84.9 psig), respectively as shown in Figure 13. Further, the VSP2 tests show that the severity of the runaway would have resulted in conditions that exceeded the PRVs setpoint pressure of 4.4 barg (63.8 barg), violently discharging DVB80 from the ISO-tank containers. In addition, Figure 14 shows the reaction resulted in pyrolyzed blackened DVB polymer similar to what was observed after the incident.

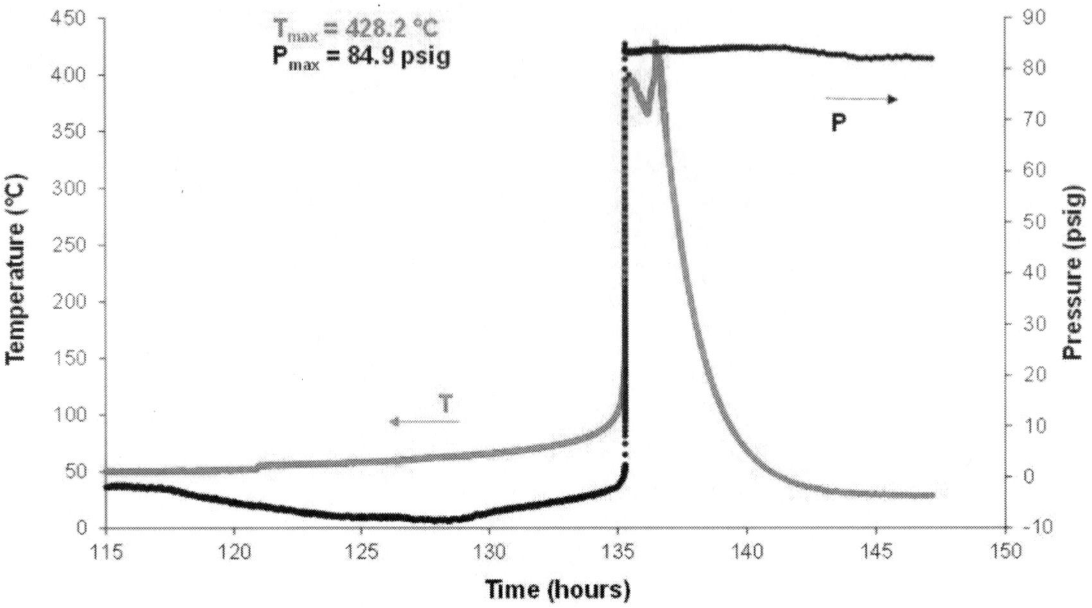

Figure 13: Temperature and pressure of DVB80 after the onset of polymerization and runaway in a VSP2 test. Initial conditions were 55 °C and 1.01 bar.

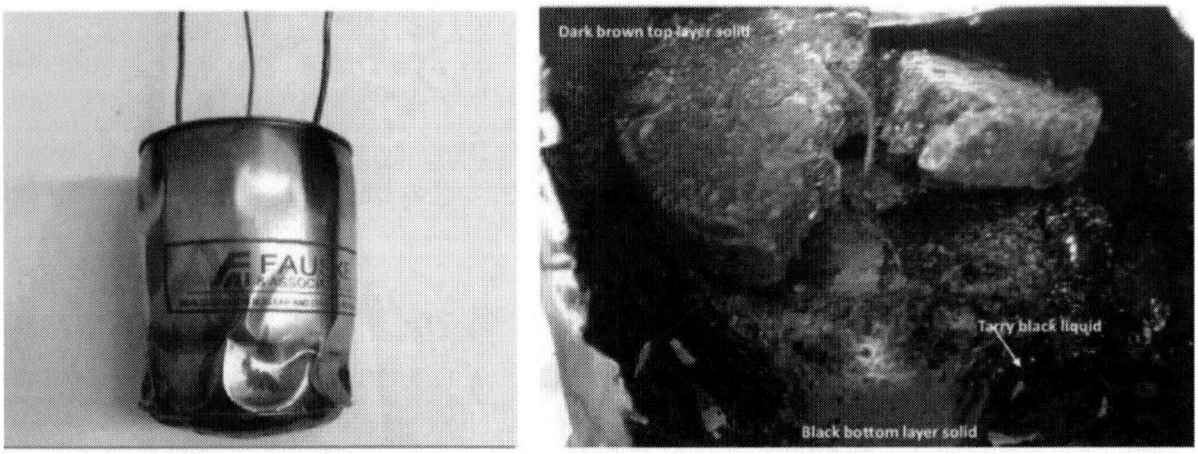

Figure 14: Image of the test cell after VSP2 test in air (left) and polymer sample after cell was cut open (right).

4 Full scale testing

In order to demonstrate the power and effectiveness of utilizing the appropriate lab-scale tests to predict the actual consequences at large scales, full-scale testing was performed to determine when DVB80 would reach the initial exothermic auto-polymerization during shipment; when DVB80 would reach thermal runaway type conditions for the actual ISO-tank container geometry; and would a runaway of DVB80 exceed the containers PRVs set to 4.4 barg (63.8 psig), resulting in a violent discharge from the container.

More specifically, numerous lab scale tests and theories were presented regarding what occurs during thermal exposure of DVB80 in an ISO-tank container until the product auto-polymerizes. The analyses included: the induction times necessary to deplete the inhibitor and oxygen during thermal exposure both with and without air headspace, resulting in auto-polymerization; the temperatures and pressures that would occur within the ISO-tank container during the auto-polymerization event; and whether the runaway would result in venting of the pressurized gases via the PRV during auto-polymerization.

4.1 Test setup

The ISO-tank container test was conducted at a remote site. Refrigerated DVB80 was transported to the test site and an exemplar 25,000 liter insulated ISO-tank container was provided. An insulated enclosure was built around the ISO-tank container so that it could be exposed to temperature conditions expected on the dock and in the hold (see Figure 15). The enclosure and ISO-tank container were instrumented with thermocouples and pressure transducers to measure the temperatures and pressures during the test (see Figure 16). DVB was added to the ISO-tank container at 80% fill and the oxygen levels were measured to be 95% saturation, consistent with the levels estimated at the dock after delivery.

Figure 15: Temperature controlled enclosure (left image) housing the ISO-tank container (right image). Photos show the enclosure before the venting stack and flare were installed.

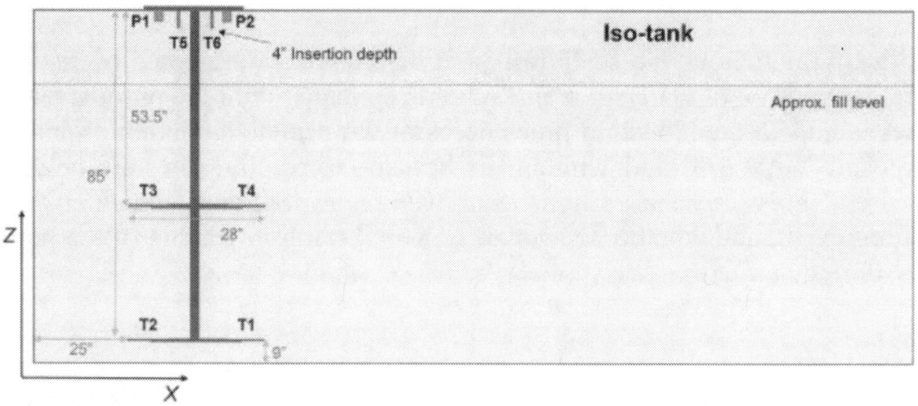

Figure 16: Location of thermocouples (red lines) and pressure transducers (blue squares) recording internal temperatures and pressure.

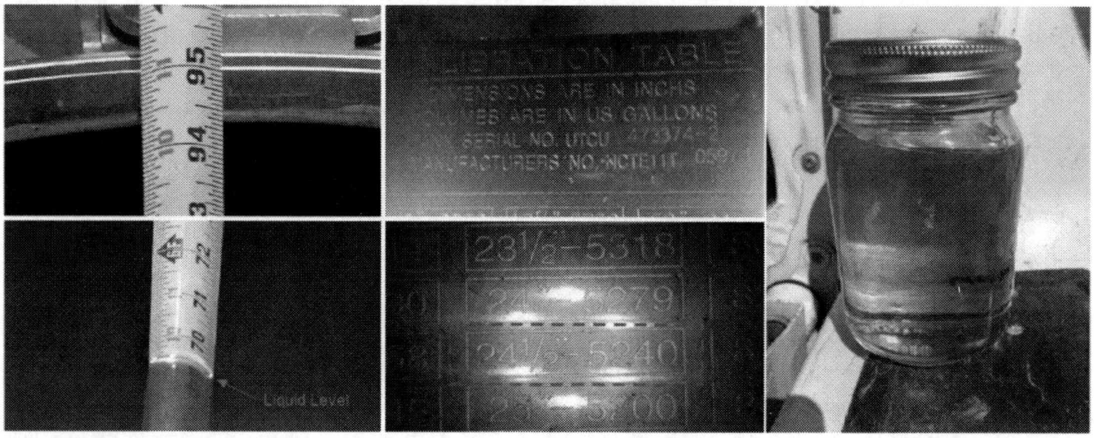

Figure 17: Liquid level measurement for determination of fill capacity and DVB sample.

4.2 Testing Phase 1 and 2

During phases 1 and 2 of the test, DVB80 starting at 6.7 °C in an ISO-tank container was exposed to the ambient temperatures estimated for the Port of New Orleans when the subject ISO-tank containers were stored there (see Figure 4). The ISO-tank container was then subjected to the hold temperatures estimated for a duration equivalent to the elapsed time from loading onboard the Flaminia to the time of the incident, which were taken as the conservatively high estimate (see Figure 5). The ISO-tank container was then exposed to a constant 25 °C condition for three more days, which was estimated to be the remaining transport time to the port had the incident never occurred. Figure 18 provides the proposed test temperature during phase 1 and phase 2, which would yield conservatively higher DVB80 temperatures and larger inhibitor consumption rates, as well as the actual test temperatures.

Figure 19 compares the average enclosure temperature to the proposed test temperature. During the on-dock portion of the test, the enclosure temperature matched very well with the proposed exposure temperature, with slightly lower lows on a few occasions and higher highs during many days of the test. The enclosure temperature also tracked well during the in-hold portion of the test, which the exception of two dips, a power outage and heater outage. In order to compensate for the lower temperatures, the new temperature profile proposed was verified by comparing the predicted final DVB temperatures and inhibitor lifetime levels at the time the incident occurred and when the ship reached the port. To view the final DVB and inhibitor levels, the new time profile was time shifted by eight hours to more directly compare the DVB80 temperatures and inhibitor lifetime levels.

The testing showed that auto-polymerization did not occur in the alleged temperature exposure due to storage at the dock and onboard the Flaminia.

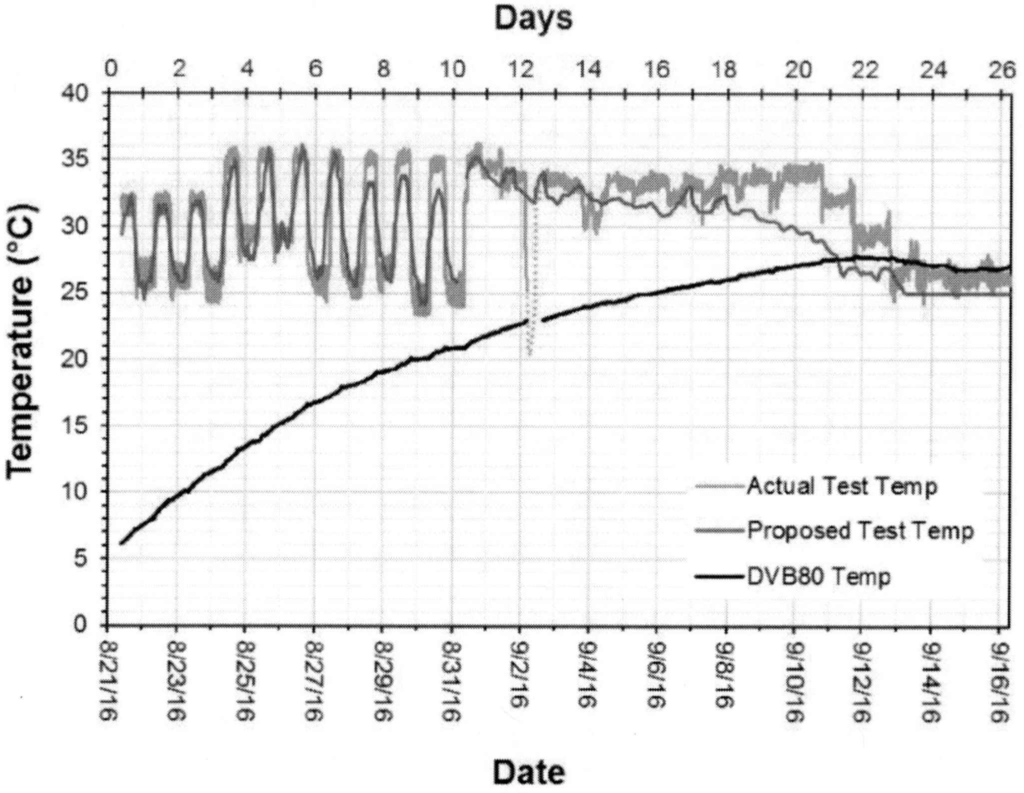

Figure 18: Original proposed temperature profile (blue) versus the actual test temperature profile (orange).

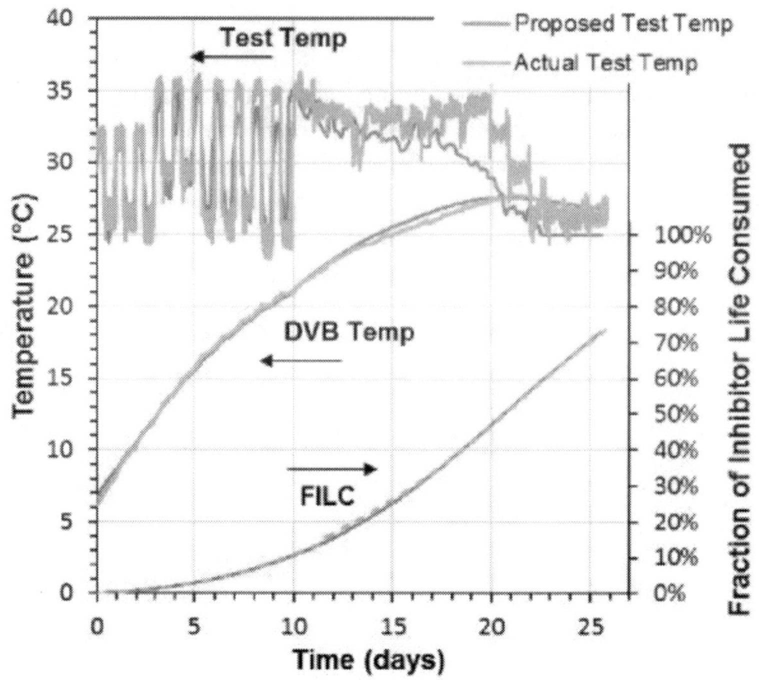

Figure 19: DVB80 temperature and inhibitor lifetime levels predicted in actual time shifted 8 hours for the original proposed temperature exposure (orange) modified to account for the power outage and heater malfunction (blue).

4.3 Testing Phase 3

After approximately 26.5 days of exposure for phase 1 and 2 testing, the temperature within the enclosure was set to 45 °C to induce auto-polymerization within the time constraints of the deadline for this study. During the daytime, the heaters were able to maintain 45 °C and above, while in the evening the heaters could only achieve approximately 40 °C as shown in the left image in Figure 20.

The no headspace Fauske equation used to predict the onset of auto-polymerization without the influence of oxygen diffusion and the manufacturer's recommended shelf-life guidelines both conservatively predict shorter times to auto-polymerization. More specifically, the Fauske no headspace equation predicts 95% fraction of inhibitor consumed at approximately 28.17 days, which is extremely close to the 28.5 days observed during the full-scale tests (see Figure 21). The guidelines conservatively predicted an even shorter time of 26.89 days (see in Figure 21).

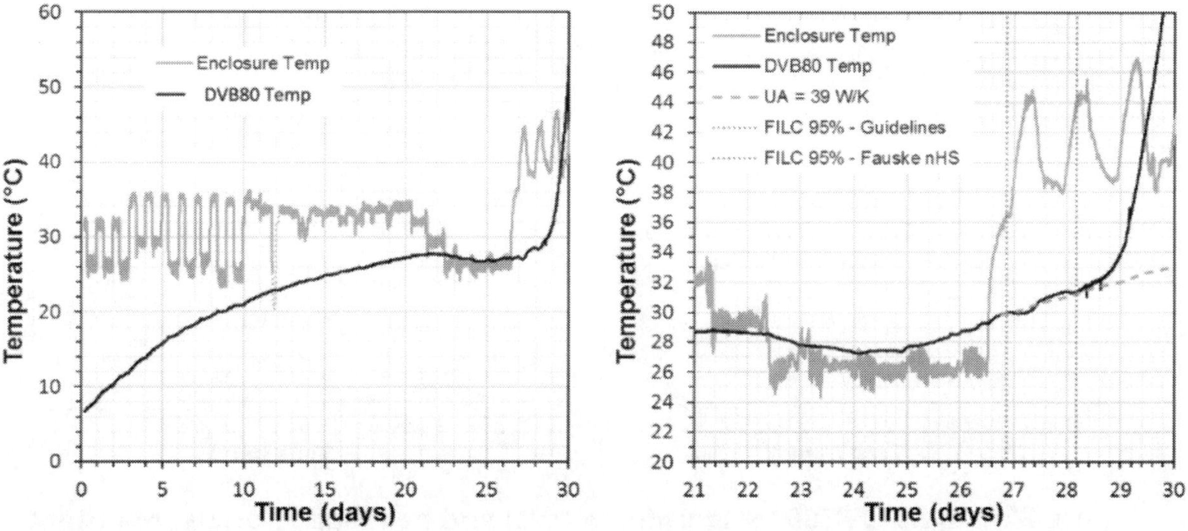

Figure 20: Comparison of actual versus predicted DVB80 temperature. Also shown are the predicted time to auto-polymerization by the Fauske no-headspace equation and by the manufacturer's recommended guidelines.

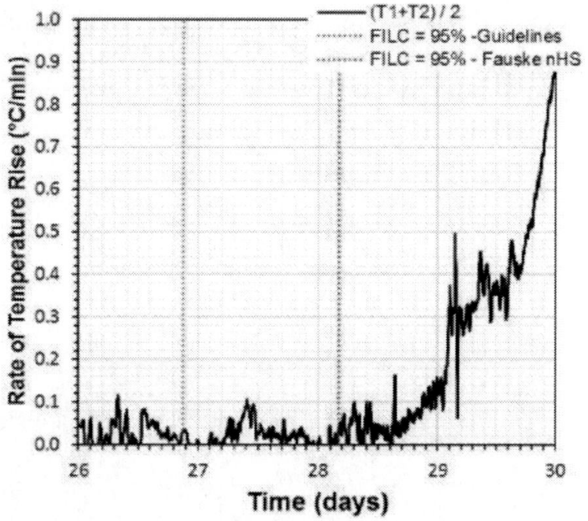

Figure 21: Rate of temperature rise as a function of time (using average measured temperature).

While the auto-polymerization initiated at approximately 28.6 days, the temperature and pressures within the ISO-tank container did not "runaway" (e.g., temperature and pressure conditions that would result in uncontrolled reaction and initiate the PRV) until approximately 1.5 days after the exothermic auto-polymerization was first detected.

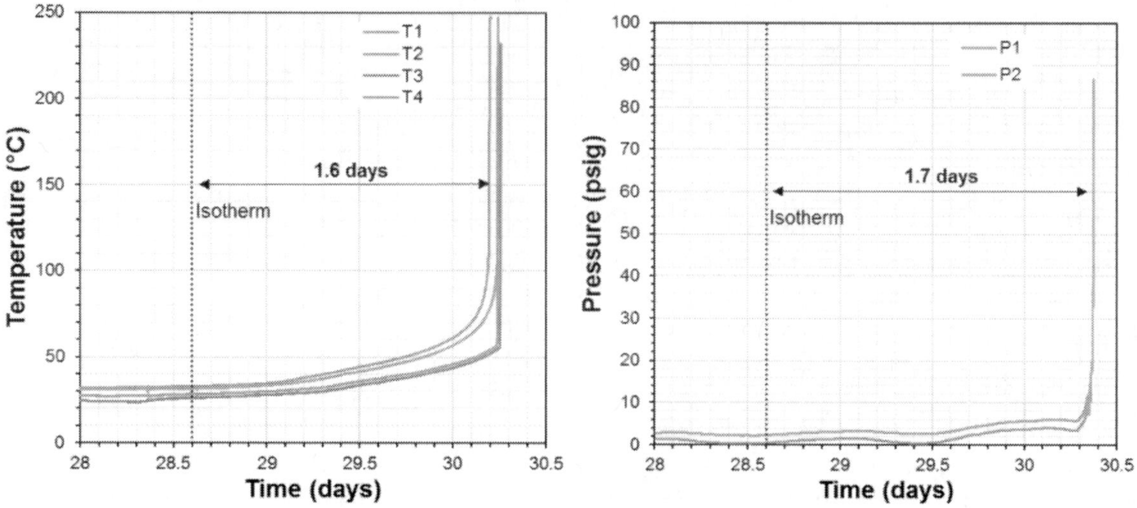

Figure 22: Liquid DVB80 temperatures (left) and headspace pressures (right) leading up to run-away.

Once the runaway reaction was achieved, the camera outside the window of the vent box captured the initial moments of the vapors discharging via the pressure relief valve (PRV). Figure 23 shows the initial vapor being discharged via the PRV. The images show only vapor and there appears to be no initial evidence of liquid discharging from the valve. The discharge was so violent that it failed the observation window on the vent chamber and knocked the camera to the ground, resulting in DVB vapor venting both to the flare stack and within the wooden enclosure. It should be noted that burning DVB results in a very black and sooty flame, while venting DVB has the appearance of a white vapor as shown in Figure 23.

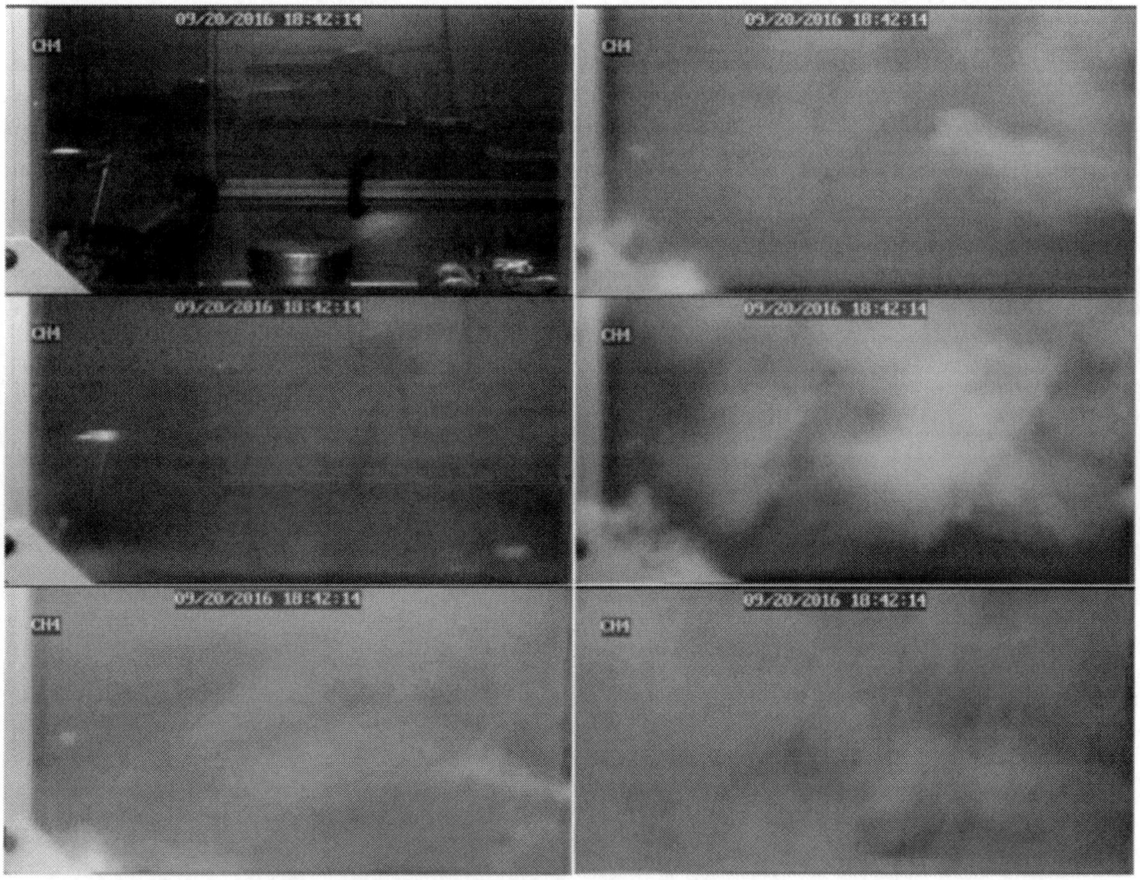

Figure 23: Camera view of the PRV during the release. Note the viewing window failed and knocked the camera to the ground.

Figure 24: White vapor from the enclosure.

5 Summary

The present analysis shows how appropriate lab-scale calorimetry tests can be used to accurately predict the induction time to auto-polymerization or shelf life of DVB80 in full-scale iso-tank containers. Further, this paper also demonstrated how certain investigators used incorrect test methods, which appeared justified, and erroneously concluded that DVB80 could not violently runaway, and that the reaction would only result in "mild" exotherms. This paper confirms the importance of increased heat loss due to the large surface-area-to-volume ratio inherent in smaller test apparatus, which is well-documented in scaling experiments. Testing with appropriate lab-scale calorimetry tests demonstrated that violent thermal runaway was accurately predicted. In fact, full-scale runaway tests were also performed with an actual ISO-container and agreed very well with the lab-scale calorimetry tests, demonstrating the importance of taking into account scaling issues when investigating runaway conditions.

SPRING24+20th GCPS
A Joint AIChE and CCPS Meeting

DEKRA

Thermal Hazard Evaluation and Safety Assessment for 1-Hydroxybenzotriazole Hydrate by DSC, ARC and Calvet Calorimeter (C80)

Jiliang He, Zhizhong Dong, Michael F. Carolan, Mike Snyder

DEKRA Process Safety, 113 Campus Drive, Princeton, New Jersey 08540

Holistic Safety Assessment

Contents

- Background about 1-Hydroxybenzotriazole Hydrate
- Thermal Hazard Evaluation by DSC, ARC, and Calvet C80
- Thermal Safety Assessment by T_{D24} and T_{SADT}
- Conclusions

1-Hydroxybenzotriazole (HOBT)

A compound with a vast spectrum of applications in chemical, agricultural and pharmaceutical industries. Containing thermally unstable triazole group.

(crystalline solid)

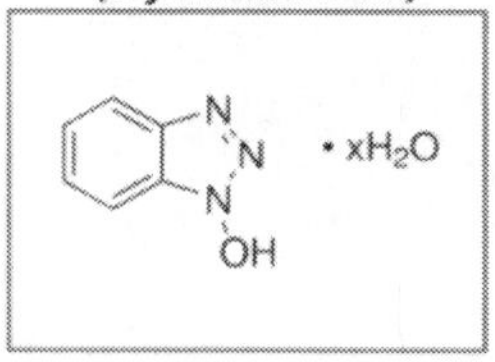

UN/DOT Classification (current)

HOBT anhydrous: UN0508, Class 1.3C explosive
HOBT hydrate (10-20% water): UN3474, Class 4.1 desensitized explosive/self-reactive

Accident Reports

Lacamas Laboratories Inc, USA, a reactor explosion, 3 employees severely injured, 2005
(OSHA ACC. Nr# 201635109)

Chemical Plant, Jiangsu, China, a distillation column explosion, complete destruction of production plant, 1 employee injured, 2005
(Ma, J.M., 2007. Cause analysis of an explosion accident of benzotriazole distillation column. *Chem. Saf. Environ.* 20, 1)

Thermal Hazard Evaluation in Literature

Only one publication for HOBT anhydrous, no report for HOBT hydrate.

Why needs to assess thermal hazard for HOBT hydrate?
Are the thermal behaviors between HOBT anhydrous and hydrate the same?

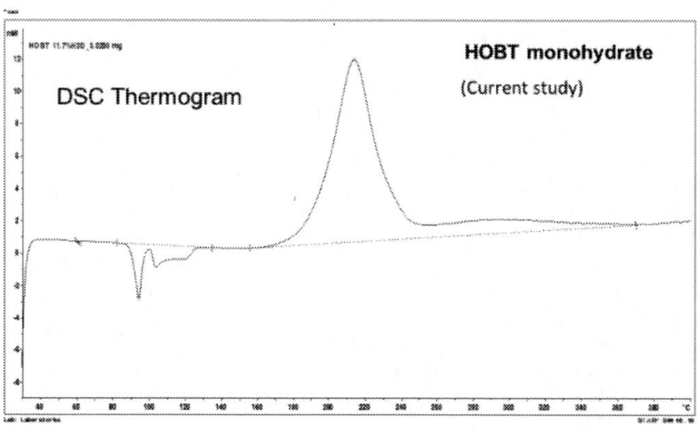

Thermal behaviors are different between HOBT anhydrous and hydrate (e.g., lower melting point and onset exotherm, broader exotherm)

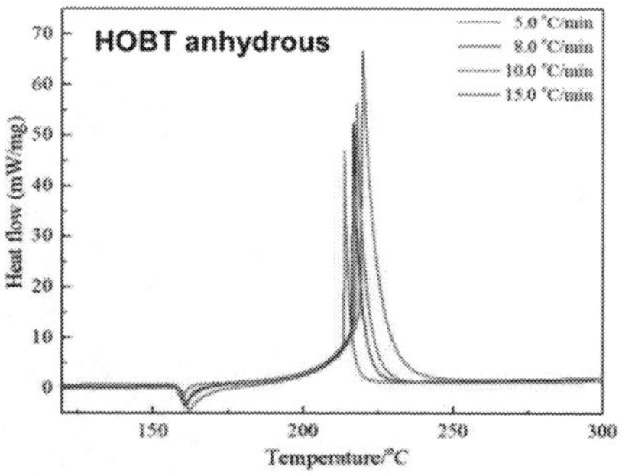

Zhang H., et al, Thermal hazard characteristics and essential mechanism study of 1-hydroxybenzotriazole: Thermodynamic study combined DFT simulation, *Process Saf. and Environ. Prot.*, 2022, 168, 713-722.

DSC (Differential Scanning Calorimetry) Results

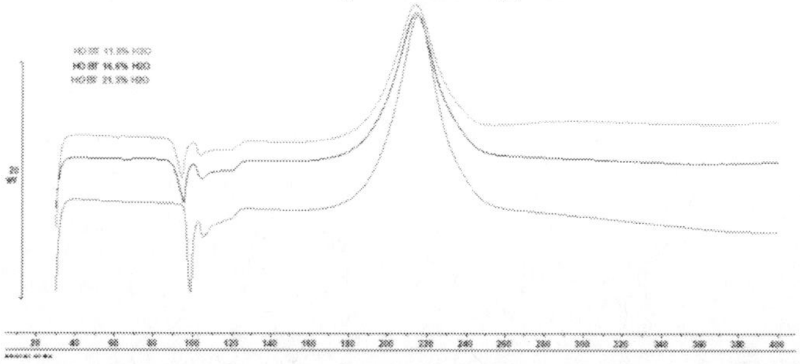

Severity: catastrophic

The more water is added, the more energy is released.

No linear relationship between energy release and water content.

Energy released from hydrate is higher than that from anhydrous.

Sample	Start Temperature °C	Peak Temperature °C	End Temperature °C	Energy J/g	HOBT Based Energy J/g
HOBT 11.8% H_2O	152	214	359	-1895	-2149
HOBT 16.6% H_2O	150	215	364	-1847	-2214
HOBT 21.5% H_2O	152	216	362	-1966	-2504

ARC (Accelerating Rate Calorimetry) Results

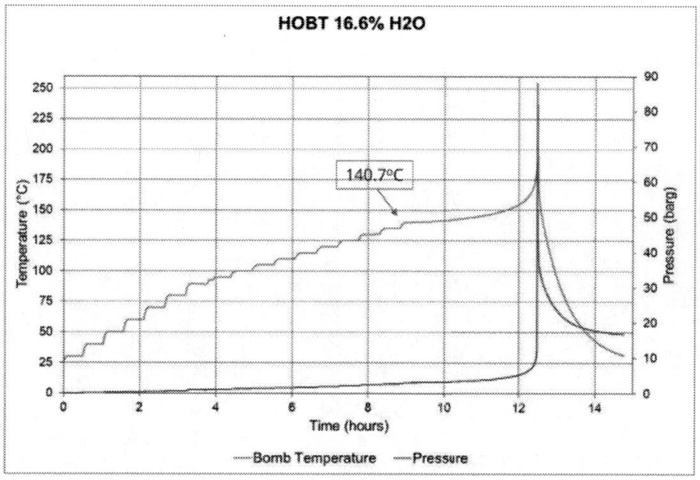

Temperature and Pressure *vs.* Time for HOBT 16.6% H_2O

Onset temperature of hydrate is much lower than that of anhydrous (170.1°C).

Onset temperature is lower than the melting point (~160°C) of anhydrous HOBT!

ARC Results

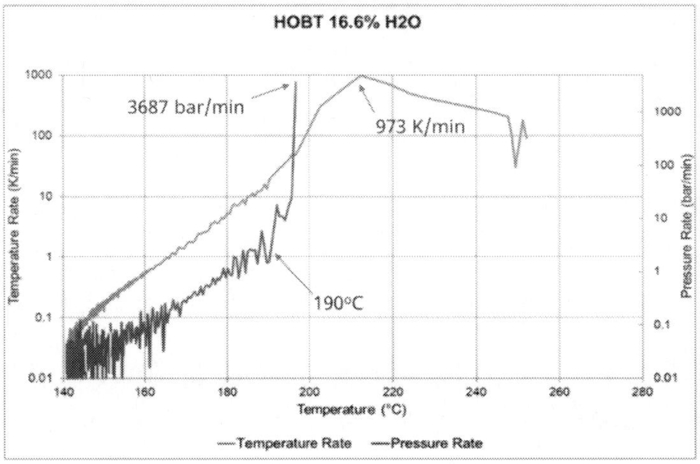

Temperature Rate and Pressure Rate Plot for HOBT 16.6% H_2O

ARC Results

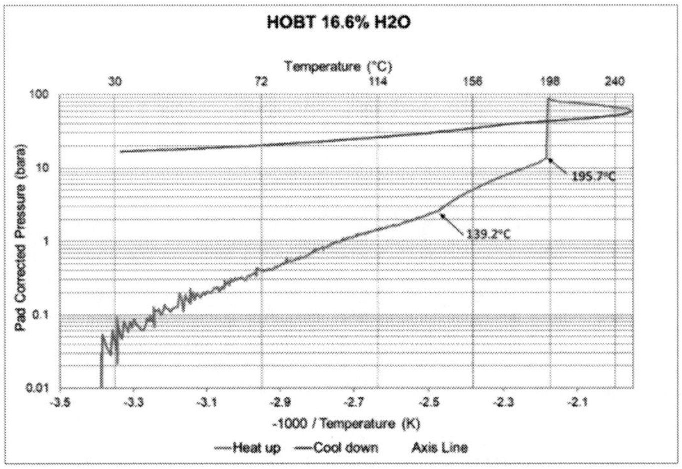

Antoine Plot for HOBT 16.6% H_2O

Kinetic Modelling of Liquid (based on ARC data)

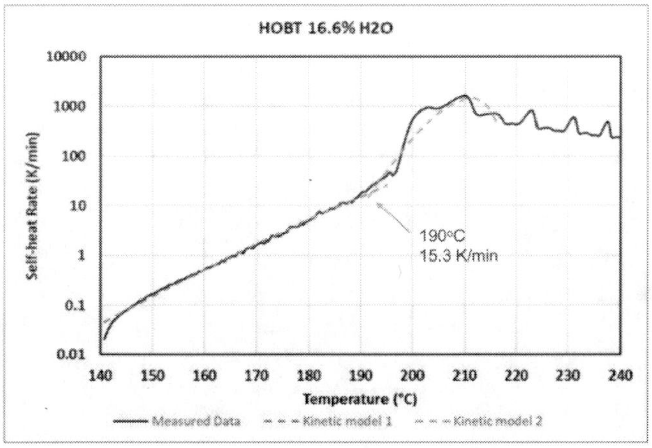

Self-heat rate *vs.* temperature for kinetic modelling for HOBT 16.6% H_2O

$$\frac{dT}{dt} = AC_0^{\,n-1}\exp\left(-\frac{E_a}{RT}\right)\left[\frac{(T_f - T)^n}{(\Delta T)^{n-1}}\right]$$

$$ln\left[\frac{(\Delta T)^{n-1}}{(T_f - T)^n}\frac{dT}{dt}\right] = ln\left(AC_0^{\,n-1}\right) - \frac{E_a}{RT}$$

C_0: initial concentration (mole/L)
T: temperature (K)
T_f: exothermic final temperature
ΔT: $T_f - T_0$
R: 8.314 J/mol.K

$E_a = 184\ (\frac{\text{kJ}}{\text{mole}})$

$A = 1.24 * 10^{20}(\text{S}^{-1})$

$n = 0$

(Zero-order reaction before runaway)

Summary of ARC Results for HOBT 16.6% H_2O

Parameters	HOBT 16.6% H_2O
Thermal Inertia (Phi factor)	7.62
Initial Temperature (°C)	141
Final Temperature (°C)	252
Adiabatic Temperature Rise (K)	850
Heat of Reaction (J/g)	-1700
Max Self-heat Rate (K/min)	973
Max Pressurization Rate (bar/min)	3688
Max Rate of Gas Generation (mL/s.g, stp)	606
Quantity of gas generated (mL/g, stp)	275
Activation Energy (KJ/mol)	184
Pre-exponential Factor (S^{-1})	1.24E+20

Due to large quantity of gas evolving instantaneously, there is no practical emergency relief system (ERS) that can be utilized to mitigate the hazard.

Appropriate thermal monitoring and temperature control become viable choices for preventing the hazard in HOBT hydrate production, storage, transport, and use.

Calvet C80

Resolution: 0.1 μW, scanning rate: 0.05 K/min
Setaram, KEP Technologies

To study if there is any exotherm before melting.

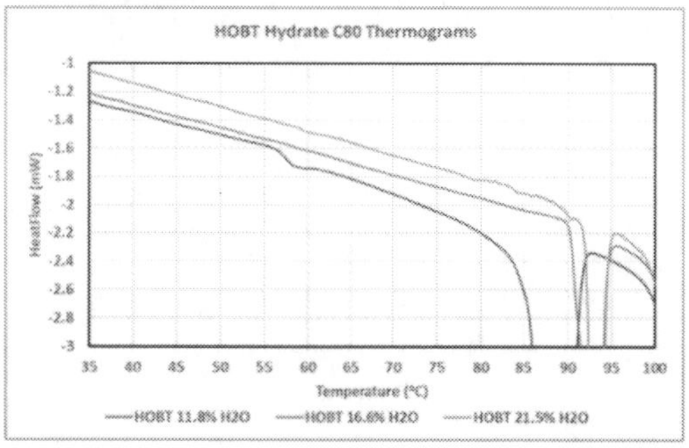

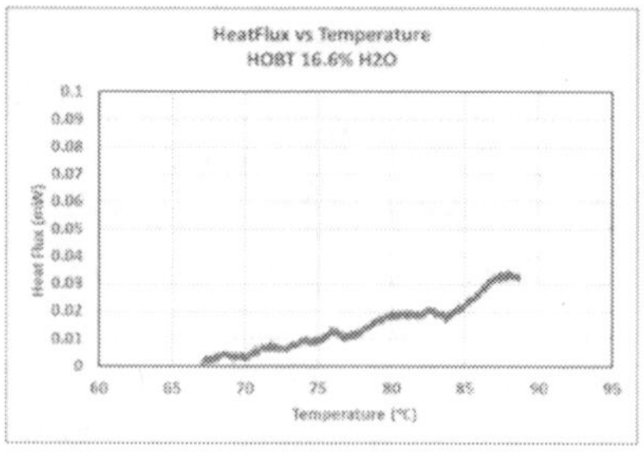

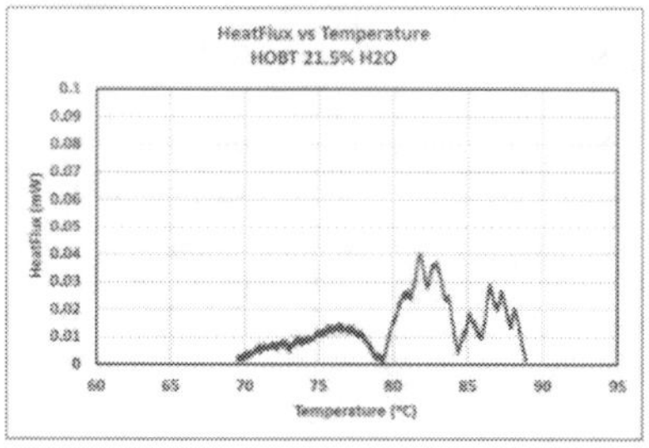

Slow exotherm detected HOBT 16.6% H_2O and 21.5% H_2O before melting.

11 |

Kinetic Modelling of Solid (based on C80 data)

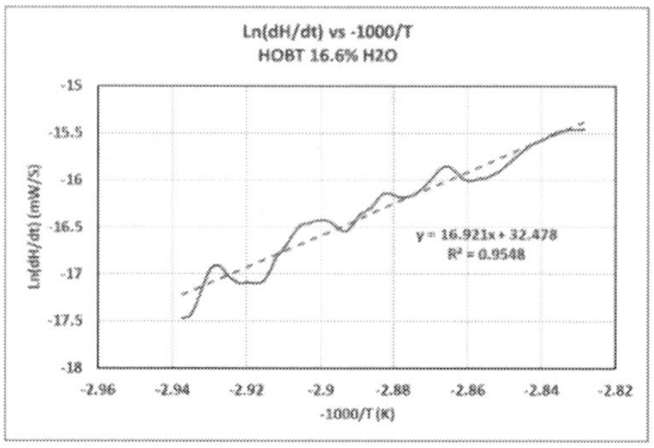

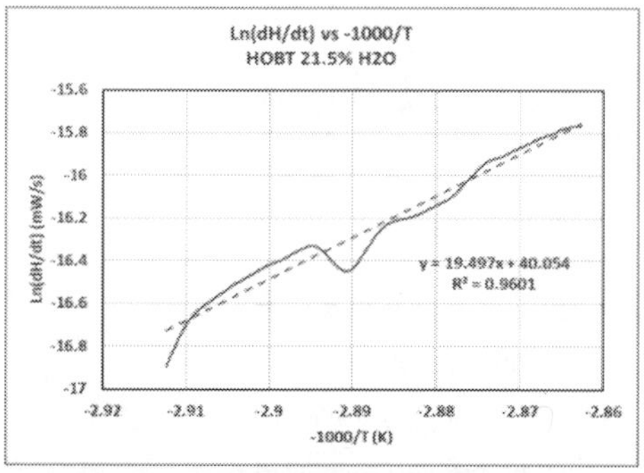

$$\frac{dH}{dt} = A\Delta H \exp\left(-\frac{E_a}{RT}\right)$$

$$\ln\left[\frac{dH}{dt}\right] = \ln(A\Delta H) - \frac{E_a}{RT}$$

$\Delta H = M_0 \Delta H_d$
M_0 *is the initial mass*
ΔH_d *is heat of reaction in J/g*

	HOBT 16.6% H_2O	HOBT 21.5% H_2O
E_a (KJ/mol)	141	162
A (S^{-1})	4.72E+8	6.698E+11

12 |

THERMAL SAFETY ASSESSMENT

Time to maximum rate (TMR_{ad})
from the kinetics of liquid based on ARC data

$$TMR_{ad} = \frac{C_p R T^2}{q E_a}$$

Low probability of thermal runaway if TMR > 24h
High probability of thermal away if TMR < 8h

Safe process temperature T_{D24}: 108.9 – 110.6°C
30°C lower than that of HOBT anhydrous.

TMR (hours)	Temperature (°C)		
	HOBT 11.8% H_2O	HOBT 16.6% H_2O	HOBT 21.5% H_2O
24	108.9	110.6	109.4
8	116.7	118.0	117.5

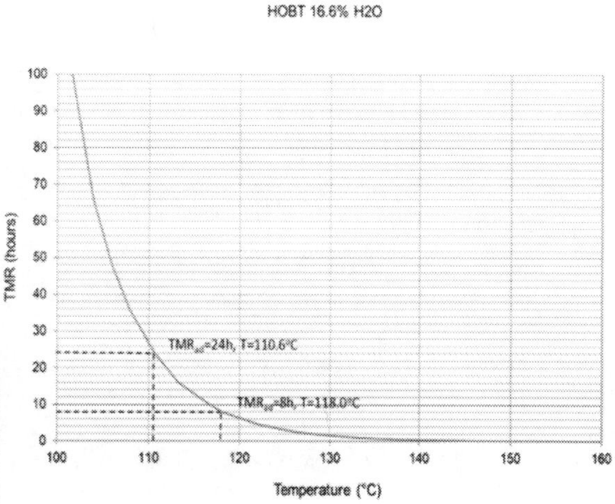

Variation of TMR as a function of temperature for HOBT 16.6% H_2O

TMR_{ad} from the kinetics of solid based on C80 data

$$TMR_{ad} = \frac{C_p R T^2}{A exp(-\frac{E_a}{RT}) \Delta H E_a}$$

(ΔH: *total heat of reaction*)

TMR (hours)	Temperature (°C)	
	HOBT 16.6%H_2O	HOBT 21.5%H_2O
24	172.2	127.9
8	190.8	137.6

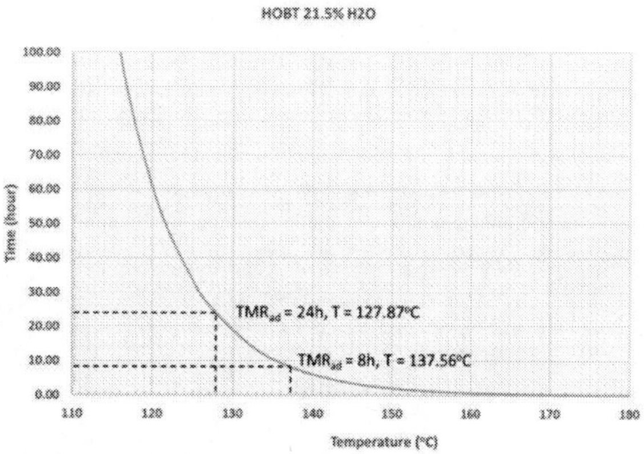

Variation of TMR as a function of temperature for HOBT 21.5% H_2O

T_{D24} and T_{D8} for HOBT 16.6% H_2O and HOBT 21.5% H_2O from ARC and C80 data

TMR (hours)		Temperature (°C)	
		HOBT 16.6% H_2O	HOBT 21.5% H_2O
ARC	24	110.6	109.4
	8	118.0	117.5
C80	24	172.2	127.9
	8	190.8	137.6

T_{D24} and T_{D8} values derived from ARC data are smaller and more conservative.

DEKRA

Self-accelerating decomposition temperature (T_{SADT})

The lowest temperature at which a self-heating of 6°C or more within 7 days or less may occur for a certain substance in the commercial packaging used in transport (by UN manual)

$$T_{SADT} = T_{NR} - R\,(T_{NR} + 273)^2/E_a$$

Semenov model

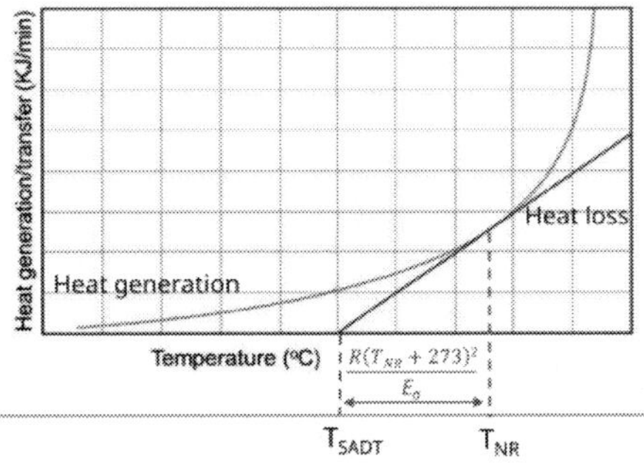

T_{NR} (temperature of no return) is the temperature corresponding the thermal time constant τ of system (package) in the curve of TMR_{ad} *vs* Temperature:

$$\tau = \frac{mC_p}{US}$$

m: mass of substance in a package
C_p: specific heat capacity of substance
U: heat transfer coefficient
S: heat exchange area

DEKRA

SADT and T_{NR} data for HOBT hydrates from ARC and C80 (50 Kg drum package)

Material	ARC		C80	
	T_{NR}, °C	SADT, °C	T_{NR}, °C	SADT, °C
HOBT 11.8% H_2O	116.0	110	-	-
HOBT 16.6% H_2O	117.4	110	189.2	175
HOBT 21.5% H_2O	116.8	110	136.8	130

T_{SADT} of HOBT hydrate is about 50°C lower than anhydrous (compared with literature).

T_{SADT} values derived from ARC data are smaller and more conservative.

The results are not as expected but reasonable:

1) only slow exothermic activity is detected in solid.
2) the melting alters kinetic behavior, thermal decomposition appears faster in liquid than solid.

17 |

DEKRA

Conclusions:

1. Slow exothermic activities are detected in both HOBT 16.6% H_2O and HOBT 21.5% H_2O in solid by C80.

2. Exothermic decomposition of HOBT hydrate takes place in liquid (after it melts), with large energy release (1.7 - 2.0 KJ/g). On HOBT basis, the more water is added, the more energy is released.

3. The self-heat rate and pressure rise rate of the decomposition reaches as high as 1000 K/min and 3700 bar/min, respectively. Nearly 1000°C adiabatic temperature rise along with rapid pressurization can cause catastrophic consequences of fire and explosion.

4. T_{D24} for HOBT hydrates is calculated to be in the range of 108.9°C - 110.6°C, and T_{SADT} is estimated to be about 110°C for 50 Kg package.

5. Compared with literature data, HOBT hydrates demonstrates lower onset temperature, higher energy released, lower T_{D24} and lower T_{SADT} than those of HOBT anhydrous. Thermal hazard of HOBT hydrate may be more severe than HOBT anhydrous, hence, they shall be handled with extra cautions.

18 |

DEKRA

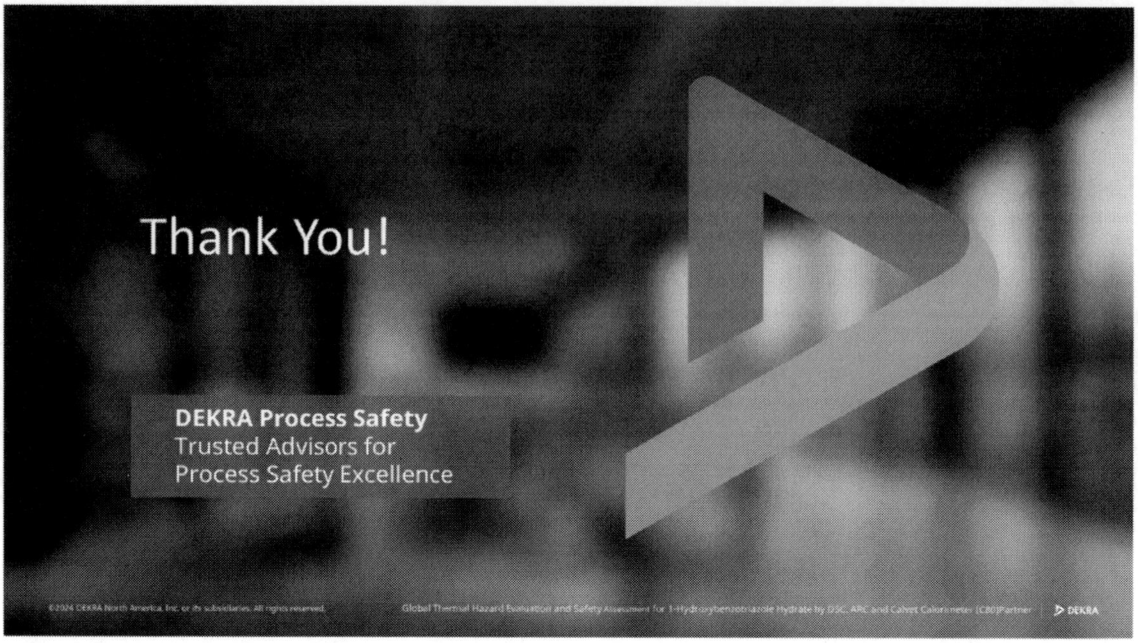
Thank You!
DEKRA Process Safety
Trusted Advisors for
Process Safety Excellence
©2024 DEKRA North America, Inc. or its subsidiaries. All rights reserved.
Global Thermal Hazard Evaluation and Safety Assessment for 1-Hydroxybenzotriazole Hydrate by DSC, ARC and Calvet Calorimeter (C80)Partner
DEKRA

A comprehensive pipeline source term model for pressurized/flashing liquids in Phast/Safeti (PIPEBREAK-II)

Adeyemi Oke
DNV
5th floor Vivo Building 30 Stamford Street London SE1 9LQ
Adeyemi.Oke@dnv.com

Jan Stene
DNV

David Worthington
DNV

Michael Harper
DNV

Frank Hart
DNV

Keywords: discharge modelling, long pipelines, risk analysis, MAH pipelines, Phast/Safeti

Abstract

PIPEBREAK is an integral model in Phast/Safeti that simulates flow conditions at the outlet and upstream ends of a pipeline as well as the attendant mass flow rate out of a breach. However, a potential limitation of PIPEBREAK is that it assumes instantaneous depressurization to saturated conditions at the onset of a breach.

This paper presents an updated model, PIPEBREAK-II, the outcome of modelling work to address potential limitations within long pipeline source term models commonly employed for major accident hazard analysis in industry. PIPEBREAK-II caters for initial rapid depressurization effect modelling, drainage flow for sub-cooled inventories, pumped inflow, flow reversal, valve closure effects as well as the occurrence of punctures of any size at any downstream location with flow contributions (disjoint or interactive) from both ends of the pipeline.

Predicted results from PIPEBREAK-II have been verified and validated against published results from selected pipeline/vessel blowdown models and field tests involving long pipeline release scenarios covering a broad range of aperture ratios, respectively. Simulated results based on PIPEBREAK-II compare very well against other pipeline release models and field data with PIPEBREAK-II results tending to be reasonably conservative (i.e., higher flowrates, initial inventories, quicker depressurization, and lower flow temperatures).

1 Introduction

PIPEBREAK is an integral model in Phast/Safeti that simulates the evolution of pressure and temperature at the outlet and upstream ends of a pipeline as well as the attendant mass flow rate out of a breach. PIPEBREAK supports the modelling of breaches at any location along the pipeline length, pumped input as well as valve closure effects. Furthermore, PIPEBREAK has been shown to compare very well against field data on propane releases from a 100m pipe with aperture ratios >20% (the Shell "Isle of Grain" trials) at least as well as more complicated models [1]. However, the current application to huge CO2 transport pipelines have highlighted some weaknesses in the model – this has been acknowledged also by UK HSE [2] and this has been the motivation for a major improvement program. A potential limitation identified is that PIPEBREAK assumes instantaneous depressurization to saturated conditions along the entire pipeline length at the onset of a breach [1]. This assumption is generally tenable for loss of containment events from long pipelines with large aperture ratios (>20%) and modest lengths (<100m) up to ca 5km. For long (>5km) pipelines, where conveyed fluids are stored at pressures significantly above their bubble point, this assumption may potentially lead to:

- Underprediction of fluid inventory contributing to the overall release process.
- Underestimation of initial source terms for release scenarios involving small breaches (e.g., with aperture ratios <20%).

Other potential limitations, identified from an inhouse review of modelling in PIPEBREAK, include:

- Assumption that flow, upon commencement of a loss of containment event, is choked, irrespective of upstream conditions and breach size. There is evidence that metastable flow conditions may prevail at the breach, particularly for smaller breach sizes and with increasing line pressures [2, 3].
- PIPEBREAK modelling is not suitable for breach sizes with aperture ratios below 20%. This is a known limitation and is based on comparison of simulated results against field data [1, 2].
- The thermodynamic modelling of liquid densities along the saturated vapor curve may be inaccurate. PIPEBREAK employs the Peng-Robinson cubic equation of state (PR-CEOS) in the modelling of liquid densities. Recent inhouse sensitivity tests indicate the PR-CEOS may underpredict saturated liquid densities by up to 25% (e.g., for pure Ammonia), particularly close to the critical point.
- Inability to model loss of containment events involving sub-cooled liquid inventories as well as any drainage flow that may ensue upon depressurization of the bulk fluid inventory to ambient pressure.

A model (PiRRaM) has been developed for the UK Health and Safety Executive (HSE), that closely resembles PIPEBREAK and attempts to address some of the key limitations in PIPEBREAK described above. However, PiRRaM was developed to satisfy HSE's requirements and focusses on the initial 30s of a release for superheated liquids. As such, PiRRaM does not support many features currently in PIPEBREAK, such as: pumped input, valve closure and heat transfer effects.

Given the above, DNV have undertaken model improvement work to address some of the above limitations. In particular, the extension of PIPEBREAK to support better modelling of:

- The initial (rapid) depressurization behavior for flashing fluids transported at pressures significantly above their saturated vapor pressure.
- Liquid densities along and at pressures above the saturated vapor-pressure curve, while ensuring predicted liquid densities away from the saturated vapor-pressure curve behave in a consistent manner (i.e., predicted liquid densities increase with increasing pressure in a non-ideal manner and in line with any relevant thermodynamic properties that may be employed by PIPEBREAK in its modelling).

This paper presents an updated model, PIPEBREAK-II, the outcome of modelling work to address the various limitations described above. PIPEBREAK-II has been developed to simulate the full range of hole sizes, from full bore ruptures to small breaches for superheated as well as sub-cooled liquid inventories, while accounting for pumped input, valve closure effects, drainage flow modelling for sub-cooled inventories, heat transfer between the fluid in the pipeline and its surroundings as well as modelling the effects from interacting flow from upstream and downstream sections of the pipeline at the breach. PIPEBREAK-II tracks both rarefaction and feedback waves, frictional effects as well as the presence of pre-existing (pumped/non-stationary) flow during the expanding liquid depressurization process. Furthermore, this paper describes work to improve the liquid density (pipeline inventory) predictions of the traditional cubic equations of state methods employed within PIPEBREAK and the results of an extensive validation exercise to support the selected approach.

2 Literature overview

A high-level review of mathematical models reported in open literature for simulating loss of containment events from long pipelines transporting pressure liquefied fluids has been undertaken. The focus was on identifying models for the initial (rapid) depressurization behavior at the onset of the release, their strengths, weaknesses, and relevance to achieving the objectives of this work. The reviewed models are summarized below in order of relevance.

2.1 Pipeline Release Rate Model (PiRRaM) for Liquefied Gases [2]

Newton [2], presented a new model (PiRRaM), developed for the Health and Safety Executive (HSE) to predict the characteristics of releases from pipelines conveying pressurized liquefied fluids from full-bore ruptures down to small holes.

PiRRaM closely resembles PIPEBREAK but includes an extension to better handle the rapid depressurization behavior at the onset of loss of containment [2]. PiRRaM assumes flow in a pipeline may be characterized in terms of three zones: a stationary zone, an expanding pressurized liquid zone and a saturated two-phase zone. Also, depending on the length of the line, all three zones may or may not coexist. The saturated flow model is very similar to PIPEBREAK, as described by Webber et al. [1], while the pressurized liquid zone is an extension of the saturated flow model to a compressible liquid. PiRRaM employs a rarefraction-wave balancing mechanism in modelling the pressurized initial conditions. This allows PiRRaM to transition smoothly from conditions where the liquid compressibility effect is less important (guillotine ruptures) to scenarios where the liquid compressibility effect is important (small holes) [2].

The potential limitations of PiRRaM, i.e., in the context of the objectives/requirements of this work, are its inability to model the impact of [2]:

- Pumped flow on initial conditions in the pipeline and during transient evolution of inventory.
- Valve-closure effects during the transient evolution process as well as any specific closure mechanisms, e.g., isolation from check valves (flow-reversal), excess flow valves (from exceeding threshold flow rate).
- Loss of containment events involving gas/vapor or sub-cooled liquid at storage conditions.
- Breaches along the length of the pipeline with flow contributions (interactive) from the upstream and downstream ends of the pipeline.

2.2 Other models

Other models exist, in varying levels of complexity, for modelling long pipeline releases transporting pressure liquefied fluids. Some of these are described in open literature ([2], [3], [4]) and are summarized below, in terms of complexity, for reference.

- Simple empirical models: These models are largely phenomenological and employ empirical curve fits to model pipeline decompression. Examples are SLURP [5] and the model proposed by Tam and Higgins [6]. The latter is based on empirical curve fits to the Isle of Grain tests and assumes an exponential decay of the mass flow rate [2]. These models should be used with care as they may be unreliable when applied to scenarios with conditions that differ from the experimental data set on which they were derived.
- Integral models: These break down the conservation equations, with the aid of simplifying assumptions, into simpler mathematical models that are easier to solve computationally. These include PIPEBREAK [1], BLOWDOWN ([7], [8], [9], [10], [11], [12]) and FRED [13]. These models typically assume instantaneous decompression to the saturated vapor pressure and generally lack a consistent method for modelling the effect of liquid compressibility [2]. These simplifications/weaknesses may result in gross underestimation of initial flowrates where the pressurized liquid flow regime is present.
- Complex models: These apply complex numerical solutions to the governing conservation equations. As a result, these models tend to be computationally expensive. Examples are OLGA ([14], [15], [2]), PROFES/PLAC ([16], [17], [18]) and PipeTech ([19], [4]).

3 Model theory and assumptions

The governing theory of PIPEBREAK-II is described in detail by Webber et al. [1] and is based on the analytical solutions of simplified forms of the conservation equations for mass, momentum, and energy for one-dimensional homogenous equilibrium flow in a pipeline. Depressurization behavior is broken down into stages/flow regimes and analytical solutions to simplified, quasi-steady, forms of the conservation equations are derived for each stage.

Following a breach in the pipeline, PIPEBREAK/PIPEBREAK-II assume there will in general be four successive flow regimes [1]: rapid depressurization (regime 1), flash front propagation (regime 2), continuing two phase flow (regime 3) and unpressurized flow (regime 4). PIPEBREAK is only applied to releases involving pressure liquefied gases (superheated/flashing liquids) and as a result, assumes instantaneous isothermal depressurization from storage to saturated conditions during regime 1, and ignores regime 4, given the potential low release rates from unpressurized flow compared to outflow during regimes 2 and 3. However, PIPEBREAK-II, to support modelling releases involving pressure liquefied gases as well as sub-cooled (non-flashing) inventories (where regime 4/drainage flow modelling is important), extends regime 1 and 4 modelling in PIPEBREAK as follows:

- **Regime 1 - rapid depressurization**: This is characterized by a series of rarefaction and reflected waves travelling back and forth from the breach to the upstream/intact end. Once the initial rarefaction wave reaches the intact end of the pipeline (and is reflected), the process is repeated until the pressure in the line drops to (or below) saturated (liquid) conditions. At this junction, regime 2 kicks in, i.e., at the downstream end of the line with a flash front travelling upstream from the breach. Outflow at the breach during regime 1 is assumed to be incompressible/metastable in nature and may be modelled using the Bernoulli equation for incompressible flow ([2], [11], [12]).
- **Regime 4 – unpressurised (drainage) flow**: This regime is only modelled for sub-cooled inventories. This regime starts once the pressure at the breach drops to ambient and there is insufficient flowing fluid/pressure upstream of the breach to maintain fully developed flow across the entire pipeline cross-section. As a result, the escaping fluid drains out of the pipeline through the breach. During this regime, flow through the breach is driven by the fluid static head upstream of the breach. The pipeline is assumed to be horizontal with the breach located conservatively such that the active flow area through the breach is the lower of the liquid cross-sectional area in the pipeline, immediately upstream of the breach and the stipulated breach size.

It is assumed that the depressurization process during regimes 1 and 4 follow an isothermal thermodynamic trajectory. In both cases, the bulk fluid remains as a liquid and the absence of any boil-off/flashing due to depressurization is assumed not to result in significant cooling of the bulk expanding liquid. This assumption is somewhat corroborated by data from the Isle of Grain trials, where, despite the rapid, modest drop in propane pressure across the line (e.g., from 21barg to ca 8barg in the P40 test) as recorded at the onset of each test, there appears to have been no significant attendant temperature decrease [1].

The governing theory, together with any underlying modelling assumptions adopted in PIPEBREAK-II are summarized in the following sections.

3.1 Rapid depressurization regime (regime 1)

The modelling of the rapid depressurization regime (i.e., regime 1) in PIPEBREAK-II has been undertaken by employing a modified/extended version of the rarefaction wave balancing model described by Newton [2] as implemented in PiRRaM. Unlike PiRRaM, the extended rapid depressurization regime modelling in PIPEBREAK-II caters for pre-existing pumped inflow, flow reversal downstream of the breach, valve closure effects as well as the occurrence of punctures at locations other than at pipeline ends with flow contributions (disjoint or interactive) from the upstream and downstream ends of the pipeline.

The rarefaction wave model implemented in PiRRaM is a phenomenological model based on observations. The model assumes there is a rarefaction wave that travels at the local speed of sound (C_s) upstream of the breach. The fluid flow between the rarefaction wave and the breach location is frictionless. The model balances the pressure drop induced flow into the frictionless zone between the rarefaction wave and the breach to the appropriate outflow model. The governing equation of the wave balancing model adopted in PiRRaM is given by [2]:

$$C_s(\rho_0 - \rho_{fric})\,A_{pipe} = G_e A_{hole} \tag{1}$$

Where:

$$G_e = \begin{cases} G_{Bernoulli} \\ \quad G_{sat\ choked} \end{cases} \tag{2}$$

A_{pipe} and A_{hole} are the pipeline and breach cross-sectional areas respectively; ρ_0 and ρ_{fric} are the fluid densities at storage and frictionless flow balancing pressure (P_{fric}), respectively; G_e is the fluid mass flux through the breach.

Note that the expression for G_e in equation (2) indicates that either an incompressible/metastable fluid flow model (Bernoulli model) is employed in calculating G_e (at the breach) or a saturated choked flow approach, i.e., where the predicted rates by the Bernoulli model is too high to balance depressurization effects from the rarefaction wave propagating into the pipeline: which is usually the case with full-bore rupture/large aperture ratio releases.

The assumption of frictionless flow between the rarefaction wave and the breach imposes a constant pressure (P) profile along the length of the line (i.e., $\Delta P = 0$) in the frictionless flow region, i.e., the fluid comes effectively to rest (stationary flow zone) throughout the region impacted by the rarefaction wave. The pressure (P_{fric}) in the frictionless region is applied as the upstream pressure in the determination of flowrate through the hole, which, in the case where P_{fric} is predicted to be greater than the saturated pressure (P_{sat}) at initial/storage temperature (T_0) (i.e., $G_e = G_{Bernoulli}$):

$$G_e A_{hole} = C_d A_{hole}\sqrt{2\rho_{fric}(P_{fric} - P_{amb})} \tag{3}$$

Where P_{amb} is the ambient pressure and C_d is the discharge coefficient through the breach. For full-bore ruptures, $C_d = 1$ at the rupture plane (as there are no constriction losses).

With PIPEBREAK-II, however, there is need to cater for pre-existing pumped flow, i.e., potential for finite pressure drop along the pipeline, prior to and following the loss of containment event.

The governing equation of the wave balancing model is extended and generalized to account for pre-existing flow in the line (e.g., due to pumped inflow) as follows:

For locations upstream/downstream (i.e., following flow-reversal) of the breach:

$$\left(C_{s,i-1} - u_{0,i-1}\right)\left(\rho_{0,i-1} - \rho_{eff,i}\right) A_{pipe} = G_e A_{hole} \quad (4)$$

Where $u_{0,i-1}$ is the local velocity through the pipeline, during the previous time step (i-1) and therefore the velocity ahead of the current rarefaction wave (i) propagating from the breach towards the intact end. The subscript "i" implies that this equation applies to the current time interval where the depressurization wave is present.

The effective velocity (u) downstream of the decompression wave front can be expressed as:

$$u_{0,i} = \frac{\left(C_{s,i-1} - u_{0,i-1}\right)\left(\rho_{0,i-1} - \rho_{eff,i}\right)}{\rho_{eff,i}} \quad (5)$$

The effective liquid density ($\rho_{eff,i}$) at the current time step is determined via a number of steps described below as follows:

As presented in Webber et al. [1], for any given pipeline, with diameter, D, the variation p(x) of pressure along the pipe can be expressed implicitly in the inverse form, i.e., in terms of an upstream distance x(p), measured from the exit point (e) to the point where the pressure and density are (p, ρ), by solving the steady state momentum equation as:

$$x(p) = \frac{D}{2f}\left[\frac{1}{G^2}\int_{p_e}^{p} \rho\, dp - ln\left(\frac{\rho}{\rho_e}\right)\right] \quad (6)$$

Where: f is the flow regime dependent fanning friction coefficient, while ρ and G are the fluid density and (quasi) steady-state mass flux of the conveyed liquid respectively: ρ_e is the fluid density at the downstream (e.g., breach) end of the pipeline for which the pressure drop is being evaluated.

The total mass of fluid (per unit cross-sectional area) in the pipe (M) within the relevant pipeline section x(p), is found based on the formular in Webber et al. [1], reproduced below:

$$M = \int_0^L \rho\, dx = \int_{p_e}^{p_0} \rho \frac{dx}{dp} dp \quad (7)$$

$$= \frac{D}{2f}\left[-(\rho_0 - \rho_e) + \frac{1}{G^2}\int_{p_e}^{p_0} \rho^2 dp\right]$$

The effective fluid density ($\rho_{eff,i}$) is determined as the averaged density over the pipeline volume (V_{pipe}) spanning the relevant length (x(p) = L) exposed to the rarefaction wave front, i.e.,

$$\rho_{eff,i} = \frac{M}{V_{pipe}} = \frac{M}{LA_{pipe}} \tag{8}$$

The objective at the end of regime 1 is to transition smoothly to regime 2. For this, there is need to determine the location of the boundary between regimes 1 and 2 such that the upstream pressure is at P_{sat}, with the flow through the line equal to "$G_{sat\ choked}$", i.e., at the commencement of regime 2. Thus, the length ($\Delta L_{pipe,i}$) of the boundary between regimes 1 and 2, measured relative to the breach is given by:

$$\Delta L_{pipe,i} = \frac{G_e A_{hole} \Delta t^*}{(\rho_{0,i-1} - \rho_{sat})\, A_{pipe}} \tag{9}$$

Where the subscript "i-1" indicates conditions during the previous time-step. Note that at the upstream boundary of the travelling wavefront, $\rho_{eff,i} = \rho_{sat}$, implying $P_{fric} = P_{eff,i} = P_{sat}$. The flow profile downstream of this boundary (i.e., towards the breach end) is based on regime 2. The rarefaction wave front at which $P_{fric} = P_{sat}$ is then tracked until it reaches the upstream (or intact) end of the pipe at which point regime 2 fully kicks in across the entire pipeline.

Alternatively, the time elapsed (Δt^*) for the final rarefaction wave front (i.e., at which $P_{fric} = P_{sat}$) to reach the upstream end of the pipeline of length, L_{pipe}, is given by:

$$\Delta t^* = \frac{(\rho_{0,i-1} - \rho_{sat}) L_{pipe} A_{pipe}}{G_e A_{hole}} \tag{10}$$

In terms of interactive flow, i.e., where flow from the upstream and downstream end of the pipeline relative to the breach, interact at the breach, the continuity equation over a fixed time interval (Δt) is given by:

$$m_{pump} + m_{expelled,A} + m_{expelled,B} = G_e A_{hole} \tag{11}$$

Where:

$$m_{expelled,A} = (\rho_{0,A,i-1} - \rho_{eff,A,i}) L_{pipe,A} A_{pipe} \tag{12}$$

$$m_{expelled,B} = (\rho_{0,B,i-1} - \rho_{eff,B,i}) L_{pipe,B} A_{pipe} \tag{13}$$

Where $\rho_{0,A,i-1}/\rho_{eff,A,i}$ and $\rho_{0,A,i-1}/\rho_{eff,B,i}$ are the effective (average) fluid densities upstream (A) and downstream (B) of the breach during the current (i) and at the end of the previous (i-1) propagating rarefaction waves, respectively; $L_{pipe,A}$ and $L_{pipe,B}$ are the line lengths upstream and downstream of the breach, respectively.

The time interval for the propagation of the rarefaction wave is governed by the longer of the time it takes for the rarefaction wave to travel the lengths of the upstream and downstream sections (branches) of the pipeline. In general, Δt is given by:

$$\Delta t = \max\left(\frac{L_{pipe,A}}{C_{s,i-1} - u_{0,i-1}}, \frac{L_{pipe,B}}{C_{s,i-1} - u_{0,i-1}}\right) \quad (14)$$

However, for finite pumped flow, the rarefaction wave will originally travel down the downstream branch along the positive mach line, in this case, Δt is given by:

$$\Delta t = \max\left(\frac{L_{pipe,A}}{C_{s,i-1} - u_{0,i-1}}, \frac{L_{pipe,B}}{C_{s,i-1} + u_{0,i-1}}\right) \quad (15)$$

The objective in the interactive flow model is to determine a common boundary condition at the breach location, satisfying continuity and momentum constraints, such that the two branches experience the same pressure and temperature at the breach/exit. This process is repeated for each rarefaction wave until the pressure in the line drops to saturated conditions at which stage subsequent flow regimes kick in.

3.2 Sub-cooled modelling with drainage flow

With sub-cooled inventories, the transition to regimes 2 and 3 does not occur. Rather, the fluid pressure in the pipeline rapidly drops to ambient pressure, thereafter, the escaping fluid drains out of the pipeline through the breach. The drainage flow regime (regime 4) starts once there is insufficient flowing fluid/pressure upstream of the breach to maintain fully developed flow across the entire pipeline cross-section. During the drainage regime, flow through the breach is driven by the fluid static head upstream of the breach, relative to the base of the pipeline. The pipeline is assumed to be horizontal with the breach located conservatively such that the active flow area through the breach is the lower of the liquid cross-sectional area in the pipeline, immediately upstream of the breach and the stipulated breach size.

Drainage flow modelling is applied to any sections along the pipeline where a finite vapor space exists above the evacuating fluid and may co-exist with fully developed flow conditions upstream of the breach.

Two types of drainage flow modelling are supported:

- "Uniform Profile": This is illustrated by Figure 1 and conservatively assumes the fluid's static head is the same across the entire length of the pipeline at a given time but varies with time. Drainage flow conditions prevail instantaneously, across the entire line length as soon as the flowrate through the breach drops below the maximum flowrate through the breach based on hydrostatic liquid head.
- "Non-uniform profile": This is illustrated by Figure 2 and assumes the fluid static head varies along the pipeline with a profile that satisfies the steady-state momentum equation. In this model, drainage flow conditions gradually travel upstream of the pipeline from the breach end until drainage flow conditions prevail across the line. Drainage flow conditions commence at the breach end as soon as the flowrate through the breach drops below the maximum flowrate through the breach based on hydrostatic liquid head.

Figure 1: Uniform profile drainage flow modelling based on constant liquid level along pipeline length

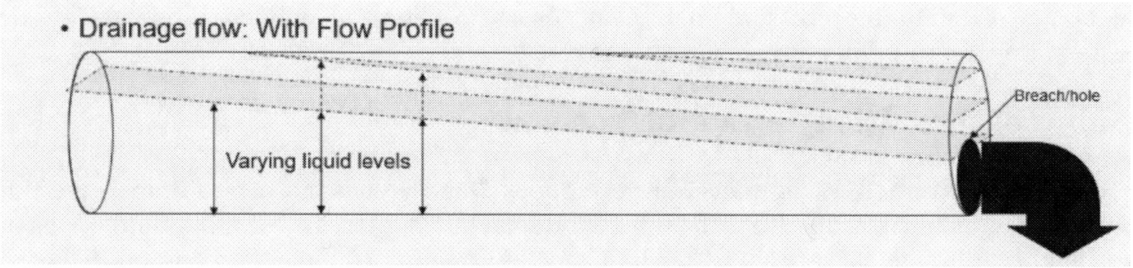

Figure 2: Non-uniform profile drainage flow modelling based on varying liquid levels along pipeline length

Details of the non-uniform profile modelling are presented below, given the uniform profile modelling is a simplification of the non-uniform approach.

The steady state momentum equation in one dimension along the length (x) of a pipeline is given by:

$$\rho u \frac{du}{dx} = -\frac{dP}{dx} - \frac{2f}{D_h} \rho u|u| \tag{16}$$

Where D_h is the hydraulic diameter, i.e., the ratio of the fluid cross-sectional area (A) and the wetted perimeter (W_p).

For a given liquid level of height, h, at any location along the length of the pipeline with diameter, D and radius, R, the fluid flow cross-sectional area, A, is given by:

$$A = R^2 \cos^{-1}\left(\frac{R-h}{R}\right) - (R-h)\sqrt{h(D-h)} \tag{17}$$

The rate of change of cross-sectional area with height, can be expressed as:

$$\frac{dA}{dh} = 2\sqrt{h(D-h)} \tag{18}$$

The wetted perimeter (W_p) for a given liquid level of height, h, at any location along the length of the pipeline is given by:

$$W_p = D\sin^{-1}\left(\frac{2\sqrt{h(D-h)}}{D}\right) \tag{19}$$

At steady state, the flowrate (m) through the line is constant, such that:

$$m = \rho u A \tag{20}$$

During drainage flow, the liquid density, ρ, is constant, hence, the rate of change of flow velocity (u) with flow cross-sectional area (A) can be simply expressed as:

$$\frac{du}{dA} = -\frac{m}{\rho A^2} \tag{21}$$

Hence,

$$du = -\frac{m}{\rho A^2}dA \tag{22}$$

In addition, the fluid pressure, P, accounting for hydrostatic head, is given by:

$$P = P_{amb} + h\rho g \tag{23}$$

Where, P_{amb} and g are the ambient pressure and acceleration due to gravity, respectively.

Given P_{amb} is constant, we have:

$$dP = \rho g\, dh \tag{24}$$

By substituting for dP, du and u, as well as dividing through by ρu^2, the momentum equation can be written as:

$$-\frac{1}{A}\frac{dA}{dx} = -g\left(\frac{\rho A}{m}\right)^2\frac{dh}{dx} - \frac{2f}{D_h} \tag{25}$$

Substituting for dA in terms of dh in the momentum equation, we can use the momentum equation to derive the variation h(x) of hydrostatic head along the pipe implicitly in the inverse form x(h) as:

$$x(h) = \frac{1}{2f}\int_{h_e}^{h} D_h\left[g\left(\frac{\rho A}{m}\right)^2 - \frac{2\sqrt{h(D-h)}}{A}\right]dh \tag{26}$$

The mass of fluid in the pipeline, M, at any given point in time can be determined as:

$$M = \rho\int A\,dx \tag{27}$$

Again, for any region within the pipeline where drainage flow conditions prevail, M can be calculated in terms of the variation in hydrostatic head along the relevant pipeline region as:

$$M = \frac{\rho}{2f}\int_{h_e}^{h} AD_h\left[g\left(\frac{\rho A}{m}\right)^2 - \frac{2\sqrt{h(D-h)}}{A}\right]dh \tag{28}$$

The time taken, dt, to evacuate the mass, dM, based on a fluid cross-sectional area change, dA at the breach, can be expressed as:

$$\frac{\int dM}{m_{ave}} = \int dt \tag{29}$$

Where m_{ave} is evaluated based on the cross-sectional area change at the breach between each quasi-steady state steps in "h", and is given by:

$$m_{ave} = \frac{\int (m_{in} - m_{out})\,dA}{\int dA} \tag{30}$$

Where m_{in} and m_{out} are the input mass flowrate (or pumping rate) and outflow mass flow rate through the breach, respectively.

3.3 Determination of liquid densities

The Peng-Robinson (PR) cubic equation of state (CEOS) is applied by default in the determination of fluid properties in PIPEBREAK. However, the PR-CEOS may not accurately predict liquid densities, particularly for non-hydrocarbon fluids.

PIPEBREAK/PIPEBREAK-II may also employ the DIPPR correlations for estimating saturated liquid densities as an alternative. This approach guarantees accurate prediction of liquid densities along the saturation curve for pure component materials. However, the DIPPR correlations are only valid along the saturation curve.

Away from the saturation curve, there is a need to predict liquid densities as varying with changes in pressure, whilst ensuring accurate predictions along the saturation curve. Various approaches have been investigated as part of this study and a simple "offset" method, based on the Soave-Redlich-Kwong (SRK) CEOS has been adopted in PIPEBREAK-II. The SRK "Offset" method is employed in estimating the fluid liquid density ($\rho_{liq}(P,T)$) away from the saturation curve whilst employing DIPPR correlations along the saturation curve and is given by:

$$\rho_{liq}(P,T) = \rho_{liq,DIPPR}(P_{sat},T) + \left(\rho_{liq,CEOS}(P,T) - \rho_{liq,CEOS}(P_{sat},T)\right) \quad (31)$$

Where $\rho_{liq,DIPPR}(P_{sat}, T)$, $\rho_{liq,CEOS}(P,T)$ and $\rho_{liq,CEOS}(P_{sat},T)$ are the liquid densities based on DIPPR correlations, SRK-CEOS at the desired pressure (P) and temperature (T) and SRK-CEOS at the saturated pressure (P_{sat}) corresponding to the desired temperature (T).

The SRK "Offset" method is akin to "volume-translation" methods, which are widely applied in improving liquid density prediction capabilities of CEOS [20]. The SRK "Offset" method simply modifies the DIPPR calculated density by an "offset", i.e., the difference in predicted densities based on the SRK-CEOS at the desired (P, T) relative to SRK-CEOS predicted liquid density along the saturation curve.

4 Model verification and validation

4.1 Model verification/intercomparison

Predicted results from PIPEBREAK-II have been verified/sense-checked against representative case studies, sensitivity scenarios and results from selected pipeline/vessel blowdown models.

Newton [2] presents results from the comparison of PiRRaM against SLURP and PROFES model predictions for a hypothetical full bore rupture release of Propane from a 250mm diameter pipeline of infinite length initially at 25°C and 45barg (Case 1), 70barg (Case 2) and 20barg (Case 3), respectively. Simulated flowrate profiles with time are reported for the first 20min (1200s) from the onset of the release and have been digitized for the model verification/intercomparison exercise.

For the verification exercise, a nominal pipeline length of 5.5km has been stipulated. This corresponds approximately to the minimum pipeline length at which the propagating flash front from the rupture end may not reach the intact end within the 20min elapsed time. Essentially, beyond 5.5km, discharge behavior at the rupture end, during regime 2, is insensitive to pipeline length over the 20min interval of interest (i.e., infinite pipeline length behavior based on regime 2).

Figure 3 shows transient flowrate profiles predicted by PIPEBREAK-II (PBRK-II) compared against PiRRaM, PROFES and SLURP data as extracted from [2]. From Figure 3:

- PiRRaM and PIPEBREAK-II results show, as expected, very good agreement. The flowrate profiles are largely insensitive to initial pressures.
- PIPEBREAK-II results show a slight delay in commencement of regime 2 with the delay increasing with increasing initial pressure. This is down to the modelling of the rapid depressurization process (regime 1). PIPEBREAK-II results suggest that Regime 1 effects are not too important in these cases, being full bore rupture releases. In the worst case (i.e., the 70barg scenario), regime 1 is predicted to last for just under 4.3s and in the 45barg and 20barg cases, under 2.7s and 1.1s, respectively.

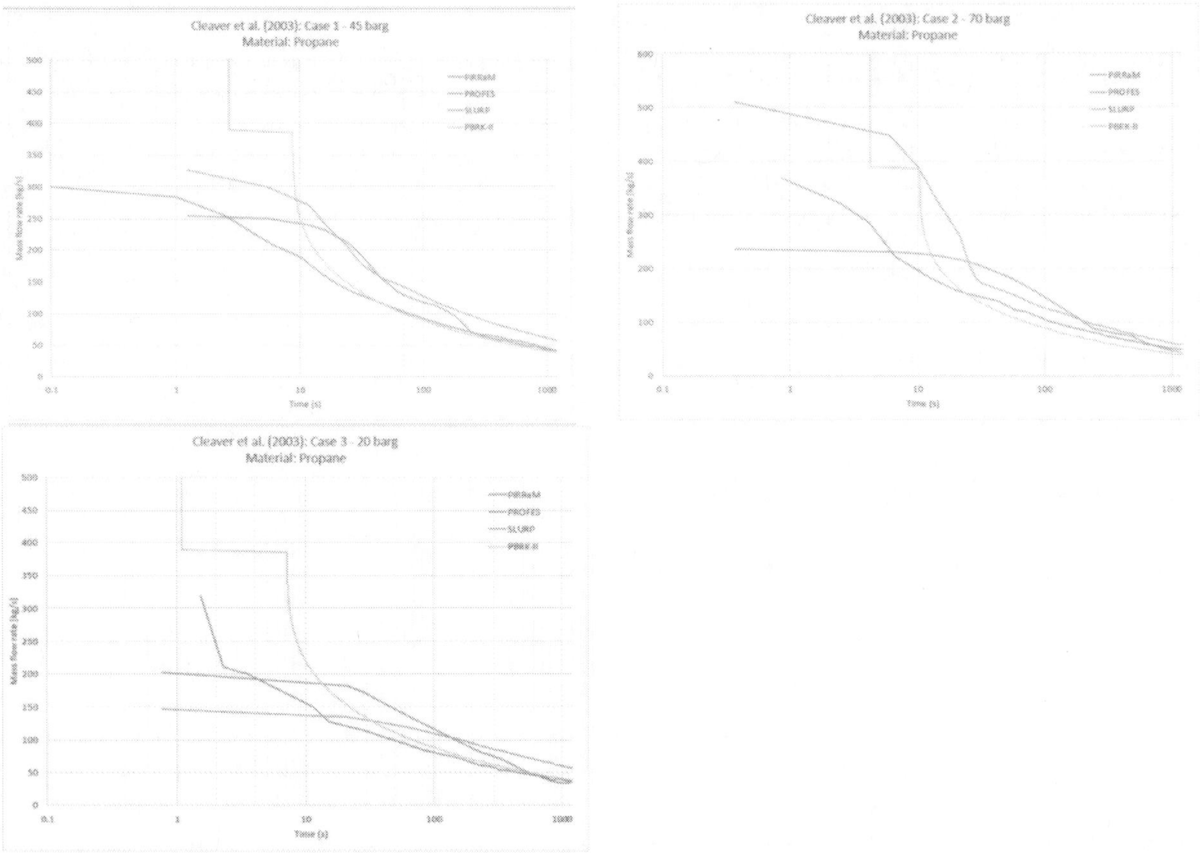

Figure 3: Comparison of mass flowrate profiles with time for PIPEBREAK-II versus predictions reported by Newton [2] for SLURP, PROFES and PiRRaM for a release of Propane from a 250mm diameter pipeline at 25°C and 45barg (Case 1), 70barg (Case 2) and 20barg (Case 3)

The drainage flow modelling for releases involving sub-cooled inventories has been verified by comparing against results from the Phast/Safeti time-varying (vessel) discharge model (TVD2). The pipeline and breach characteristics are summarized below:

- Inventory: 50% n-Octane (n-C8) and 50% n-Decane (n-C10) (mol/volume basis)
- Storage pressure and temperature: 0.01barg and 25 °C
- Pipeline diameter: 152mm
- Aperture ratio, i.e., ratio of breach to pipeline cross-sectional areas (%): 5%
- Pipeline inventory storage state: Liquid
- Pipeline length: 200 m
- Breach location: 200 m
- Release direction: horizontal
- Pumped throughput: 0 kg/s
- Maximum duration of interest: 7200 s
- Minimum pool thickness for termination of drainage flow calculations: 1mm

The TVD2 vessel is set up as a horizontal cylinder atmospheric storage tank with vessel diameter and length equal to the pipeline diameter and length, respectively, as well as a vessel liquid inventory of 2599.21 kg (corresponding to the total inventory in the pipeline at the onset of the release).

Figure 4, Figure 5 and Figure 6 show, respectively, the transient profiles of:

- Upstream/intact and downstream/breach end pressures: Pipeline releases only
- Total flowrate
- Total expelled mass

Each plot shows simulated results based on:

- PIPEBREAK-II with drainage flow modelling based on uniform flow profile: "Regime1-C8_C10_Mix_5%_Disjoint_NoProf_DISC".
- PIPEBREAK-II with drainage flow modelling based on non-uniform flow profile: "Regime1-C8_C10_Mix_5%_Disjoint_Prof_DISC".
- TVD2 results: "5%_33.9882mm_TVD2".

From Figure 4, Figure 5 and Figure 6:

- For sub-cooled inventories, a significant proportion of outflow could occur during the drainage flow regime, and it would be overly optimistic to ignore drainage flow in pipeline loss of containment modelling in these cases.
- TVD2 and PIPEBREAK-II, with uniform profile drainage flow, modelling results (flowrate and expelled mass) match from the onset of the release and only diverge at ca 3000s. The divergence in results at ca 3000s, is due to PIPEBREAK-II predicting the liquid level in the pipeline to drop below the representative hole height (i.e., the aperture is only partly covered by discharging fluid from this time onwards). TVD2 modelling conservatively assumes the aperture is fully covered by the discharging fluid for the entire release duration.
- As expected, the fluid inventory is predicted to drain out quicker with uniform as compared against non-uniform flow profile modelling. The uniform flow profile modelling results are conservative and indicate the minimum pool thickness criterion is reached across the entire length of the pipeline at ca 5100s, while non-uniform flow profile modelling indicates otherwise, with drainage flow predicted to continue beyond the duration of interest (i.e., 7200s).

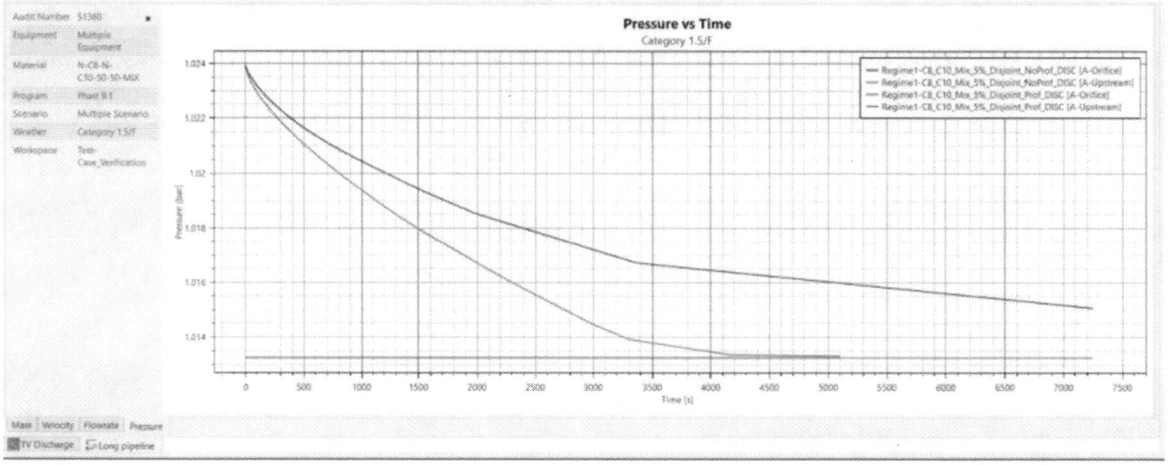

Figure 4: Variation of upstream end pressure with time for sub-cooled inventory: PIPEBREAK-II with drainage flow modelling based on uniform and non-uniform profiles.

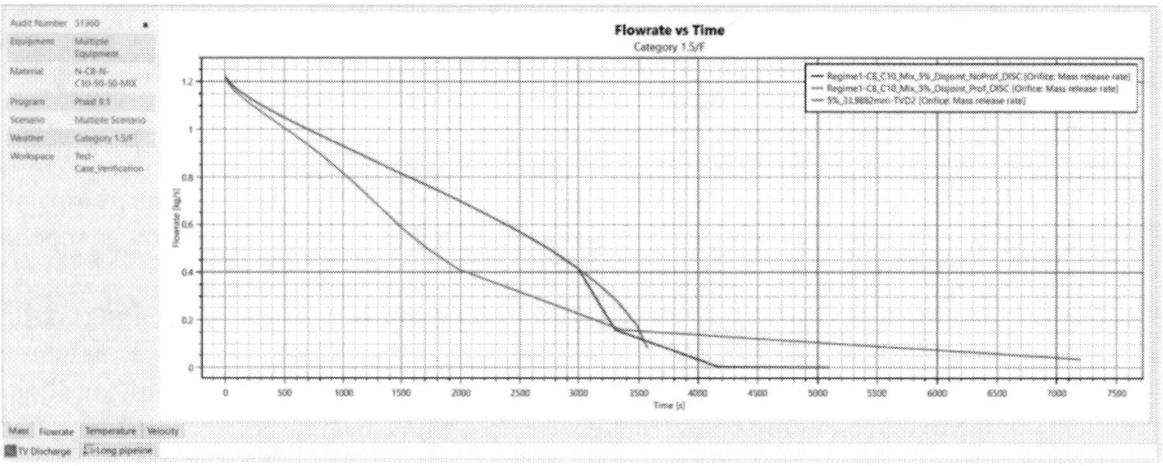

Figure 5: Variation of total mass flowrate with time for sub-cooled inventory: PIPEBREAK-II with drainage flow modelling based on uniform and non-uniform profiles compared against TVD2 results.

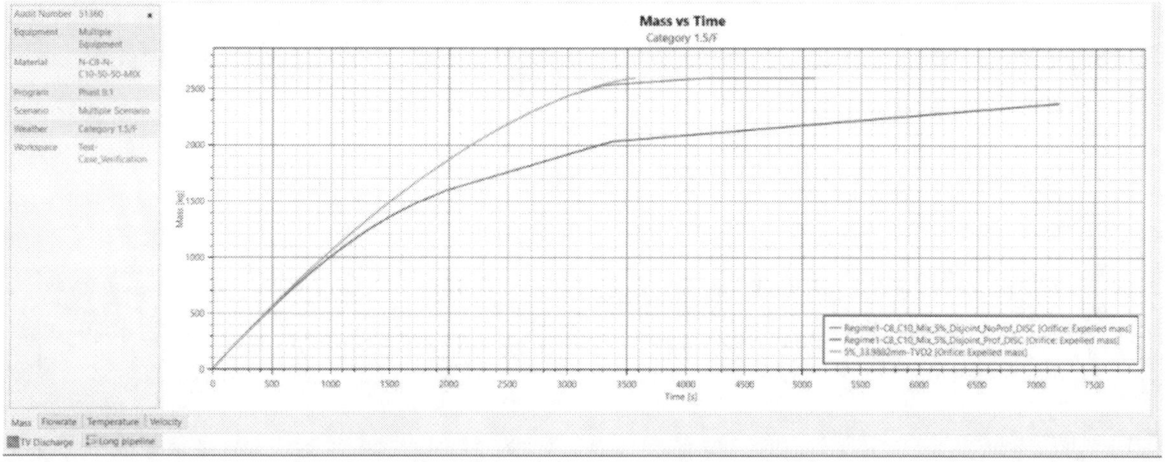

Figure 6: Variation of total expelled mass with time for sub-cooled inventory: PIPEBREAK-II with drainage flow modelling based on uniform and non-uniform profiles compared against TVD2 results.

The SRK "Offset" liquid density method has been verified by comparing against NIST [21] reference data for a selection of key materials. The materials are listed here with the seven different temperatures for which the properties were evaluated. The temperatures selected ranged from the lowest temperature with a saturation pressure of 1barg up to a temperature close to the fluid's critical point. For each temperature, the pressure varied from saturation pressure up to four times the critical pressure of the material in question. In total ca 2800 data points were evaluated as part of the verification exercise.

- Carbon dioxide (T: -50°C, -35°C, -20°C, -5°C, 10°C, 25°C, 30°C)
- Ammonia (T: -30°C, -10°C, 10°C, 30°C, 50°C, 70°C, 75°C)
- Propane (T: -35°C, -15°C, 5°C, 25°C, 45°C, 65°C, 75°C)
- Hydrogen (T: -252°C, -250°C, -248°C, -246°C, -244°C, -242°C, -241°C)
- Methane (T: -160°C, -145°C, -130°C, -115°C, -100°C, -90°C, -85°C)
- Ethane (T: -80°C, -60°C, -40°C, -20°C, 0°C, 20°C, 30°C)
- Butane (T: 5°C, 15°C, 25°C, 35°C, 45°C, 55°C, 65°C)
- Octane (T: -45°C, -25°C, -5°C, 15°C, 35°C, 55°C, 75°C)

Figure 7 shows the ratio of SRK "Offset" predicted liquid densities to NIST reference data compared to NIST reference data. From Figure 7, the SRK "Offset" method is observed to compare very well with NIST data, with predicted liquid densities tending to be higher (conservative) but generally within +/- 10% of NIST data.

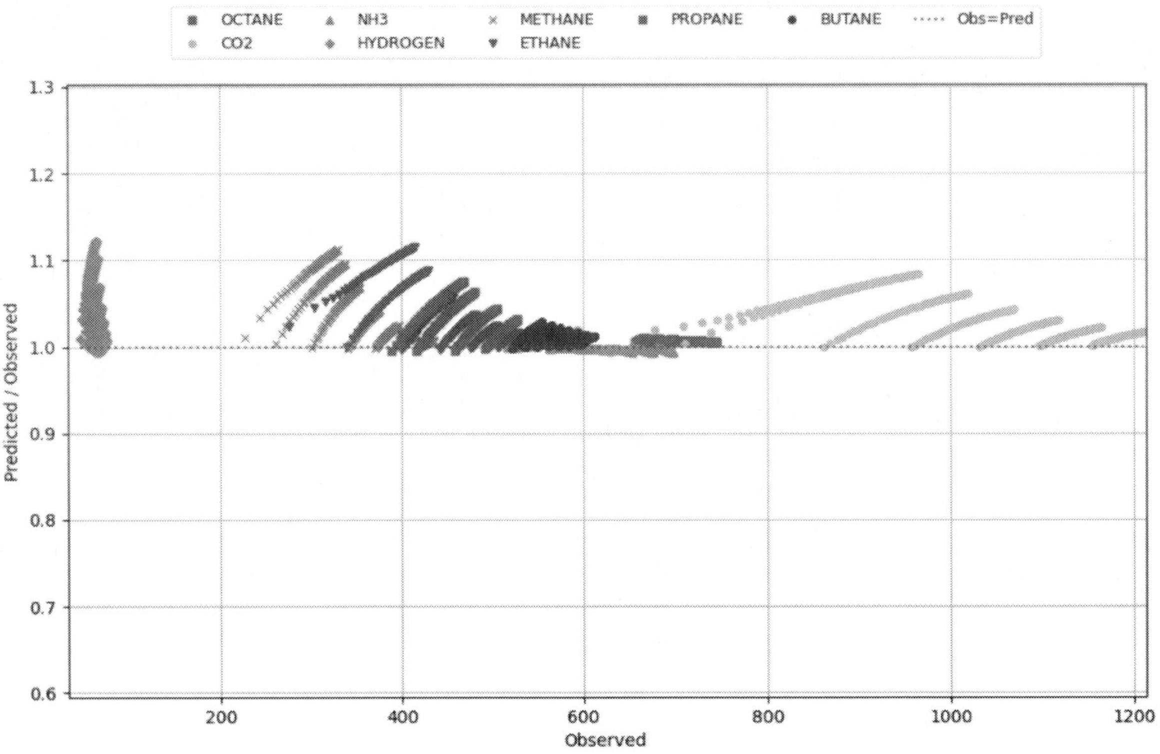

Figure 7: Liquid density predicted by SRK "Offset" method against NIST reference data.

4.2 Model validation

Simulated results from PIPEBREAK-II have been validated against published data for a wide range of experiments/field tests identified in open literature. The results of the validation of PIPEBREAK-II against selected data from the following tests are summarized below:

- The Isle of Grain LPG pipeline rupture/puncture experiments [11]
- The CO2PIPETRANS pipeline rupture/puncture experiments [22]

Table 1 presents a summary of key field test data for the selected validation cases.

Table 1. Summary of field test data for selected validation cases

Test Set	Trial	Pipeline length (m)	Pipeline Diameter (mm)	Aperture (mm)	Aperture ratio (% pipe cross-section)	Initial Temperature [°C]	Initial Pressure (barg)
Isle of Grain (Propane tests)	P40	100	154	154	100	20	21.6
	P45	100	154	75	23.72	17.3	11.4
	P47	100	154	50	10.54	15.6	21.3
CO2PIPETRANS (CO_2 tests)	Test 4	200	52.4	50	91	4.9	100.5
	Test 5	200	52.4	20	15	6.4	100.9
	Test 6	200	52.4	10	3.6	5.8	101.2

4.2.1 Isle of Grain tests

Figure 8, Figure 9 and Figure 10 compare predicted PIPEBREAK and PIPEBREAK-II results for mass inventory, closed and open end pressure and temperatures to Isle of Grain data for tests P40, P45 and P47, respectively.

From Figure 8, Figure 9 and Figure 10, it can be observed that PIPEBREAK and PIPEBREAK-II results are very similar. The minor differences in results are down to the use of different liquid density prediction methods, i.e., PR-CEOS in PIPEBREAK and SRK "Offset" in PIPEBREAK-II.

For all the tests the initial rapid depressurization regime does not play a significant role in the overall picture of things given the relatively short (100m) pipeline length. As pointed out by Newton [2], literature relating to the Isle of Grain tests indicate that the P47 trial appear to have been subject to a higher level of uncertainty, due to unaligned readings from the load cell and neutron beam scattering (NBS) measurements [11], [12].

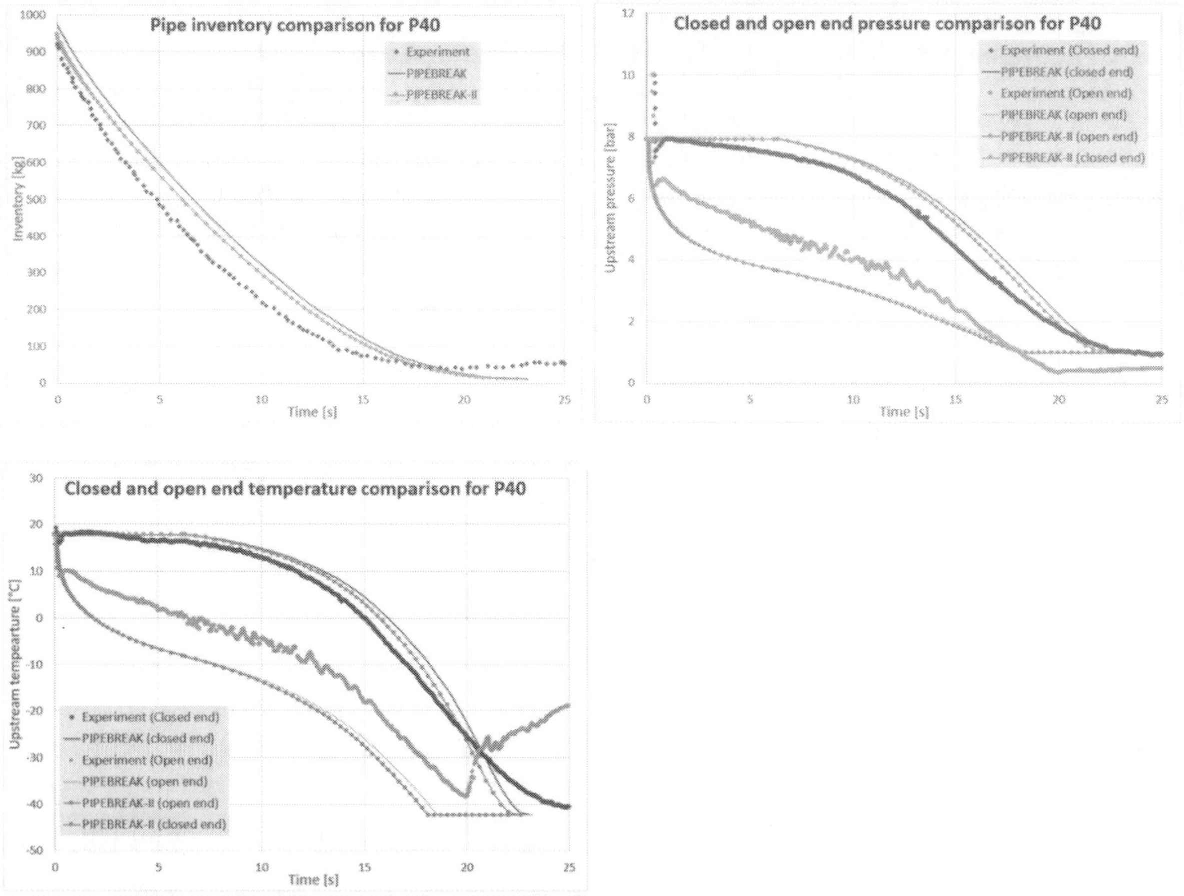

Figure 8: Comparison of PIPEBREAK and PIPEBREAK-II predicted Inventory mass, Pressure and Temperature to Isle of Grain experiment P40.

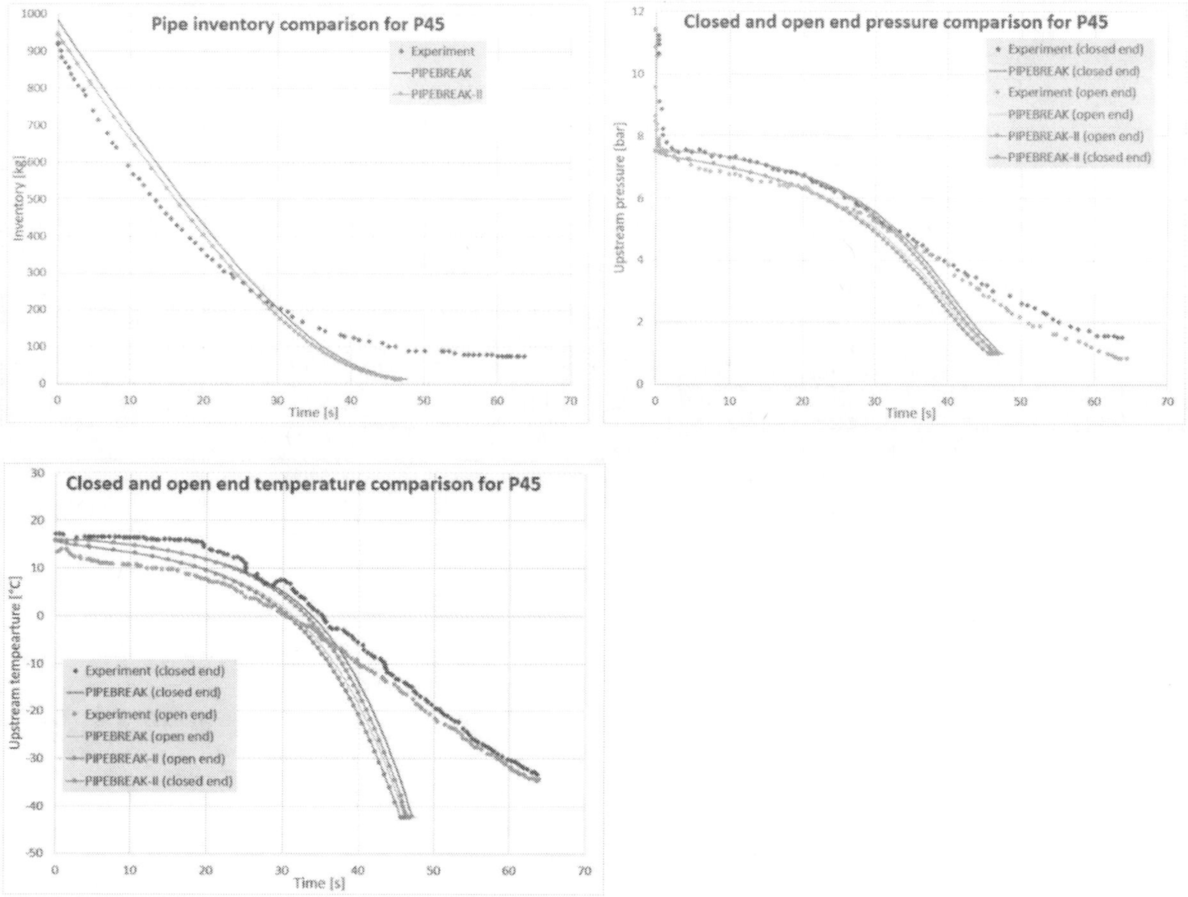

Figure 9: Comparison of PIPEBREAK and PIPEBREAK-II predicted Inventory mass, Pressure and Temperature to Isle of Grain experiment P45.

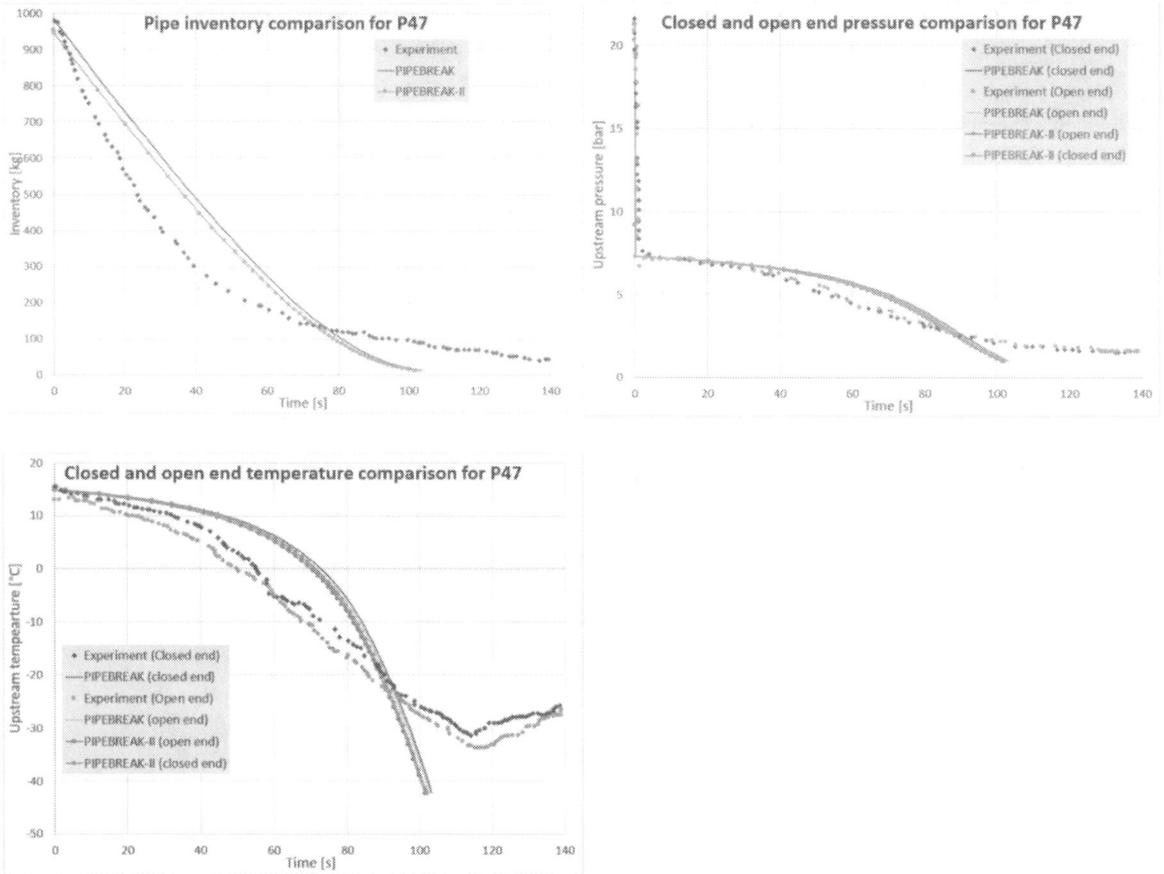

Figure 10: Comparison of PIPEBREAK and PIPEBREAK-II predicted Inventory mass, Pressure and Temperature to Isle of Grain experiment P47.

4.2.2 CO2PIPETRANS tests

A series of eight releases of CO_2 from a 200 m long shock tube with inner diameter 52.44 mm were conducted at Spadeadam as part of the CO2PIPETRANS JIP between November 2012 and April 2013 [22]. Orifice sizes varied from 10 mm to 50 mm. To help assess PIPEBREAK-II, we here select three of these tests with very different hole sizes for validation. Key model inputs for the three selected tests are given in Table 1.

4.2.2.1 Expelled mass and density

Accurate prediction of the total mass expelled when modelling a release is important. Figure 11 shows a comparison of the total expelled mass for Test 4, 5 and 6 as derived from experiments (blue bars) and as predicted by the PIPEBREAK (grey bars) and PIPEBREAK-II (yellow bars) models. The two shades of the experimental bars represent an uncertainty in the measured data due to some load sensors not behaving as expected. We observe that:

- The PIPEBREAK model significantly under-predicts the total expelled mass when compared to the experiments.
- The updated PIPEBREAK-II model predicts the expelled mass very well, with a slight under-prediction for Test 5.
- The improved predictions in PIPEBREAK-II is due to the new Regime 1 modelling (initial depressurization regime) instead of dropping to saturation pressure

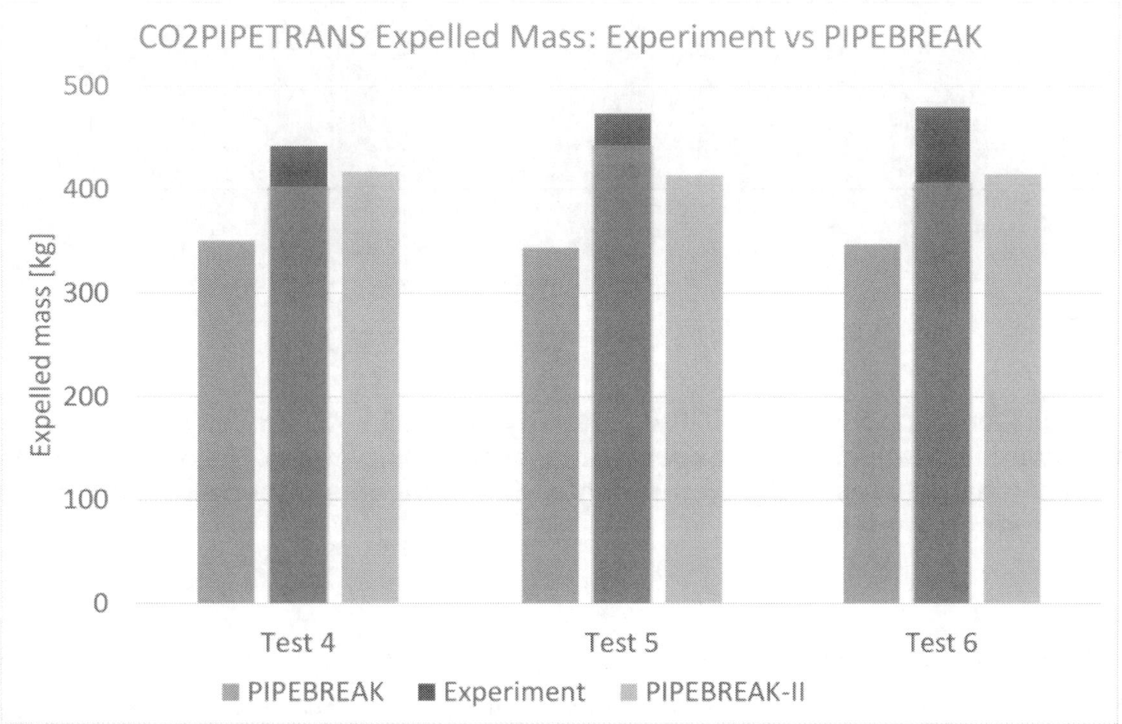

Figure 11: Expelled mass as derived from load cells in the experiment vs predictions from PIPEBREAK and PIPEBREAK-II. The shaded area in the experimental data indicates uncertainty in the data due to measurement issues (i.e., excluding potentially faulty load cells).

Getting the density predictions accurate is crucial to get an accurate prediction of the total expelled mass. We here look at some comparisons of the initial fluid density in the pipe immediately prior to the onset of the release – see Figure 12. Reference density of CO_2 at the given initial conditions given in Table 1 is shown by the green bars and was obtained from the NIST website [21] (using the Span-Wagner equation of state). The experimental density in blue is derived from the experiments by dividing total expelled mass by the initial pipe volume, with the shades again representing experimental uncertainty. The yellow bars is the density as predicted by the new hybrid DIPPR-SRK offset method in PIPEBREAK-II. The following can be seen:

- The new hybrid DIPPR-SRK offset density method predicts densities remarkably close to the Span-Wagner reference values, giving confidence to the new PIPEBREAK-II modelling.
- The back-calculated density values for Test 5 seem somewhat higher than expected.

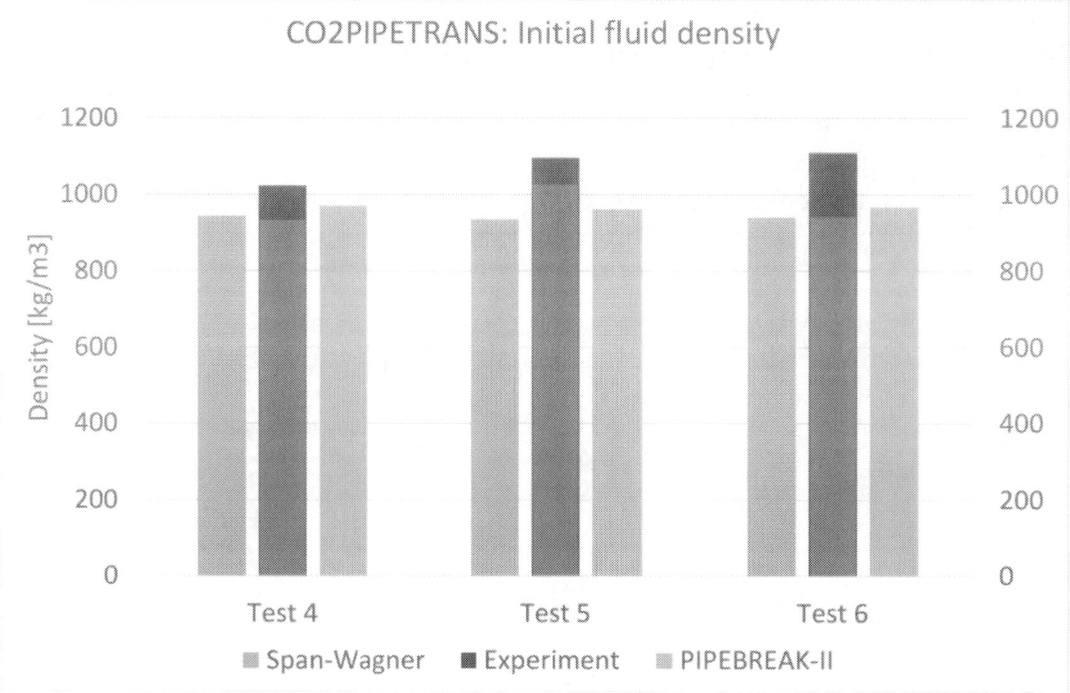

Figure 12: Comparison of initial fluid density of CO2 – Span-Wagner reference density, density derived from experiments and density from the PIPEBREAK-II model. The shaded area in the experimental data indicates uncertainty in the data due to measurement issues.

4.2.2.2 Expelled mass vs time

The expelled mass over time during the tests was also derived from the load cells and compared against the expelled mass predicted by PIPEBREAK and PIPEBREAK-II as shown in Figure 13, Figure 14 and Figure 15. We can see that:

- PIPEBREAK-II predictions tend to be a bit conservative in the initial stage of the release, but over time the predictions fall within the band of uncertainty for the experimental data. The higher rate of expelled mass predicted with PIPEBREAK-II at the onset of the release is due to PIPEBREAK-II employing onerous release rate modelling assumptions in these cases, such as: ignoring any constriction losses at the breach (i.e., $C_d = 1$) as well as metastable liquid flow during the rapid depressurization regime and its transition to the flash front propagation regime.
- PIPEBREAK predictions match quite well in the early stage of the release, but over time a significant under-prediction occurs.
- The higher rate of expelled mass predicted with PIPEBREAK-II compared to PIPEBREAK at the onset of the release is due to modified PIPEBREAK-II release rate modelling assumptions such as metastable liquid flow and higher line pressures.

- Overall, PIPEBREAK-II predictions are significantly better than PIPEBREAK predictions from a consequence modelling point of view.

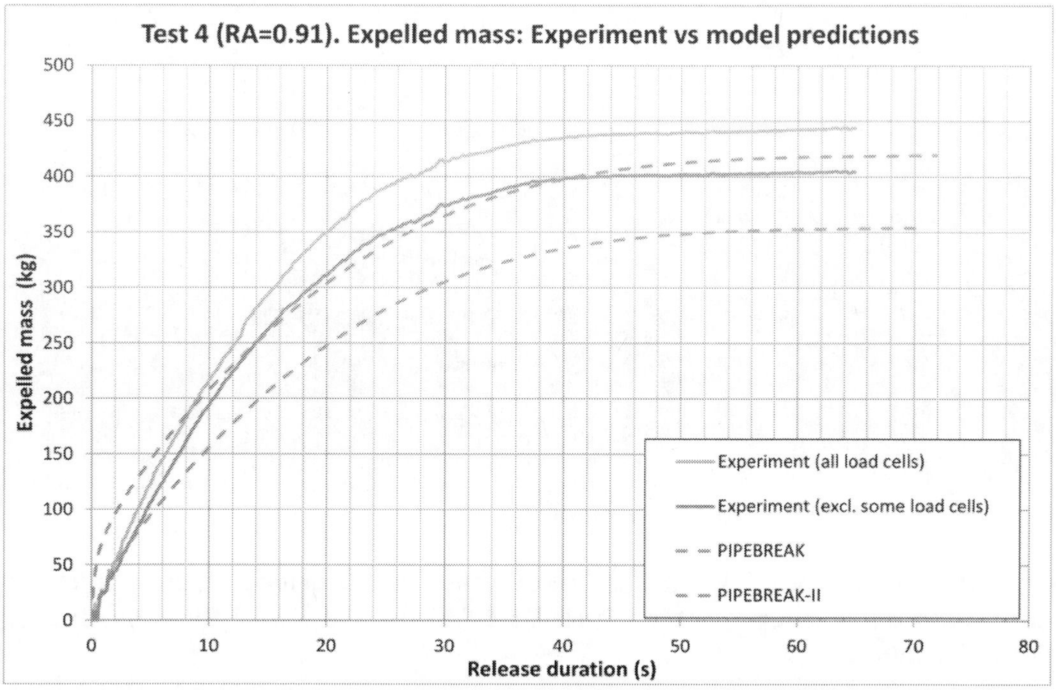

Figure 13: Comparing expelled mass over time as derived from load cell measurements and as predicted by the PIPEBREAK and PIPEBREAK-II models for CO2PIPETRANS Test 4.

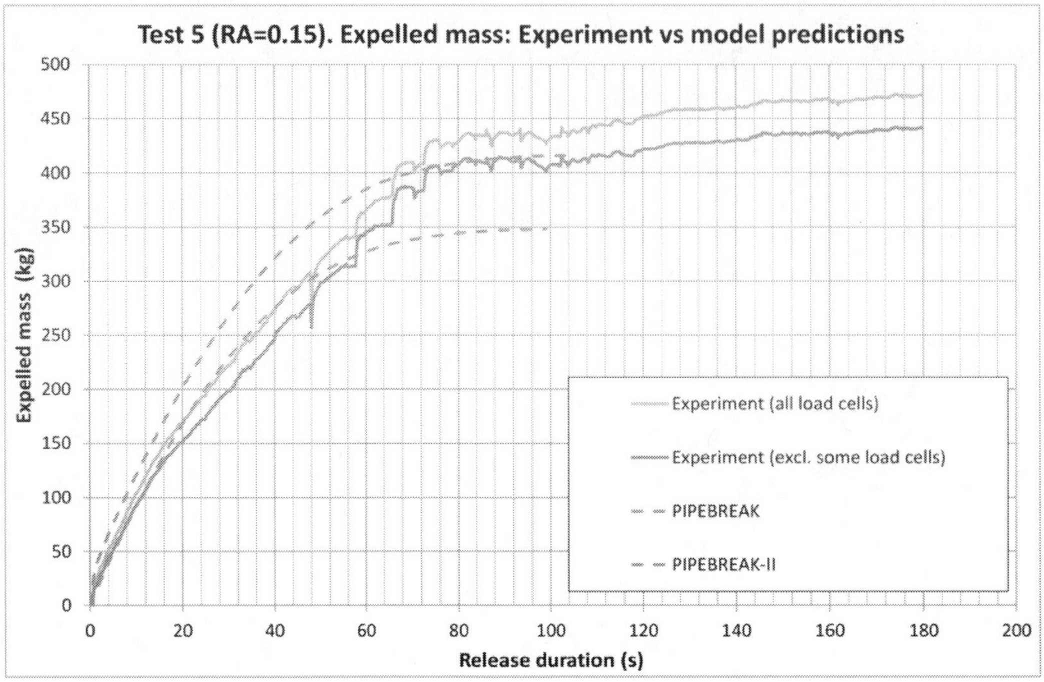

Figure 14: Comparing expelled mass over time as derived from load cell measurements and as predicted by the PIPEBREAK and PIPEBREAK-II models for CO2PIPETRANS Test 5.

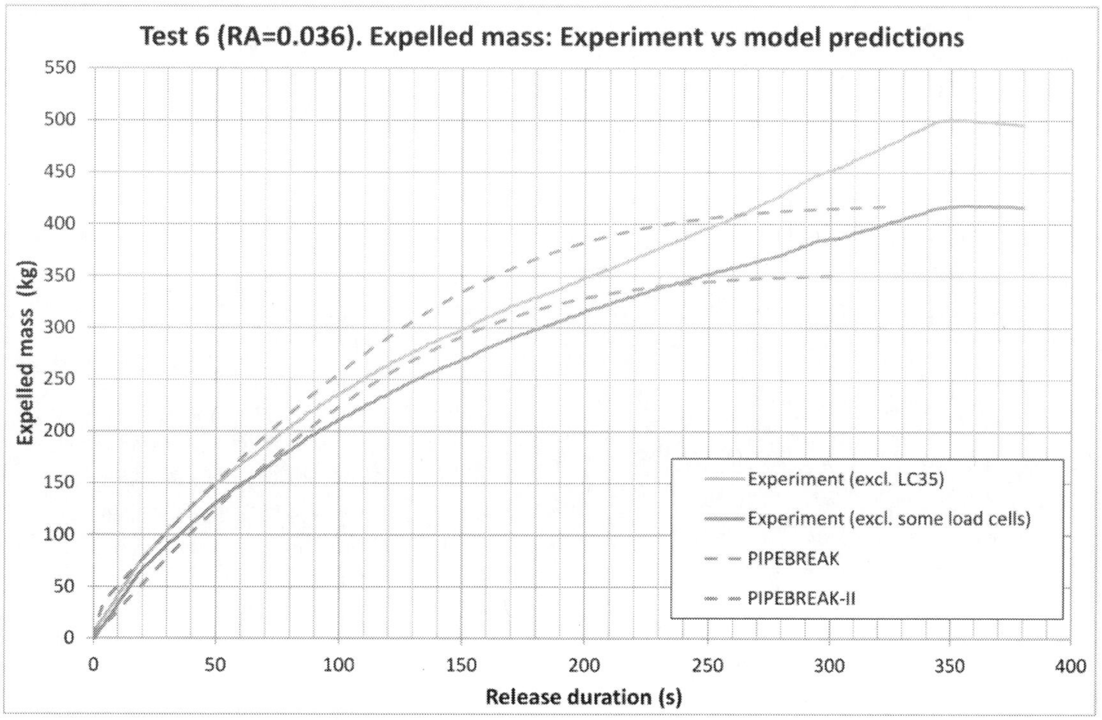

Figure 15: Comparing expelled mass over time as derived from load cell measurements and as predicted by the PIPEBREAK and PIPEBREAK-II models for CO2PIPETRANS Test 6.

4.2.2.3 Pressure and temperature evolution at pipe ends

Figure 16, Figure 18 and Figure 20 compare predicted PIPEBREAK and PIPEBREAK-II results for closed and open-end pressure to CO2PIPETRANS measured data for tests 4, 5 and 6, respectively.

From Figure 16, Figure 18 and Figure 20 it can be observed that:

- PIPEBREAK-II results show quite good agreement with measured data particularly at the onset of the release (i.e., during the rapid depressurization regime) compared to PIPEBREAK, in each case.
- PIPEBREAK-II results also show the step-like effects of propagating rarefaction and reflected waves from the breach to the intact end, in alignment with reported trends in measured data, with the total number of propagating wave pulses between regime 1 and the onset of regime 2 decreasing with increasing aperture size.
- PIPEBREAK-II underpredicts the total duration of regime 1 in test 4 (91% aperture ratio), i.e., ca 0.2s as against 0.4s and overpredicts the duration in test 5 (14.6% aperture ratio), i.e., ca 0.8s as against 1.2s, and test 6 (3.6% aperture ratio), ca 2s as against 4s.

Figure 17, Figure 19 and Figure 21 compare predicted PIPEBREAK and PIPEBREAK-II results for closed and open-end temperature to CO2PIPETRANS measured data for tests 4, 5 and 6, respectively.

From Figure 17, Figure 19 and Figure 21 it can be observed that PIPEBREAK-II and PIPEBREAK results show quite good agreement with measured data with PIPEBREAK-II predicting a longer delay in the commencement of cooling at the onset of the release (i.e., during the rapid depressurization regime) and higher temperatures overall compared to PIPEBREAK, in each case.

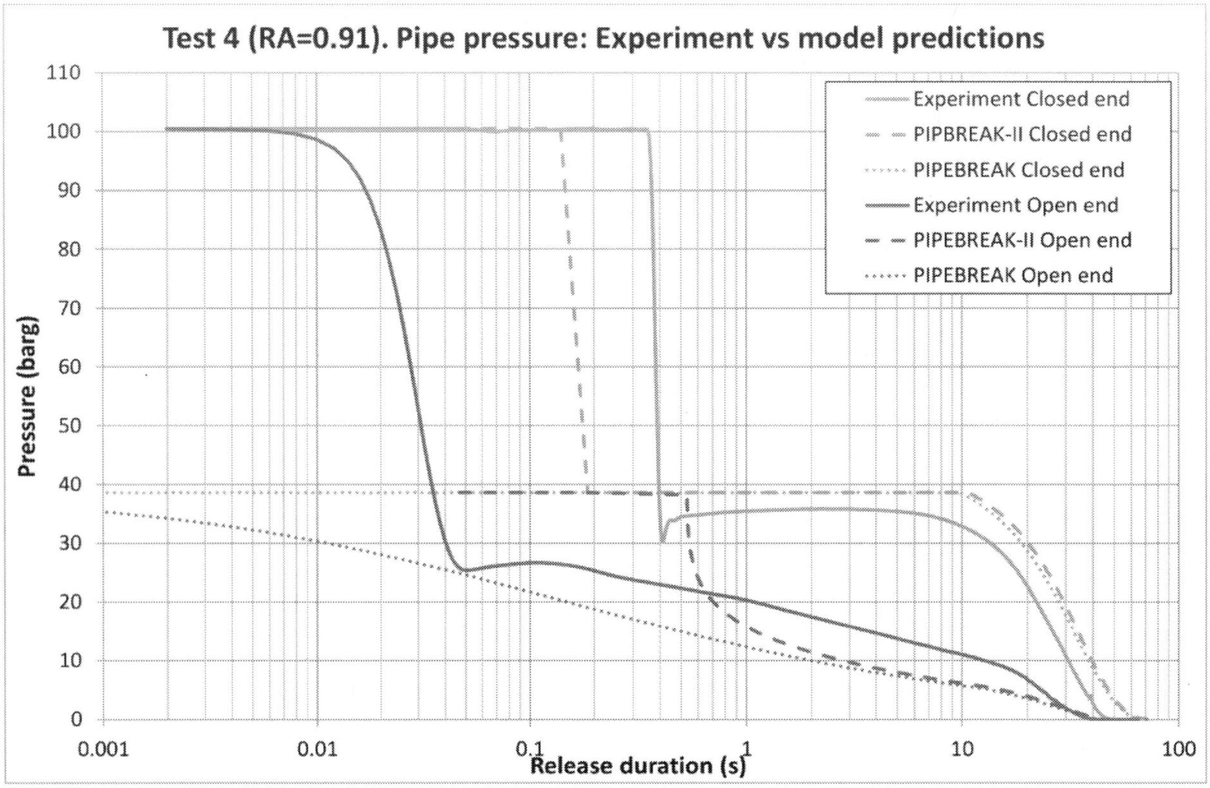

Figure 16: Experimentally observed and PIPEBREAK predicted pressure evolution at the pipe ends for CO2PIPETRANS Test 4.

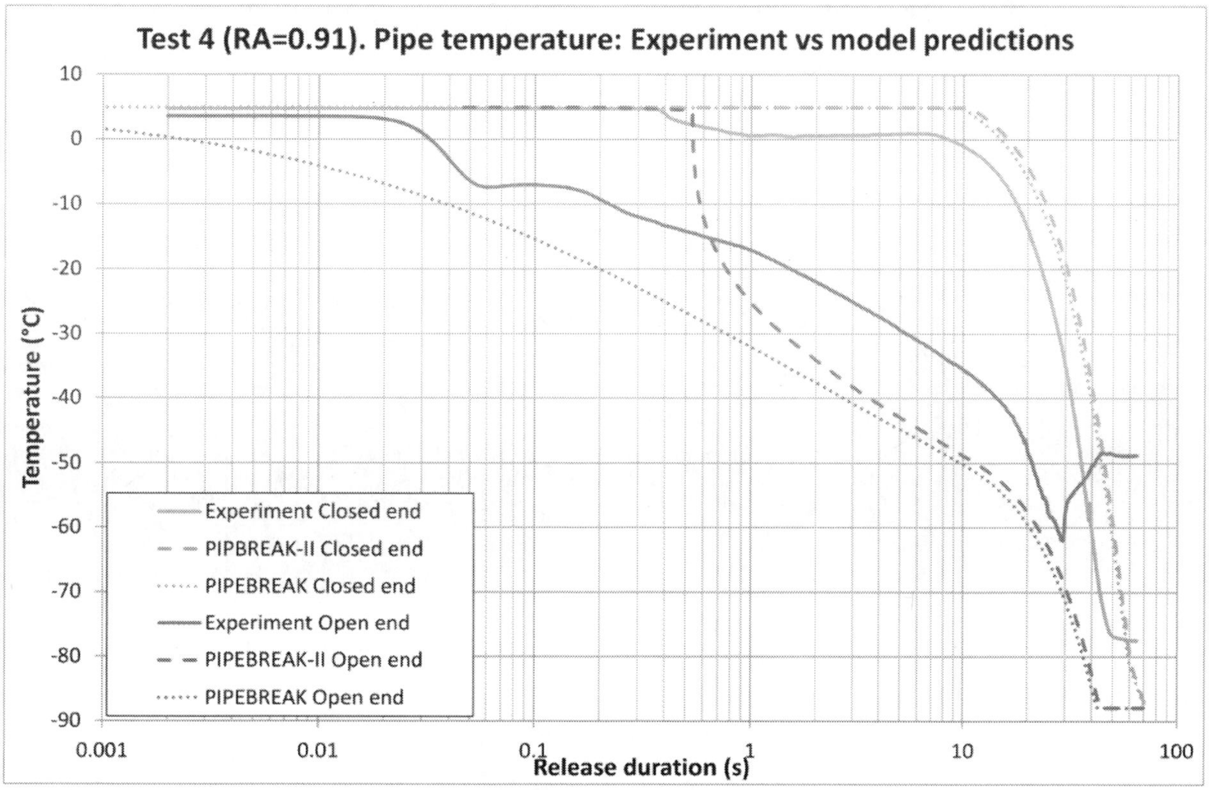

Figure 17: Experimentally observed and PIPEBREAK predicted temperature evolution at the pipe ends for CO2PIPETRANS Test 4.

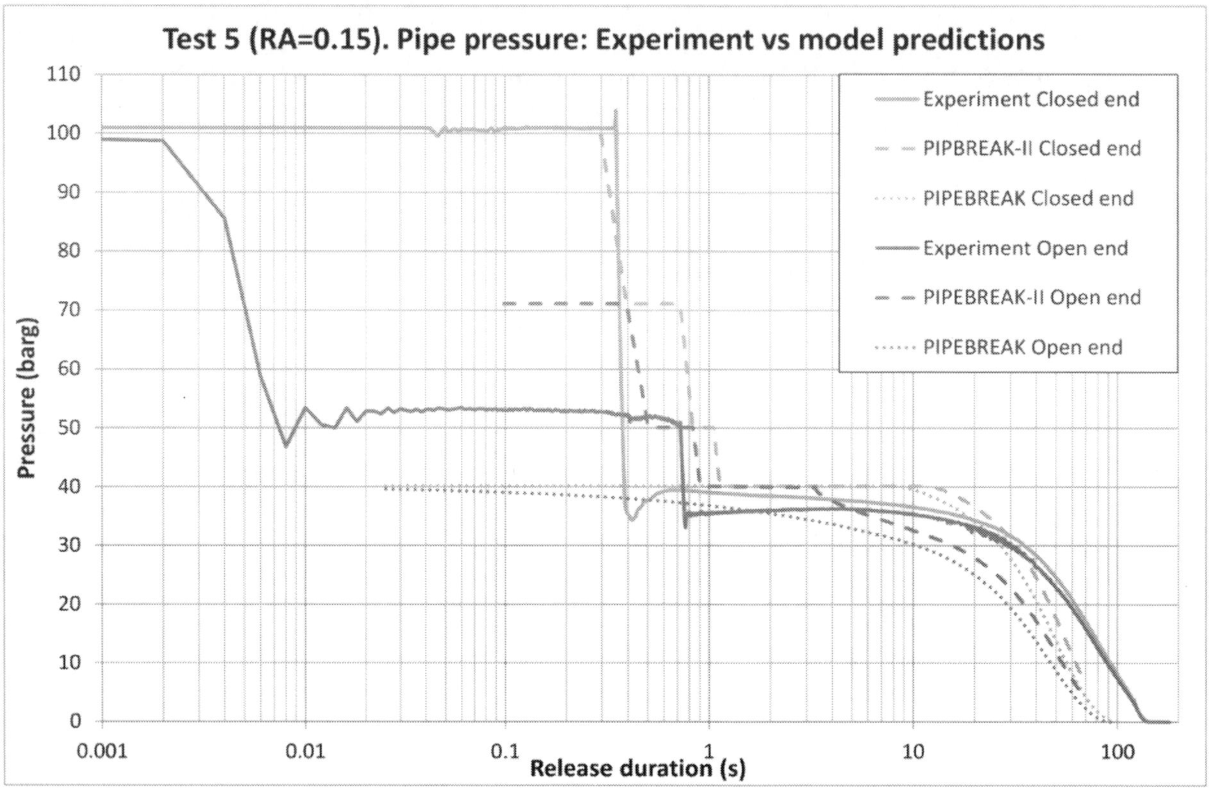

Figure 18: Experimentally observed and PIPEBREAK predicted pressure evolution at the pipe ends for CO2PIPETRANS Test 5.

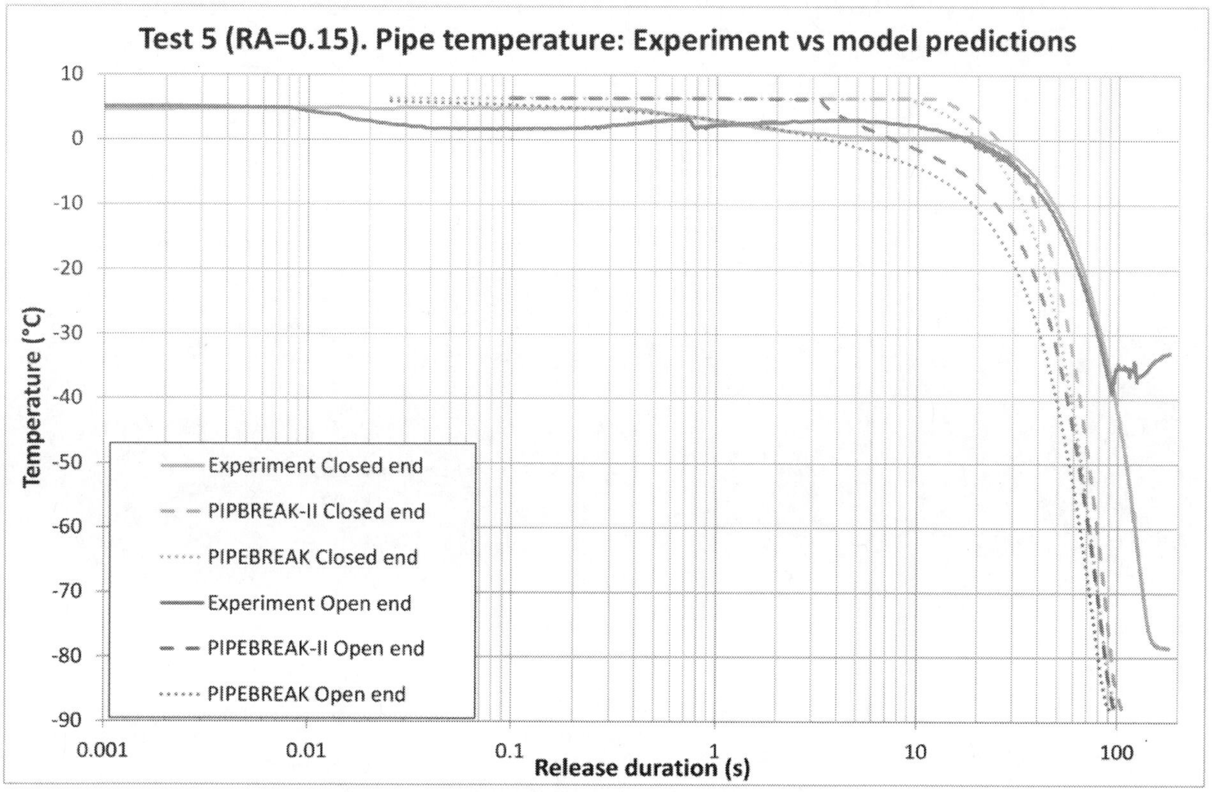

Figure 19: Experimentally observed and PIPEBREAK predicted temperature evolution at the pipe ends for CO2PIPETRANS Test 5.

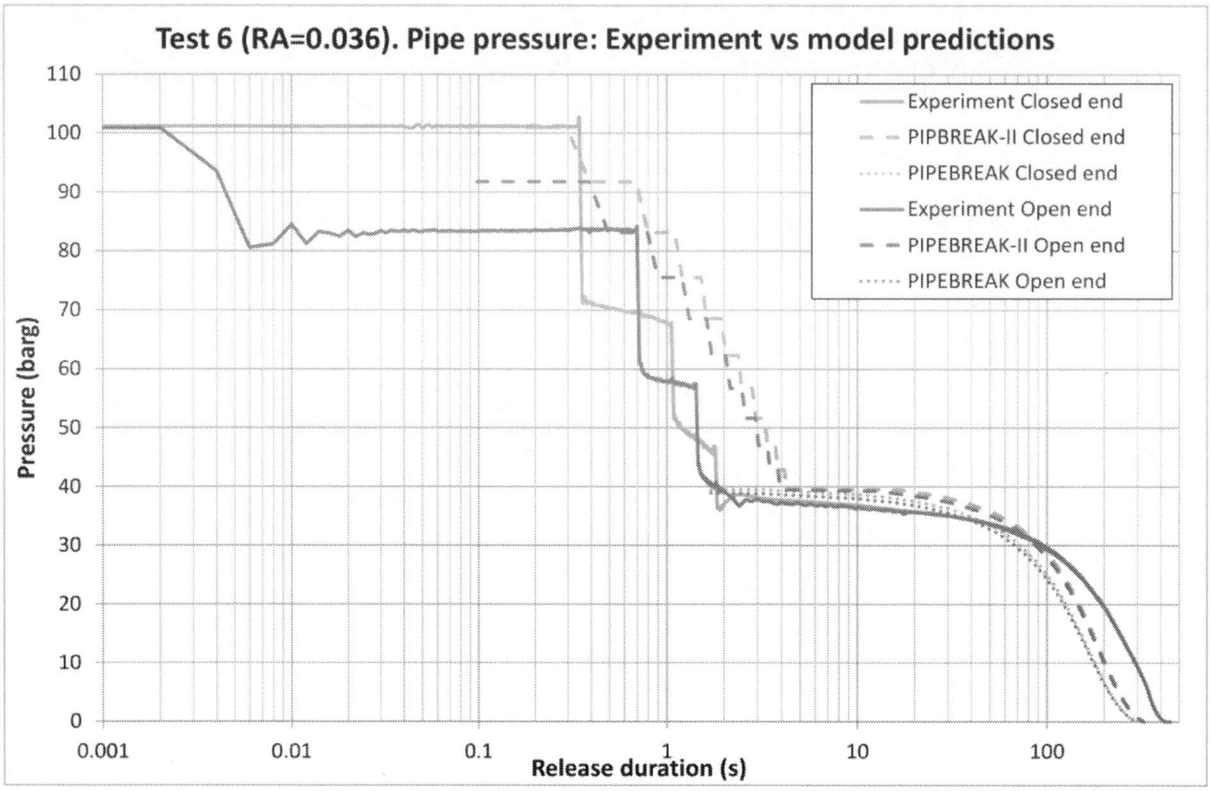

Figure 20: Experimentally observed and PIPEBREAK predicted pressure evolution at the pipe ends for CO2PIPETRANS Test 6.

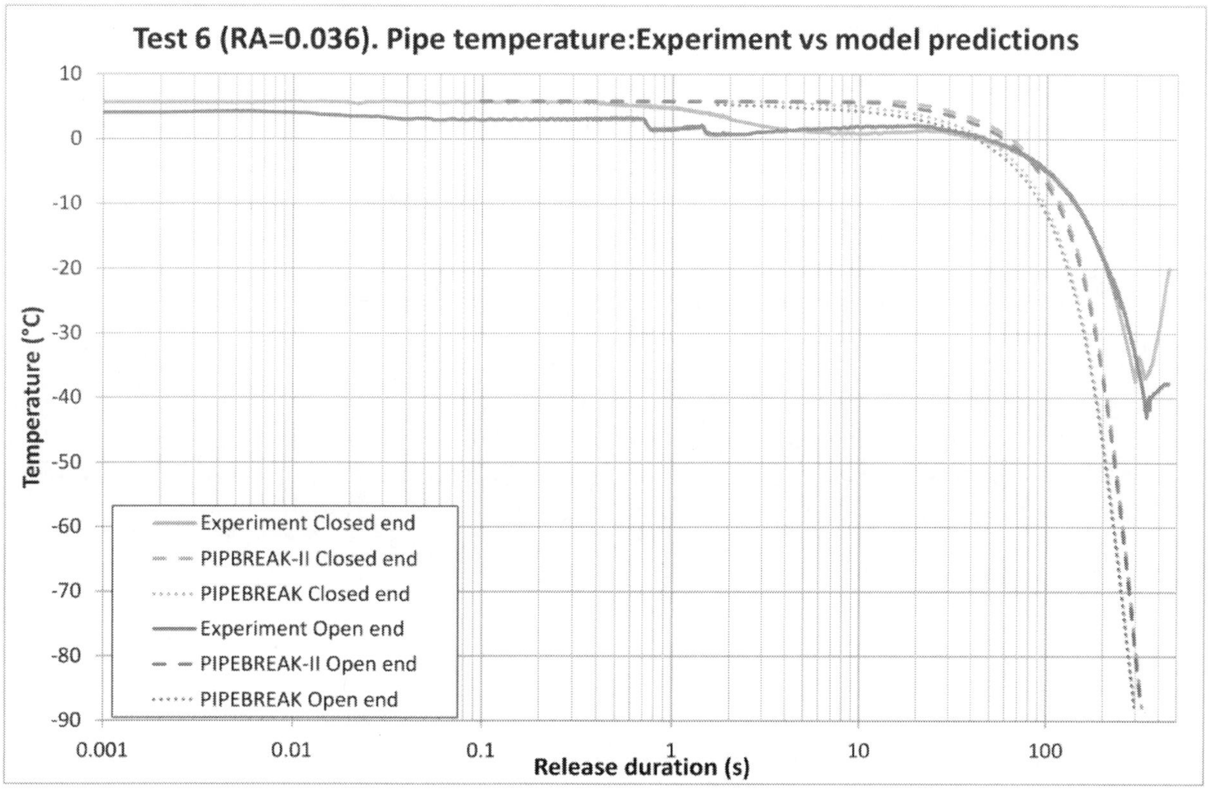

Figure 21: Experimentally observed and PIPEBREAK predicted temperature evolution at the pipe ends for CO2PIPETRANS Test 6.

5 Conclusion

This paper presents an updated model, PIPEBREAK-II, the outcome of modelling work to address various limitations identified within long pipeline source term models commonly employed for major accident hazard analysis in industry. In particular:

- Overly simplistic (non-conservative) modelling of the initial depressurization behavior for flashing fluids transported at pressures significantly above their saturated vapor pressure.
- Inability to model breach sizes with aperture ratios below 20%.
- Inaccurate modelling of liquid densities at pressures above and along the saturated vapor pressure curve.
- Inability to model loss of containment events involving sub-cooled liquid inventories as well as any drainage flow that may ensue upon depressurization of the bulk fluid inventory to ambient pressure.

The modelling of the rapid depressurization regime in PIPEBREAK-II has been undertaken by employing a modified/extended version of the rarefaction wave balancing model described by Newton [2] as implemented in PiRRaM. The modelling in PIPEBREAK-II caters for initial rapid depressurization effect modelling, sub-cooled inventories, drainage flow modelling for sub-cooled inventories, pre-existing pumped inflow, flow reversal downstream of the breach, valve closure effects as well as the occurrence of punctures at locations other than at pipeline ends with flow contributions (disjoint or interactive) from the upstream and downstream ends of the pipeline.

Predicted results from PIPEBREAK-II have been verified/sense-checked against representative case studies, sensitivity scenarios and results from selected pipeline/vessel blowdown models. In addition, PIPEBREAK-II has been validated against published results from field tests involving long pipeline release scenarios covering a broad range of aperture ratios, ranging from <1% to full-bore rupture. Simulated results based on PIPEBREAK-II compares very well against other pipeline release models and field data with PIPEBREAK-II simulated results tending to be conservative in general (i.e., higher flowrates, higher initial inventories, quicker depressurization, and lower flow temperatures).

Potential improvements to PIPEBREAK-II include the need to extend current modelling to support for multi-component flashing liquid and vapor inventories as well as modelling of elevated triple point effects for substances like CO_2, i.e., employ vapor-solid in place of vapor-liquid equilibria thermodynamics where the triple point pressure is above ambient pressure.

6 Acknowledgements

The authors would like to acknowledge the RIVM (Dutch National Institute for Public Health and the Environment) for the financial support provided towards the work reported in this paper. The contents of this paper including any opinions and/or conclusions expressed, are those of the authors alone and do not necessarily reflect the policy of the RIVM.

7 References

[1] Webber DM, Fanneløp TK, Witlox HWM. Source terms from two-phase flow in long pipelines following an accidental breach. "International Conference and Workshop on Modeling and Mitigating the Consequences of Accidental Releases of Hazardous Materials", *AIChE* 1999; **ISBN 0-8169-0781-1**: 145-167

[2] Newton A. Pipeline release rate model (PiRRaM) for pressure liquefied gases. *Hazards 32* 2022; symposium series No. 169.

[3] Hammer M, Deng H, Austegard A, Log AM, Munkejord ST. (2022). Experiments and modelling of choked flow of CO2 in orifices and nozzles. *International Journal of Multiphase Flow* 2022; **156**: 104201.

[4] Oke A, Mahgerefteh H, Economou I, Rykov Y. A transient outflow model for pipeline puncture. *Chem. Eng. Sci.* 2003; **58(20)**: 4591.

[5] Morrow TB, Bass RL, Lock JA. An LPG pipeline break flow model. 1983.

[6] Tam VHY, Higgins RB. Simple transient release rate models for releases of pressurized liquid petroleum gas from pipelines. *Journal of Hazardous Materials* 1990; **25(1-2)**: 193-203.

[7] Haque MA, Richardson SM, Saville G, Chamberlain G. Rapid depressurization of pressure vessels. *J. Loss Prev. Process Ind.* 1990; **3**: 4.

[8] Haque MA, Richardson SM, Saville G. Blowdown of pressure vessels. I. Computer model. *Trans IChemE Part B: Process Safety Environmental Protection* 1992a; **70 (BI)**: 1.

[9] Haque MA, Richardson SM, Saville G, Chamberlain G, Shirvill L. Blowdown of pressure vessels. II. Experimental validation of computer model and case studies. *Trans IChemE Part B: Process Safety Environmental Protection* 1992b; **70 (BI)**: 10.

[10] Richardson SM, Saville G. Blowdown of pipelines. Society of Petroleum Engineers Europe 91, paper SPE 23070, Aberdeen, UK, 1991: 369.

[11] Richardson SM, Saville G. *Isle of Grain pipeline depressurization tests*. Bootle, U.K. HSE OTH 94441, HSE Books, HSE; 1996a.

[12] Richardson SM, Saville G. Blowdown of LPG pipelines. *Trans. IChemE* 1996b; **74B**: 236.

[13] Chamberlain GA. Shell FRED, Fire, Release, Explosion, Dispersion, Hazard consequence modelling package, Technical Guide. Shell technical report; 2007.

[14] Bendiksen KH, Brandt I, Fuchs P, Linga H, Malnes D, Moe R. Two-phase flow research at SINTEF and IFE. Some experimental results and a demonstration of the dynamic two-phase flow simulator OLGA. *The Offshore Northern Seas Conference*, Stavanger; 1986.

[15] Bendiksen KH, Malnes D, Moe R, Nuland S. The dynamic two-fluid model OLGA: theory and applications. *SPE Production Eng.* 1991; **6**: 171.

[16] Hall ARW, Butcher ER, Teh CE. Transient simulation of two-phase hydrocarbon flows in pipelines. *Proceedings of the European Two-Phase Flow Group Meeting*, Hannover; 1993.

[17] Philbin MT, Govan AH. Pipeline analysis code. UKAEA Report AEA-APS-0031; 1990.

[18] Philbin MT. The simulation of transient phenomena in multiphase production systems. AEA Harwell Report AEA-APS-0177; 1991.

[19] Mahgerefteh H, Saha P, Economou I G. Fast numerical simulation for full-bore rupture of pressurized pipelines. *AIChE J.* 1999; **45(6)**: 1191.

[20] Abudour AM, Mohammad S, Robinson R, Gasem K. Volume-translated Peng-Robinson equation of state for liquid densities of diverse binary mixtures. *Fluid Phase Equilibria* 2013; **349**: 37-55.

[21] National Institute of Standards and Technology. NIST Chemistry WebBook, SRD 69. https://webbook.nist.gov/chemistry/fluid/ . Accessed September 12, 2023.

[22] Holt H, Brown J, Witlox H, Armstrong K, and Allason D. Discharge and dispersion for CO2 releases from a long pipe: Experimental data and data review. *Hazards 25* 2015; Symposium Series No 160.

Advanced Consequence Modeling and Risk Analysis

Dr. John Currie
Dow
4770 US Highway 42 E
Carrollton, KY 41008
john.s.currie@dow.com

Karen Study, P.E.
Dow
332 S Hwy 332 E
Lake Jackson, TX 77566
kstudy@dow.com

Paul Delanoy
Dow- Retired
pdelanoy@live.co.uk

Ken First
Dow- Retired
kefirst@aol.com

Keywords: Process Hazard Analysis; Consequence Modelling; Pseudo probit coefficients; Probit

Abstract

When performing Process Hazard Analyses (PHA), basic consequence evaluation methods are often sufficient. However, sometimes these approaches lead to overly conservative estimates, and we need to conduct more thorough dispersion modeling to refine the potential impact.

This paper will evaluate different methods for determining the consequence severity of a potential release, starting with a basic approach, and ultimately using probit functions. Probit functions are used to predict the lethality of a potential release following acute inhalation exposure from a toxic release. Developing probits to use in modeling can be challenging, for obvious reasons. In the absence of human or animal toxicological data, estimates of "pseudo probits" can be used in lieu of values derived from toxicological data.

The analysis makes use of the PHAST^{TM} program that uses probits as part of its basic tool set. While probits are not new to the industry this paper provides a simple method for estimating pseudo probits for materials that do not have readily available probit data.

1 Background

Chemical plants provide materials that are used in endless products that have shaped the world that we live in, including fertilizers, medicines, personal care products, and materials integral to the technology of today's world. To produce these products, it is necessary to mix, react, distill, and package chemicals that have the potential to impact the environment and the communities in which the manufacturers operate. In some cases, the final products that are used in downstream processing originate from hazardous raw materials that need to be managed and treated with respect. Regulations govern the requirements for the chemicals used, often requiring companies to complete process hazard analyses to estimate the risks of their operations and to implement controls to prevent chemical releases.

There are many methodologies used to conduct risk analysis, but key to using any one of them is the need to determine the potential consequence severity due to a loss of containment. In Dow, modelling tools have been developed and improved over the years as computing capabilities have developed. Early tools used for consequence assessment were the Chemical and Exposure Index[1] for toxic evaluations and the Fire and Explosion Index[2] for flammables. Over time, these tools are being superseded by more sophisticated tools.

People unfamiliar with the conducting of risk assessments would probably believe that the ideal risk assessment strategy would involve a HAZOP (HAZard and OPerability study) and QRA (Quantitative Risk Assessment) on every chemical process. They may view that as the only way to ensure the best possible understanding and quantification of the risks. However, in reality this is not always a practical proposition. Such in-depth studies can be more than what is necessary for lower hazard facilities, and enthusiasm for the studies would likely suffer as a result. Instead, many companies have adopted a tiered approach, whereby distinct levels of assessment depth are applied in accordance with the process hazards present at the facility. There is one particularly important rule to such an approach that must never be forgotten:

The simpler the risk study, the more conservatism must be applied.

Failure to include sufficient conservatism in a reduced accuracy study will, almost certainly, lead to risks being underestimated.

This approach and the way it can fit with the maximum tolerable risk specified by a company or authority is shown graphically in Figure 1, where there are four levels of risk analysis available with each level being more comprehensive.

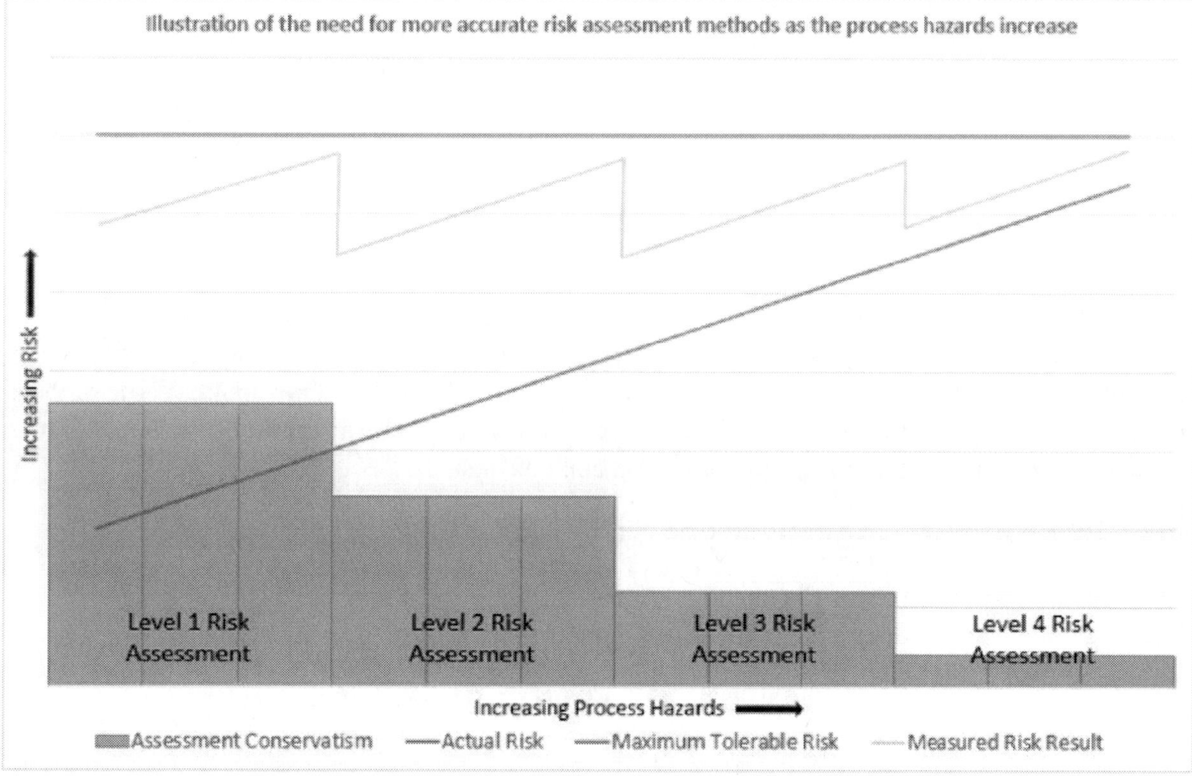

Figure 1: Comparison of risk based on process hazard.

Whilst accepting the necessity of a tiered approach to risk assessment, it is still desirable to obtain the most accurate results possible at each tier level. This accuracy is largely determined by the tools and techniques applied. The yellow line on Figure 1 shows how the estimated risk typically takes a step change down as more through a risk assessment is performed. In all cases, the measured risk is below company threshold limits.

1.1 History in Dow

The Chemical and Exposure Index [1] and the Fire & Explosion Index [2] are tools developed in Dow that date from the 1960s. These indices provided a means to rank the relative fire, explosion, and chemical impacts associated with a process. Later, Dow developed a chemical table approach that categorized materials based on their physical properties and then estimated the potential offsite impact based on the magnitude of the release.

Midway through the 2000s, other consequence modeling tools were developed. Dow developed the Risk Assessment Screening Tool (RAST) [3], a powerful tool which performs simplified dispersion modelling and evaluated potential impacts based on the ERPG (Emergency Response Planning Guidelines) extents. Around 2010, Dow began to use probits in consequence analysis to incorporate the concept of dose. While they improve modeling accuracy, probits are generated from costly toxicology studies, and probit values are only available for a select few toxic materials. To fill the gap, an approach was developed which calculates a conservative 'pseudo-probit' for toxic chemicals, such that the material concentration and duration of exposure could both be considered in the consequence severity assessment. This novel approach has allowed Dow to re-evaluate toxic impacts for several studies without requiring the extensive work needed to develop probits.

This paper will look at approaches that Dow has used to estimate the consequences of a release to improve the assessment of offsite consequence severity.

2 Consequence severity estimation

When completing a Process Hazard Analysis, an estimate of the consequence severity is required so that the appropriate protective layers can be installed for each scenario. There are different approaches that can be taken to make this estimation. The following sections look at some of the methods available to estimate the consequence severity, starting with a simplified method and then refining the estimates with the introduction of probit constants.

2.1 Chemical table consequence severity estimation

The first method presented is the chemical table method, in which some basic property information is used to determine a hazard category for a material. These parameters are shown in Table 1.

Boiling Point (°C)
ERPG-3 (ppm)
NFPA Health Rating
NFPA Flammability Rating
NFPA Reactivity Rating

Table 1. Chemical property data used to determine Hazard Category.

The ERPGs are a measure of the potential toxic impact that can affect a human, depending on the extent of the exposure. They are separated into three levels; ERPG-3 the one of most interest. The definition of ERPG-3 is:

> The maximum airborne concentration below which nearly all individuals could be exposed for up to 1 hour without experiencing or developing life-threatening health effects. [6]

The National Fire Protection Association (NFPA) [7] developed a hazard communication system that considers the main hazards associated with a chemical and rates them on a scale of 0 - 4. Considering these parameters in conjunction with the ERPG-3 value and the material boiling point allows an estimate of a material hazard category to be made. This is achieved by a scoring system whereby each parameter is given a number based on its properties and the NFPA ratings, see Table 2 and Table 3. The scores added together give a hazard score and the corresponding category is picked based on Table 4.

		Boiling point (°C)			
		<30	30-60	60-150	>150
ERPG-3 (ppm)	0-5	162	54	18	6
	5-100	54	18	6	2
	100-1000	18	6	2	2
	>1000	6	2	2	2

Table 2: Boiling point and ERPG-3 score.

NFPA rating	0	1	2	3	4
Health	2	2	6	18	54
Flammability	2	2	6	18	54
Reactivity	2	6	18	54	162

Table 3: NFPA rating score

Hazard score	0-10	10-50	50-100	100-200	200>
Hazard category	E	D	C	B	A

Table 4: Relationship of hazard score to hazard category.

Hazard Category	Quantity involved in the undesired consequence severity, pounds					
	< 10	10 to 100	100 to 1,000	1,000 to 10,000	10,000 to 100,000	> 100,000
A	1 to 2 impacts	Multiple impacts	Multiple impacts	Multiple impacts	Multiple impacts	Multiple impacts
B	Serious irreversible injury	1 to 2 impacts	Multiple impacts	Multiple impacts	Multiple impacts	Multiple impacts
C	Reportable Medical Treatment	Serious irreversible injury	1 to 2 impacts	Multiple impacts	Multiple impacts	Multiple impacts
D	N/A	Reportable Medical Treatment	Serious irreversible injury	1 to 2 impacts	Multiple impacts	Multiple impacts
E	N/A	N/A	Reportable Medical Treatment	Serious irreversible injury	Serious irreversible injury	Serious irreversible injury

Table 5: Chemical Specific Safety Target Factor Table

Once the hazard category has been determined and the extent of a release is understood, the consequence severity of a release can be determined. Table 5 provides a means to categorize the offsite impacts from a release. The table is a screening method for assessing consequence severity, and better methods may be required to obtain a more accurate understanding of the potential consequence severity.

2.2 Probits concept

2.2.1 Lethality based on ERPG-3 surrogates

Refinements to the consequence estimate can be made based on an assessment of the material concentration and the duration of the exposure. These two factors, in combination, reflect the potential dose to which an individual may be exposed.

A high-level estimate of toxic impact study can simply be based on the quantity of material released. However, the release quantity is not a good estimate of the potential individual exposure dose. Probits certainly allow this estimate of dose to be made; however, often a probit is not available for the chemical of interest. To address such scenarios without available probit values, a conservative method was developed that uses multiples of ERPG-3 values to develop a concentration/time dose relationship. This relationship can be expressed as:

$LC(x) = y_x * ERPG\text{-}3$

Where:

x = % lethality (e.g. LC-01 or LC-50)

y_x = multiple of ERPG-3 corresponding with the specific % lethality

When it is assumed that twice ERPG-3 approximately corresponds to a 1% lethality and five times ERPG-3 approximately corresponds to 50% lethality for a one-hour exposure, we can derive an anticipated probability of severe toxic impact over one hour. In Figure 2, the resulting ERPG multiple based response curve compared conservatively to internal company probit response curves with only one probit model being less conservative.

An estimate of the multiples of ERPG-3 that can be used to represent lethal impacts from a chemical release can be taken from Figure 2. The chart is a conservative representation of probability of severe toxic impact against multiples of the ERPG-3 [14]. For example, a 1% probability of severe toxic impact (LC-01) is roughly two times the ERPG-3 value (using the upper limit of the model line). Similarly, 50% probability of lethality equates to approximately five times the ERPG-3 value.

The chart is an approximation, but to quote George Box:

"All models are wrong, but some are useful." [12]

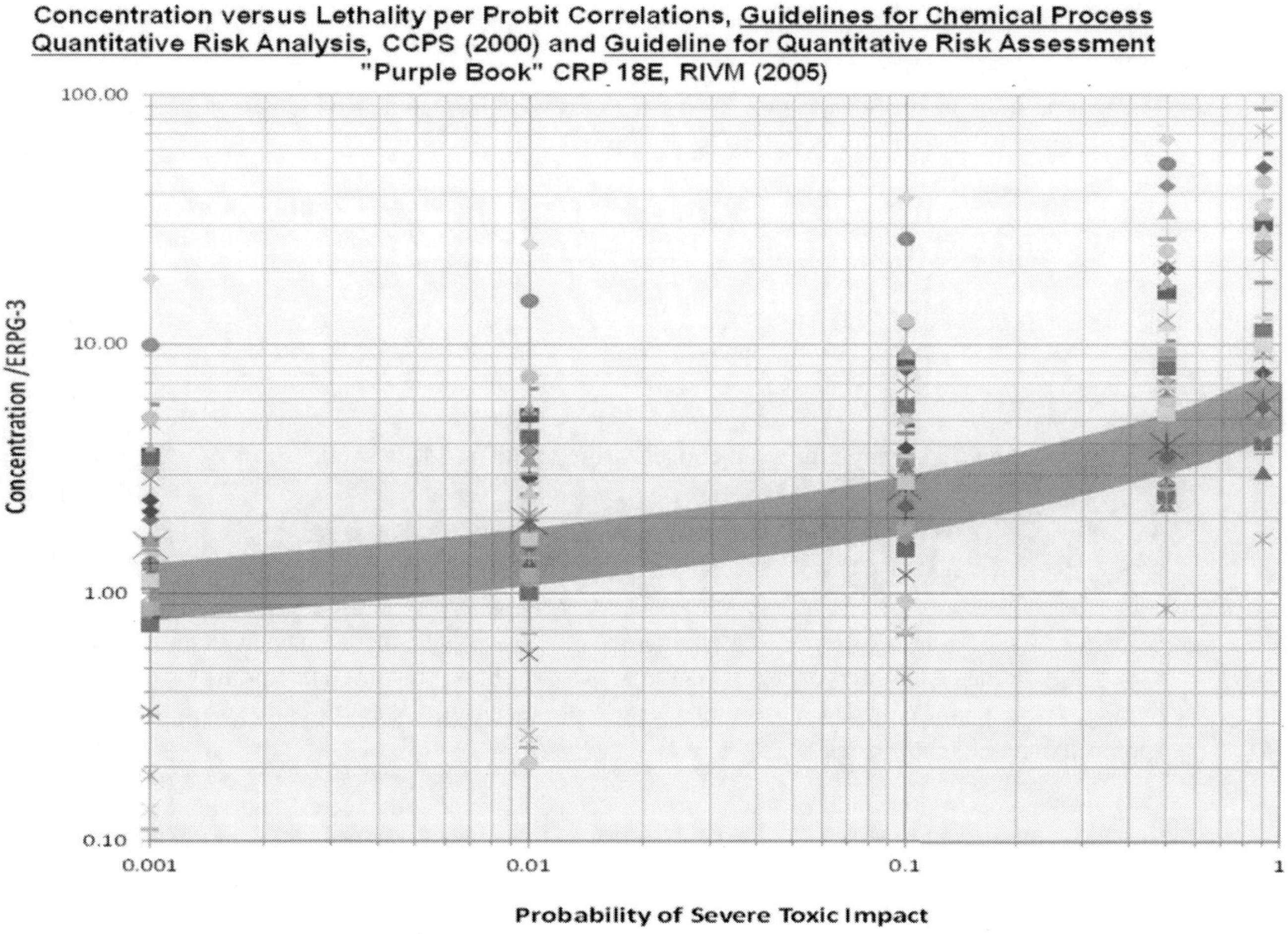

Figure 2: Probability of severe toxic impact at 60-minute exposure.

In RAST, estimates of the lethality from a toxic release are based on ERPG-3 surrogates. For a lethal concentration of 50% (LC-50) five times the ERPG-3 was used, and two times ERPG-3 is then used as a surrogate for a lethal concentration of 1% (LC-01).

2.2.2 Pseudo probit estimation

The pseudo-probit approach is a refinement of the ERPG-3 multiples approach. Similar to the ERPG-3 surrogate approach, the pseudo-probit assumes a correlation based on ERPG-3 multiples. The pseudo-probit approach also uses the same assumptions of two times the ERPG-3 concentration over one hour correlating with a one percent probability of lethality and five times the ERPG-3 concentration correlating with a 50% probability of lethality.

Probits represent the set of values of a, b, and n as outlined in Equation 1. This correlation can be found in many literature references [13].

$$Probit = a + b * \ln(C^n * t)$$

Equation 1: Probit function

A probit function expresses the probability of fatality in an exposed population as a function of both exposure concentration and exposure time. The constants used for estimating the dose response are 'a,' 'b' and 'n.'

The estimate of the probit constant 'a' incorporating multiples of ERPG-3 is detailed in Equation 2. Constant "b' is similarly detailed in Equation 3.

$$a = 2.67 - b * (4.094 + 2 * \ln(2 * ERPG3))$$

Equation 2: Probit constant 'a' assuming ERPG-3 multiple correlation.

$$b = \frac{1.165}{(\ln(5 * ERPG3) - \ln(2 * ERPG3))}$$

Equation 3: Probit constant 'b' assuming ERPG-3 multiple correlation.

When the value of 'n' is unknown a conservative value of 2.0 is used. [8]

2.2.3 Probit estimation

In the previous section, we discussed pseudo probits for cases when there are no available constants for a material. A refinement of the concept is shared in the "Methods for the determination of possible damage (Green book)" [8], which extrapolates the constants based on the LC-50 values similar to the approach outlined in this paper.

To generate probits, toxicological studies are performed, typically on mice or rats. The time/concentration/impact data gathered during the study is transformed into a linear relationship that can be analyzed to generate the constants 'a,' 'b,' and 'n.' Probit data is typically required when performing highly detailed studies such as Quantitative Risk Analysis. Probit equations represent the most accurate means we have for evaluating the potential impact of an exposure. For a small subset of chemicals, toxicological studies have been completed and probit data is available [8]. Some data is available in the public domain but many times it is company specific and as such is confidential.

3 Case study

To illustrate the process of refining the consequence severity of a given scenario, we will apply each of the methods mentioned above to an example problem. For the example problem, assume that the objective of this study is to determine if there are offsite consequences and how extensive the impacts might be using these different approaches. The event to be modelled involves the release of acrylonitrile from a vessel with an inventory of 25,000lb/ 11,340kg. In this scenario, the vessel has a partially open drain valve with an equivalent diameter of 0.4"/ 1cm. The operating conditions are:

- Temperature: 122°F/ 50°C.
- Operating pressure: 100psi/ 6.89 bar.
- Diked area: 328ft/^{2} 100m^2 with a height of 3ft/ 0.91m.
- Nearest fenceline is 656ft/ 200m away.

The consequence severity associated with the scenario will be estimated with the following methods and will illustrate a possible refinement approach as the methods become more thorough:

- Chemical table
- ERPG-3 Dispersion
- Probit surrogates
- Pseudo probits
- Probits

3.1 Chemical table

The first method to be considered will be the chemical table.

The material release is acrylonitrile, which has been determined to have Hazard category C, see Table 6. A release rate of 262 lb. minute^{-1} (1.98kg s^{-1}) equates to a total release of 15,720lb/ 7,130kg in a one-hour period. The chemical table indicates the potential for multiple impacts.

	Value	**Score**
Boiling point	171°F/ 77.2°C	6
ERPG-3	75ppm	
NFPA Health rating	4	54
NFPA Flammability rating	3	18
NFPA Reactivity rating	2	18
	Total score:	96
	Hazard Category	**C**

Table 6: Acrylonitrile hazard category determination.

3.2 RAST with ERPG-3 based lethality

The next method to be applied is the surrogate probits approach in RAST. For this model, the scenario of interest is an open drain with a 0.4 in / 1cm hole size. The release rate in this case is estimated as 228lb $minute^{-1}$ (1.72kg s^{-1}).

- Max ERPG-3: 883ft/ 269m
- 1% LC: 768ft/ 234m
- 50% LC: 335ft/ 102m

The results of this approach indicate that there is a potential for multiple offsite impacts as the 1% lethal concentration does extend over the fenceline, although the actual population locations relative to the fenceline would need to be considered when making this determination.

3.2.1 PHAST™ dispersion

PHAST™ is another consequence modeling tool that can be utilized to determine the consequence severity of a release. Employing the same model basis used in RAST, the tool estimates a comparable release rate of 254 lb. $minute^{-1}$ (1.92kg s^{-1}). The dispersion results for this scenario are depicted in Figure 3.

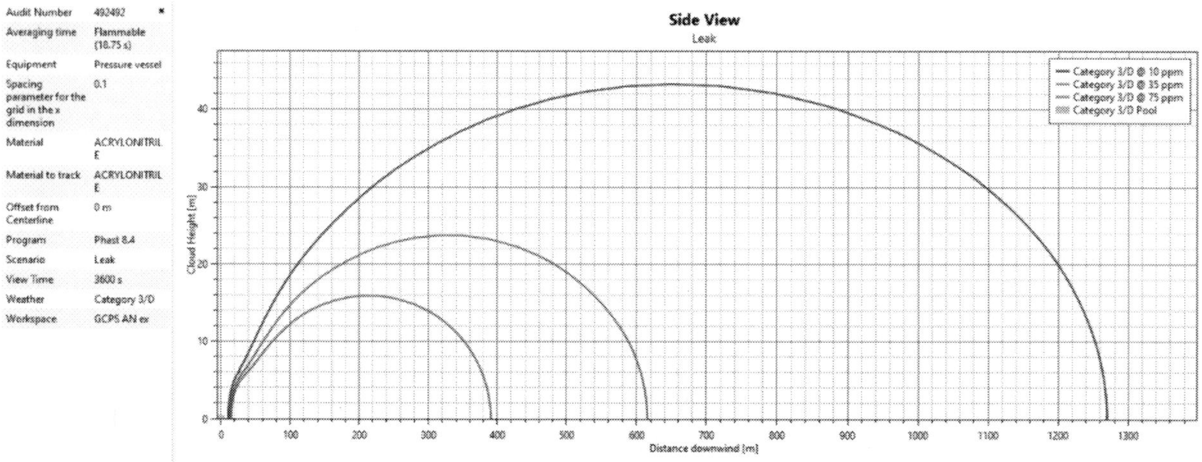

Figure 3: Acrylonitrile release. 3D weather. Side view.

Looking at the simple dispersion extents, the distance to ERPG-3 is 1,312ft/ 400m. Comparing RAST vs PHAST, the distances to toxic impact are estimated as:

Toxic Endpoint	RAST	PHAST
Release Rate	228lb. min^{-1} (1.72 kg. s^{-1})	254 lb. min^{-1} (1.92 kg. s^{-1})
ERPG-3	883ft/269m	1,312 ft/400 m

Table 7: Dispersion result comparison for RAST and PHAST™.

In both cases the ERPG-3 levels extend before the 200m fenceline and would indicate that there is the potential for offsite impact- similar to the result obtained using the simple chemical table. More comprehensive analysis is needed.

3.2.2 PHAST™ pseudo probits

The next step to refine the analysis would be to estimate the potential lethality impact based on pseudo probits for acrylonitrile. (PHAST™ does include probit constants for this material, but an initial estimate will be made using pseudo probits that are based on the ERPG-3 values.)

The constants used as listed in

ERPG-3	75ppm
a	-15.277
b	1.271
n	2

Table 8: Acrylonitrile pseudo probit constants.

The first graph, Figure 4, provides an estimate of the probability of a fatality. It can be seen that the 0.1% probability of fatality extent distance is around 640ft/ 195m, which is close to the plant fenceline.

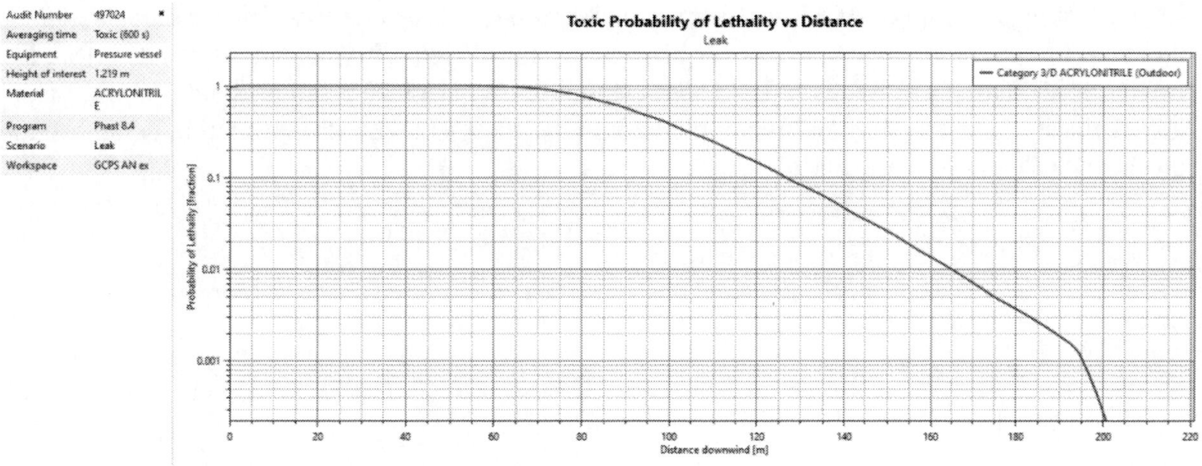

Figure 4: Acrylonitrile probability of lethality using pseudo probits.

A closer look at the lethal footprint in Figure 5 shows the 1% probability of fatality at a distance of 541ft/ 165m, which is less conservative than the768ft/ 234m estimated using RAST. The 50% lethality distances are 305ft/ 93m for PHAST™ and 335ft/ 102m for RAST – fairly comparable. Therefore, using PHAST™ with pseudo probits based on the ERPG-3 values has refined the results obtained from the surrogate probit approaches in RAST and PHAST™.

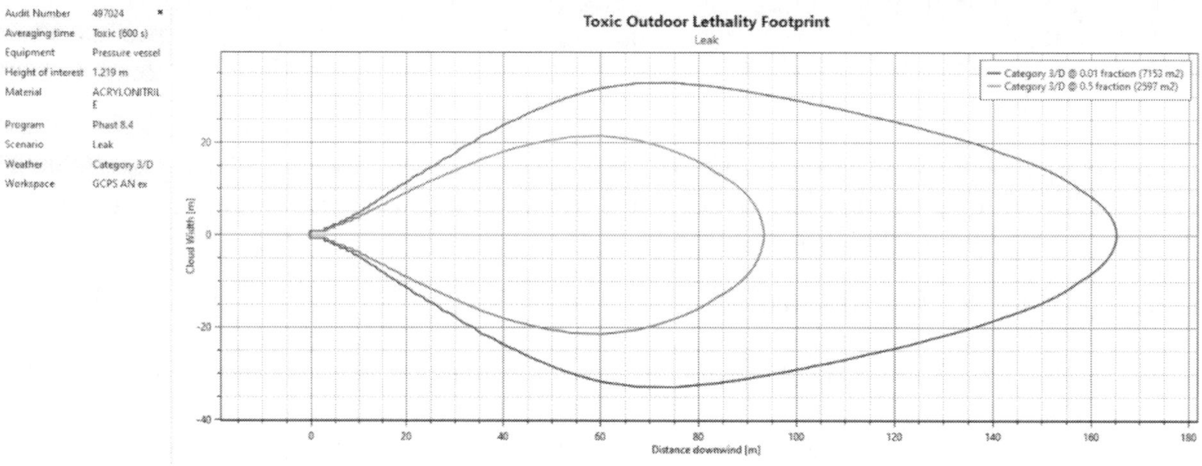

Figure 5: Acrylonitrile outdoor lethality using pseudo probits:

3.2.3 PHAST™- Actual probits

Lastly, the PHAST™ model was run using the probits for acrylonitrile taken from SAFETI™ Version NL8.8, used by companies to comply with the Dutch regulations [11].

a	-15.63
b	1.69
n	1.19

Table 9: Acrylonitrile PHAST probit constants.

Figure 6 shows a distance of 453ft/ 138m to 0.1% probability of fatality, which indicates that there are no severe offsite consequences predicted. The 1% lethality distance is 361ft/ 110m, less conservative than the previous 541ft/ 165m estimation in PHAST ™ using pseudo probits. The 50% lethality extent is 203ft/ 62m compared with the PHAST™ LC-50 estimation using pseudo probits.

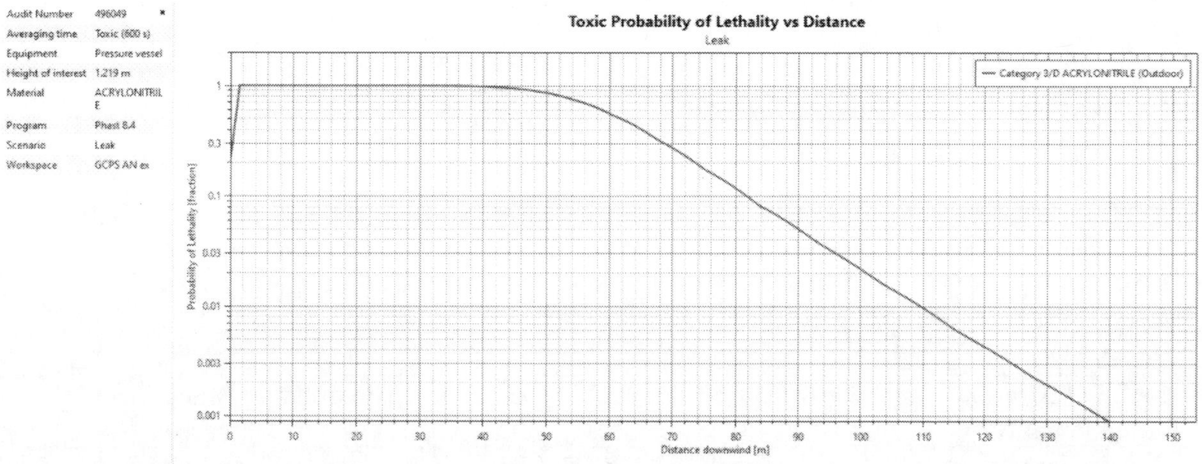

Figure 6:Acrylonitrile probability of lethality using probits.

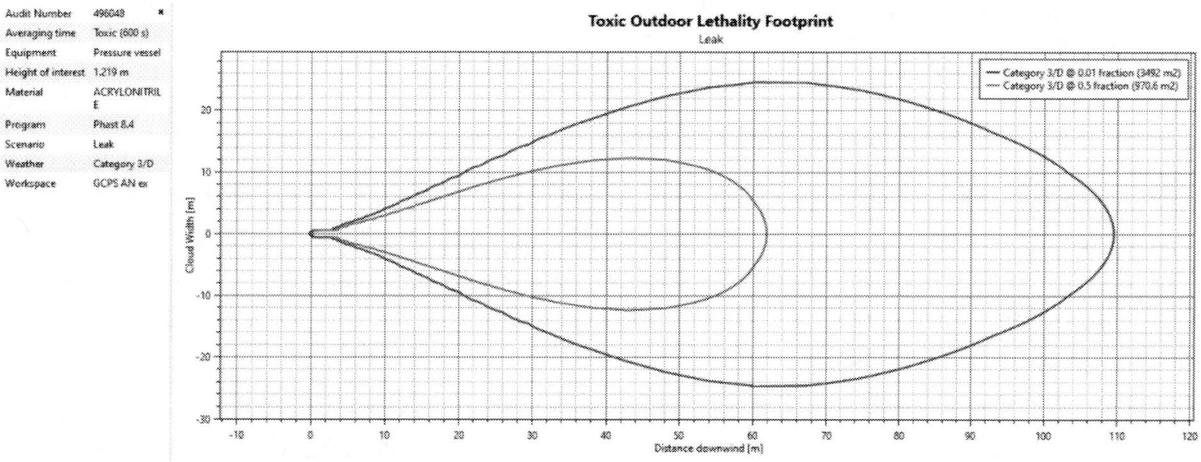

Figure 7:Acrylonitrile outdoor lethality using probits.

3.2.4 Summary table

Comparing the various model results, it can be seen in Table 10 how the results are refined for each of the techniques used.

Method	LC01	LC50	Consequence Severity
Chemical table	N/A	N/A	Multiple impacts
ERPG-3 extent distance (RAST or PHAST^{TM})	N/A	N/A	Multiple impacts
RAST- ERPG-3 based lethality	768ft/ 234m	335ft/ 102m	Potential offsite impact (*)
Pseudo probits	541ft/ 165m,	305ft/ 93m	No offsite consequence
Probits (SAFETI^{TM} Version NL8.8)	361ft/ 110m	203ft/ 62m	No offsite consequence

(*: Offsite impact depends on the proximity of the impacted populations. A buffer zone around a site minimizes any potential impact.)

Table 10: Summary of consequence severity analysis

3.2.5 Method selection considerations

A key point to consider as any consequence model is being developed is to ensure that the scenario reflects reality. The modeler or a team need to determine what will actually happen if there is a loss of containment. There may be a need to verify physical data in the field, such as distance to the fenceline and the size of a diked area, to ensure that the model is accurate.

Once a method has been agreed upon and a scenario developed, it is also recommended to complete a sensitivity analysis on the model to determine the robustness of the results. A key parameter to adjust in a dispersion model would be the weather conditions. To take a single result as the absolute answer may be wide of the mark and result in the plant adding additional layers of protection to protect against an event when it is not required.

If the scenario being analyzed can readily be managed with available protection layers using the crudest consequence method, then using a less refined model may be a logical approach. However, sometimes it is advantageous to further refine consequence estimates to properly risk-rank projects and allocate risk reduction efforts within an organization. In such instances, consequence severity can be more accurately predicted with additional analysis using more advanced techniques.

4 Conclusion

There are a number of tools available for modelers to use when completing hazard analyses and estimating potential impacts. This paper described four consequence modeling approaches containing varying degrees of rigor. The case study presented in the paper applied those methods to an example acrylonitrile release scenario and demonstrated how increasing the rigor of a modeling technique can remove over conservatism and possibly even change the conclusions of an analysis. Regardless, it is important to use a tool appropriate to the situation being reviewed and focus resources on higher consequence events to ensure the appropriate layers of protection are in place to protect our communities, employees, and the environment.

5 References

1. American Institute of Chemical Engineers (AIChE). *Dow's Chemical Exposure Index Guide* (1st Edition).
2. AIChE. *Dow's Fire and Explosion Index Hazard Classification Guide* (7th Edition).
3. AIChE. *RAST Manual and CHEF Aid*
4. Occupational Safety and Health Administration. *1910.119 - Process safety management of highly hazardous chemicals.*
5. Directive 2012/18/EU. EUR-Lex - 32012L0018 - EN - EUR-Lex (europa.eu)
6. Office of Response and Restoration. *Emergency Response Planning Guidelines (ERPGs)*
7. National Fire Protection Association *NFPA 704: Standard System for the Identification of the Hazards of Materials for Emergency Response*; 2022.
8. Instituut voor Milieu en Energietechnologie Tno Prins Maurit Tno. *Methods for the determination of possible damage (Green book) CPR16E*; 1992
9. Director-general for social affairs and employment (Pays-Bas). *Guideline for Quantitative Risk Assessment (Purple book) CRP 18E*; 2005
10. AIChE. *Guidelines for Chemical Process Quantitative Risk Analysis* (2nd ed); 1999 American Institute of Chemical Engineers (AIChE)
11. Overheid.nl. *BEVI (Besluit Externe Veiligheid Installaties). Decree on External Security Establishments.*
12. Box, G. E. P.; Draper, N. R. *Empirical Model-Building and Response Surfaces*; 1987
13. Crowl, D. A.; Louvar, J. F. *Chemical Process Safety* (4th ed); 2019
14. CCPS *Chemical Hazard Engineering Fundamentals* (V2.0); Mar 2019

Fire Impairment Analysis for Designing Offshore Depressurization System

Soraya Tamara A Malik
PETRONAS Malaysia
soraya.malik@petronas.com

Shaiful Khalid[1], M Ashraf M Ibrahim[1],
Nur Farhanah Nadhirah M Adam[1], M Sanusi N Azman[1]
[1] PETRONAS Malaysia

Keywords: ***Blowdown, ALARP, FERA, BLEVE***

Abstract

Emergency depressurization system in an offshore facility is a safety critical system to reduce pressure of a protected system to prevent escalation of a fire scenario. The design of the depressurizing system shall be adequate to ensure the stress of the protected vessel or system is not compromised. According to API 521, depressurizing a system to 7 barg in 15 minutes is commonly considered to reduce the consequences from vessel failure. This can be challenging for a brownfield facility where the vent or flare capacity is limited. Moreover, with most offshore facilities aiming for remote operations with minimal manning, the design of emergency system calls for practical design of fit for purpose.

The design of depressurizing system whether to depressurize within 15 mins or longer is dependent on risk evaluation supported with analysis to demonstrate that the risks can be mitigated to as low as reasonably possible (ALARP). This paper presents the detailed approach on how a brownfield project implemented fire impairment analysis to assess the impact to sensitive receptors. These results provide a basis to decide a practical design of the offshore depressurization system.

1. Introduction

An emergency depressurization system is installed to reduce the pressure of a protected system during fire exposure, lowering the material stress and minimizing the risk of rupture. As per API 521 standard, the design of the depressurization system should be capable of reducing the pressure to 7barg or 50% of the design pressure of the system (whichever is lower) within 15 minutes.

As part of a process simplification project, a Production Separator with a flammable inventory of 124.4 m^3 and operating conditions of 22.7barg and 27.9°C is installed on an unmanned Platform

A. Flare adequacy study revealed that the additional relief exceeds the existing flare system capacity. Process optimization was then conducted to keep the total relief within flare capacity; however, the resulting depressurization duration still exceeds 15 minutes.

According to COMPANY requirements, the design of the depressurization system on an unmanned platform depends on risk evaluation to ensure risk is mitigated to ALARP. Fire Impairment Analysis is selected to assess the potential fire impairment to the protected system, and the results will serve as the basis for the practical design of the offshore depressurization system.

2. Methodology

The methodology for Fire Impairment Analysis is conducted in accordance with the generic methodology of Fire and Explosion Risk Analysis (FERA), as used for offshore installations in the oil and gas industries. The steps involved in the analysis are depicted in **Figure 2.1**.

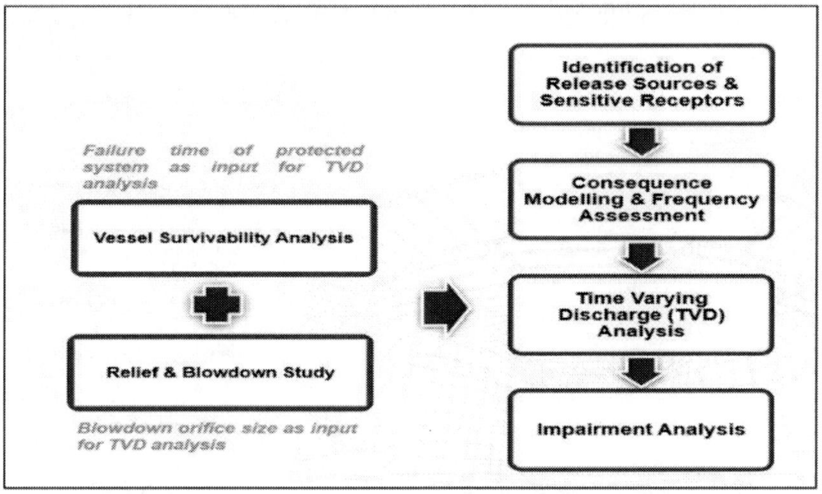

Figure 2.1: Methodology of Fire Impairment Analysis

2.1 Identification of Fire Scenarios

Fire scenario identification involves reviewing potential hazardous events resulting from loss of containment on offshore installations. Key factors determine the selection of fire scenarios for consequence modeling are type of material involved (gas or liquid), process parameters, type of releases and location of release (with or without bund).

2.2 Consequence Modelling

Potential fire scenarios are simulated using PHAST software. The simulation results will provide the specific values of the release duration, release rate, flame length, and radiation intensity level results which are crucial for carrying out the subsequent step of fire impairment analysis.

2.3 Time Varying Discharge (TVD) Analysis

TVD analysis is performed to determine the changes on consequence modelling results over time intervals. There are two (2) inputs required which are the failure time of Production Separator obtained from Vessel Survivability Analysis (VSA) and the blowdown orifice size of blowdown cases obtained from Relief and Blowdown Study. **Table 2.1** summarized the TVD cases performed for this study.

Table 2.1: Time Varying Discharge (TVD) Cases

Case	TVD Cases	Blowdown Duration	Remarks
1	Successful Isolation, No Blowdown	-	Requirement for blowdown system if impairment occurs in absence of blowdown
2	Successful Isolation, Successful Blowdown	15 minutes	Simultaneous Blowdown for all system within 15 minutes
3	Successful Isolation, Successful Prolonged Blowdown	15 minutes 30 seconds	Prolonged blowdown of more than 15 minutes for Production Separator.

2.4 Impairment Analysis

The initial step in Impairment Analysis is to screen sensitive targets with fire frequency above 1 x 10^{-04} per year. If the jet flame duration exceeds the failure time of sensitive targets or if jet flame reaches escape and evacuation facilities (EER), Risk Reduction Measures (RRM) is proposed. The complete process for conducting the impairment analysis is summarized in **Figure 2.2**.

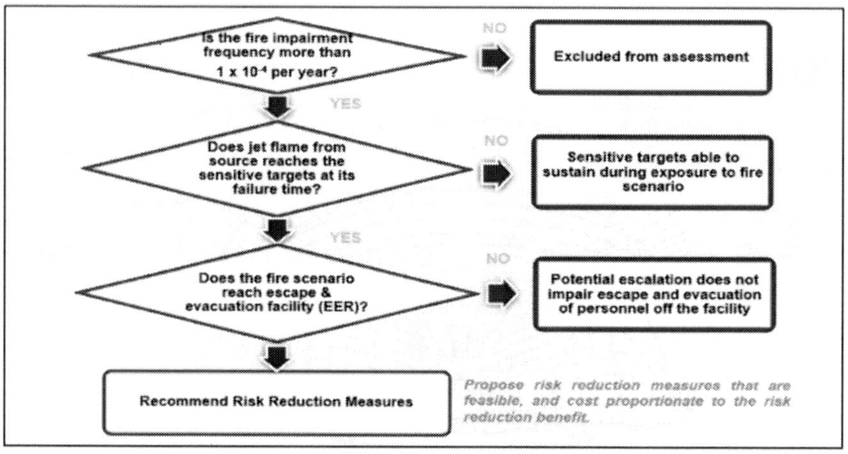

Figure 2.2: Acceptance Criteria for Impairment Analysis

3. Results and Discussion

3.1 Identification of Credible Fire Scenarios

The credible fire and explosion scenarios upon loss of containment (LOC) scenario from Production Separator is identified as follows which is in line with IOGP Consequence Modelling (Report No. 434 -7).

Table 3-1: Credible Fire and Explosion Scenarios from Production Separator

Equipment	Credible Fire and Explosion Scenario			
	Jet Fire	Pool Fire	Flash Fire	Explosion
Production Separator	√		√	√

Subsequent sections will focus on jet fire impairment due to its long duration, directions, and high heat intensity.

3.2 Fire Modelling Results

The jet fire modelling is performed for small, medium, large and rupture releases and detection time and isolation time are considered to represent the realistic jet fire profile. The fire modelling results from Production Separator for all TVD cases are shown below.

Table 3-2: Jet Fire Modelling Results from Production Separator

Time (min)	Case 1 Flame Length (m)				Case 2 Flame Length (m)				Case 3 Flame Length (m)			
	Small	Medium	Large	Rupture	Small	Medium	Large	Rupture	Small	Medium	Large	Rupture
0	4.7	19.9	36.9	74.1	4.7	19.9	36.9	74.1	4.7	19.9	36.9	74.1
5	4.7	19.9	18.9	-	4.7	19.9	21.3	-	4.7	19.9	16.8	-
10	4.0	15.3	0.00	-	3.9	14.7	-	-	4.0	14.7	-	-
15	3.9	12.9	-	-	3.0	9.3	-	-	3.1	10.0	-	-
30	3.8	10.6	-	-	2.3	6.8	-	-	2.4	7.5	-	-
60	2.0	6.9	-	-	1.4	0.00	-	-	1.5	0.0	-	-

3.3 Impairment Assessment

Impairment Assessment is conducted to evaluate whether impairment from fire source could impair nearby sensitive targets. The nearby sensitive targets identified are Crane Structure A and Crane Structure B. Time taken to cause crane structural failure upon jet flame engulfment is taken as 15 minutes and the critical distances between release source (Production Separator) and sensitive targets are 10.8 m (Crane Structure A) and 11.1 m (Crane Structure B).

Based on the results in **Table 3-2**, jet fire from small and medium releases from Production Separator could potentially impair Crane A and Crane B structures as these releases occurred more than the crane structure failure time of 15 minutes.

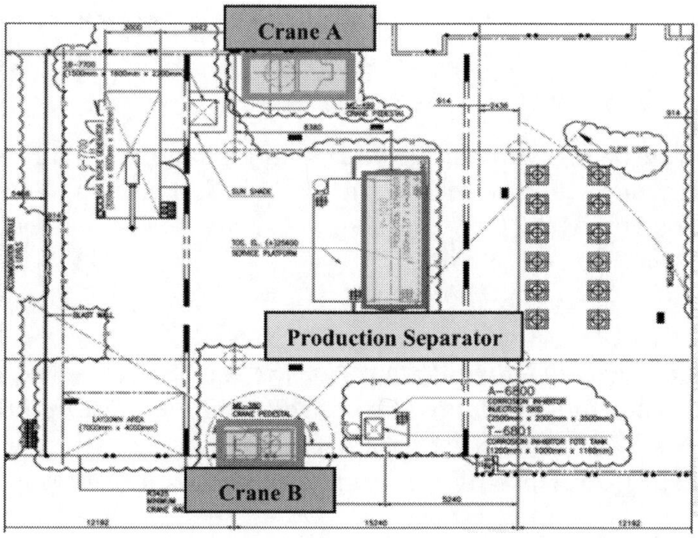

Figure 3-1: Location of Release Source and Sensitive Targets on Main Deck

The summary of impairment assessment from Production Separator to Crane Structure A and Crane Structure B is presented in **Table 3-3.**

Table 3-3: Summary of Impairment Assessment

Jet Fire Scenario		Crane Structure A		Crane Structure B	
		Flame Length (m) at Failure Time	Impaired?	Flame Length (m) at Failure Time	Impaired?
Case 1	Small	3.9	No	3.9	No
	Medium	12.9	Yes	12.9	Yes
Case 2	Small	3.1	No	3.1	No
	Medium	9.3	No	9.3	No
Case 3	Small	3.1	No	3.1	No
	Medium	10.0	No	10.0	No

The assessment above shows that jet fire could impair Crane Structure A and Crane Structure B under isolation case only. This is due to the prolong fire duration of more than 15 minutes and longer flame length than the critical distance from the Production Separator. This indicates that blowdown system is required to reduce escalation to crane structures in event of loss of containment from Production Separator.

Upon further comparison of acceptable blowdown durations, it can be observed that, with a prolonged blowdown duration of 15 minutes and 30 seconds, the jet flame did not impair the structures of Crane A and Crane B. This is due to the fact that the jet flame depleted more quickly compared to scenarios without blowdown.

Based on the assessment above, the simultaneous Prolonged Blowdown of 15 minutes and 30 seconds is able to prevent escalation to sensitive targets i.e. crane structures on Main Deck.

4. Conclusion

Outcome from the assessment found that the design of prolonged depressurization system is acceptable in terms of risk evaluation to mitigate risk to ALARP. This decision led to cost compression of approximately USD137,026 by avoiding extensive flare system modifications to cater for additional relief volume.

The method is practical for brown field facilities with limitation of flare capacity. For greenfield projects, it is recommended to provide inherent safer design at early stage of project following guideline recommended by API521.

5. References

- IOGP, *Consequence Modelling, Report No. 434 -7*, 2010.
- Spouge, J, *A Guide to Quantitative Risk Assessment for Offshore Installations*, CMPT publication No. 99/100, ISBN 1 870553 365, 1999.
- API Standard 521, *Pressure-relieving and Depressurizing Systems*, 7th Edition, 2020.

AUTHOR INDEX

AIChE
120 Wall Street, FL 23
New York, NY 10005-4020

ISBN 979-8-3313-1700-3
9 798331 317003

CURRAN ASSOCIATES INC.
proceedings.com